Contesting Space in Colonial Singapore

Contesting Space in Colonial Singapore

Power Relations and the Urban Built Environment

BRENDA S. A. YEOH

Department of Geography
National University of Singapore

NUS PRESS
SINGAPORE

Published by:

NUS Press
National University of Singapore
AS3-01-02, 3 Arts Link
Singapore 117569

Fax: (65) 6774-0652
E-mail: nusbooks@nus.edu.sg
Website: http://nuspress.nus.edu.sg

Reprint 2013
Reprint 2014
Reprint 2016
Reprint 2017
Reprint 2018
Reprint 2019

ISBN 978-9971-69-268-1 (Paper)

The first edition was published by Oxford University Press in 1996 and Singapore University Press in 2003.

Printed by: Markono Print Media Pte Ltd

For Andrew, in memory of our days at Oxford

The architect who designs, the planner who draws up master-plans, see their 'objects', buildings and neighbourhoods, from on high and from afar.... They pass from the 'lived' to the abstract in order to project that abstraction onto the level of the lived.

(Henri Lefebvre, *The Production of Space*, 1991)

As for the populace, where they were not served by the city of capital and luxury, they reacted as best as they could, according to their needs, to the extent of their provocation, and to the immediate circumstances of their everyday life. Between submission to the intolerable and outraged revolt against it, they somehow defined a human existence within the walls and along the passage of their streets.

(Anthony Vidler, 'The Scenes of the Street', 1978)

Acknowledgements

With the re-issue of this book, the trail of debt has lengthened considerably since the writing of the first acknowledgements. Research is ultimately a humanized activity, dependent first and foremost on human connectivity and ingenuity. I am thus grateful for the filaments that connect me to my colleagues and students within the research community at the National University of Singapore. I would like to renew my thanks to colleagues at the Department of Geography, particularly those whom I have had the privilege and pleasure of standing shoulder to shoulder in ploughing the research field over the last decade. At the University's Asia Research Institute, I have befitted from the liveliness of intellectual discussions among committed scholars and would also like to thank the support staff — in particular Verene Koh, Leong Wai Kit and Theresa Wong of the Asian MetaCentre — for the myriad small and not-so-small ways in which they have aided me in my work. It is also clear in my mind that without the energies of Peter Schoppert and Paul Kratoska at the Singapore University Press, this re-issue will not have happened. To Peter and Paul, I am very grateful.

Some things do not change. I wish to re-dedicate the book to my husband Andrew for always being there for me.

Singapore   BRENDA S.A. YEOH
October 2002

Acknowledgements to the First Edition

IN both the conception and materialization of this book, I have inevitably left a trail of debt in Singapore, Britain, and elsewhere. This work has its beginnings as a doctoral dissertation and I would first like to thank my supervisor at St John's College, University of Oxford, Jack Langton, for the generous amount of time spent discussing both the intricacies of archival research and broader questions concerning the nature of geography, the loan of innumerable volumes which have stimulated and shaped some of the ideas which eventually found their way into the book, as well as for efficiently supervising the 'everyday' aspects of research. I am also grateful to my advisers at the School of Geography, David Harvey and Ceri Peach, and other staff members of the University, including Alisdair Rogers and Peter Carey for advice and direction during the course of research. Research is often an isolated experience but this has been made much less so by the company of the other graduate students at the School of Geography, particularly Meshack Khosa and Margaret Byron. In as much as research and writing are contextual, the precise shape which this work takes is a reflection of the intellectual and social contexts of which I was privileged to be a part. Beyond Oxford, I am particularly indebted to Gerry Kearns for pointing me to wider debates in the realm of nineteenth-century public health, Jim Warren and Katherine Yeo for stimulating an interest in Singaporean 'subaltern' history, and Felix Driver whose own writing, while rooted in a different cultural context, has provided much inspiration and encouragement.

I am also grateful to the National University of Singapore for the award of an Overseas Graduate Scholarship and study leave to pursue research at Oxford. Two field-trips back to Singapore during the course of research were made possible through generous grants from St John's College and the Beit Fund. I am also appreciative of the efforts made by the librarians and archivists at the former Library of the Royal Commonwealth Society, the Public Record Office in Kew, the National Archives of Singapore, the National Library of Singapore, and the Chinese Resource Centre of the Singapore Federation of Chinese Clan Associations in helping me chart a path through dusty tomes and tracing the occasional 'obscure' work.

Returning 'home' to the Singapore fold after three years at Oxford meant taking on board many other dimensions of life apart from research. Immersion in a different academic community and administrative setting at the National University of Singapore, teaching commitments, and the demands and delights of ordinary Singaporean family life became other pressing engagements which competed for my time and energy. In setting aside time to refine the ideas in this book, I have depended on the support of my colleagues at the Department of Geography, particularly Shirlena Huang, Lily Kong, Peggy Teo, and Victor Savage. I would also like to thank the departmental cartographers, Mrs Lee Li Kheng and Mrs Chong Mui Gek, for their assistance in drawing the figures for the volume. Above all, I am deeply grateful to my husband, Andrew, who was with me in the beginning and has been a source of unfailing support ever since.

Some part of the present have appeared in modified or contracted form as articles in the *Journal of Southeast Asian Studies*, 22 (1991), the *Geographical Review*, 82 (1992), and *Southeast Asia Research*, 1 (1993). I am indebted to the editorial boards of these journals for permission to reproduce the material.

Singapore   BRENDA S. A. YEOH
August 1995

Contents

Tables

Figures

Plates

Abbreviations

Government Records
PLCSS   Proceedings of the Legislative Council of the Straits
               Settlements
SSGG   Straits Settlements Government Gazette

Municipal Records
ARSM   Administrative Report of the Singapore Municipality
MPMC   Minutes of the Proceedings of the Municipal
               Commissioners
MPMCOM   Minutes of the Proceedings of the Municipal Commis-
               sioners at an Ordinary Meeting
MPMCSM   Minutes of the Proceedings of the Municipal Com-
               missioners at a Special Meeting
MPMCEM   Minutes of the Proceedings of the Municipal
               Commissioners at an Emergency Meeting
MMFGPC   Minutes of Meeting of the Finance and General
               Purpose Committee
MMSYC   Minutes of Meeting of the Sanitary Committee
MMFGPSYC   Minutes of Joint Meeting of the Finance and General
               Purpose and Sanitary Committees
MGCM   Minutes of General Committee Meeting
MMSC1   Minutes of Sub-Committee Number 1
MMSC2   Minutes of Sub-Committee Number 2
MMSC3   Minutes of Sub-Committee Number 3

Singapore Improvement Trust Records
SIT   Singapore Improvement Trust

Glossary

Unless indicated, all terms are in Chinese

Bang	A dialect grouping from a more or less well-demarcated area. The main *bang* in colonial Singapore were the Hokkien, Cantonese, Teochew, Hakka, and Hainanese *bang*.
Pangkeng	Hokkien word for 'room'. It is used to refer to a rooming system where migrants with the same occupation or belonging to a particular clan sleep in and share the use of a large undivided room.
Chop	Business trade mark.
Chung-i	Doctor or medical man.
Feng shui	Literally meaning 'winds and waters' and translated as Chinese geomancy, this is defined as the art of adapting residences of the living and the dead so as to co-operate and harmonize with the local currents of the cosmic breath inherent in a particular configuration of the landscape.
Gharry	(Hindi, *gari*) A small horse-drawn carriage that plied for hire.
Gui	Ghost or evil spirit.
Jinrikisha	Also *rikisha*, *ricksha* (Japanese, *jin* = man, *riki* = power, *sha* = carriage). A small lightweight, two-wheeled vehicle with a removable hood which can be pulled by one man and which can carry one or two passengers.
Kan yu jia	Professional geomancer, or literally translated as 'specialist in the system that occupies itself with heaven and earth'.
Kapitan	(Portuguese origins) Headman of a tribe or ethnic group.
Kongsi	Chinese clan association.
Kampung	(Malay/Indonesian term) Refers to a village, but also used to describe an urban neighbourhood.

Nanyang	'Southern seas', a Chinese term for South-East Asia.
Permatang	(Malay term) A sand ridge or dune.
Purdah	Indian system of secluding women of rank from public view.
Sinkheh	'New guest or visitor', referring to newly arrived Chinese immigrants.
Samseng	Hokkien word referring to toughs, gangsters, rowdies, and unruly elements in society, many of whom are engaged by secret societies. It originates from the triad term *sanxing*, meaning 'three stars'.
Sembahyang Hantu	Malay colloquial term for the Chinese *zhong yuan jie*, also known as *gui jie* or the 'Festival of the Hungry Ghosts'.
Sultan	Islamic honorific given to a Malay/Indonesian ruler.
Temenggong	Malay chief of high rank.
Toti	Labourer employed in carrying and emptying night-soil.
Towkay	Malay/Indonesian term for rich Chinese.
Wayang	Malay term for a theatrical show.
Yamen	Government offices in China.
Yin-Yang	Complementary principles or essences (*yin* associated with the dark, wet, feminine, and absorbent and *yang* with the bright, dry, masculine, and powerful) existing on both the metaphysical and physical planes which are used in Chinese thought to explain all processes of growth and change in the natural world.

Note on Currency

UNLESS stated otherwise, the unit of currency used throughout the text is the Straits dollar. Prior to 1906, the value of the Straits dollar fluctuated between 4 shillings and 6 pence in 1874 and 1 shilling and $8\frac{1}{2}$ pence in 1902. A new Straits dollar was introduced in 1903 and the rate of exchange fixed at 2 shillings and 4 pence in 1906, remaining at this level for the rest of the period under consideration.

Note on Chinese Names and Terms

WHEREVER possible, the original spelling of Chinese names and terms as given in historical sources has been retained. Where there are no original English translations, these have been romanized in Mandarin according to the Hanyu Pinyin system.

1

Power Relations and the Built Environment in Colonial Cities

The Distinctiveness of Colonial Cities

THE morphology and development of Third World colonial cities have been enduring geographical research concerns over the last three decades, but as David Simon observed in a stock-taking effort, 'much of the published evidence is fragmentary and purely empirical, thereby rendering the important socio-scientific tasks of comparison and generalization difficult'.[1] Up to the 1960s, conceptualizations of urban development such as Sjoberg's pre-industrial/industrial dichotomy[2] failed to distinguish sufficiently between different urban dynamics in Europe and the Third World. Since then, the evolving scholarship has shown that subsuming the colonial city under such a dichotomy, or characterizing the colonial city as an 'industrializing' form articulating the transition between pre-industrial and industrial cities (as Sjoberg himself suggests),[3] is both inadequate and distorted, largely because it ignores the force and impact of colonialism and imperialism.[4]

The search for a theory of Third World urbanization which is not largely derivative of European models was spearheaded by scholars such as Ronald J. Horvath and Terence G. McGee who argue, *inter alia*, that the colonial city represents a heterogeneous but distinct urban type which does not conform to the pre-industrial, transitional, or industrial model. Instead, it has been stressed that the distinctiveness of the social, morphological, and functional features of colonial cities cannot be understood apart from their pivotal role in establishing, systemizing, and maintaining colonial rule.[5] In the main, three particular features intrinsic to the colonial process itself distinguish the colonial city. The first is its racial, cultural, social, and religious pluralism.[6] The colonial city contains a diversity of peoples, including colonialists, immigrants, and indigenes intermeshed within a social matrix comprising newly constituted relations of domination and dependence between individuals and between collectivities of people. These social groups are derived from vastly different societies, each with its own ingrained cultural behaviour, civil traditions, and institutionalized practices. The

colonial city is the archetypical exemplification of what J. S. Furnivall
has called a 'plural society', that is, one where

> different sections of the community live side by side, but separately, within the
> same political unit.... Each group holds by its own religion, its own culture
> and language, its own ideas and ways. As individuals they meet, but only in the
> market place, in buying and selling.... Even in the economic sphere there is a
> division of labour along racial lines. Natives, Chinese, Indians and Europeans
> all have different functions, and within each major group, subsections have
> particular occupations.[7]

Second, the colonial city is characterized by a social stratification
system which resembles neither the class structure of pre-industrial
cities nor that of industrial cities. Horvath distinguishes three main
components of the colonial stratification system: first, the ruling élite
comprising colonial settlers from the metropolitan country who obtain
their authority exogenously from the colonial power abroad; second,
the colonized indigenous population; and third, an 'intervening' group,
intermediate in status and power, derived either from interracial unions
or through migration from countries other than the metropolitan
country itself.[8] Not only is *race* the key mode of reference group
ascription within the colonial system of stratification, it also forms the
basis on which discontinuities between groups are institutionally organized and maintained.[9] At the core of colonial ideology is the notion that
communities are capable of being ranked on the basis of their sup-
posedly inherent racial attributes, an assumption which serves to justify
the subjugation of the 'natives' to white superiority.[10] A sociology of
ethnic pluralism comprising racially and culturally distinct groups also
serves to sustain colonialism on what D. A. Washbrook calls the 'umpire
analogy, positing the need for an independent arbiter to regulate the
affairs of naturally conflicting communities'.[11] Not only are different
'racial groups' clearly distinguished, they are also differentially
incorporated into colonial society. For example, from the perspective of
Weberian sociology, John Rex argues that colonial social structures are
based on the notion of 'estates', that is, status groups distinguished on
the basis of 'legal inequality and social separation'.[12] While recognizing
that the social relations of production are an important source of differentiation and class position in colonial societies, Rex argues that this is
often subordinated to 'other aspects of the total economic, political and
legal systems which differentiate men from one another and produce
roles which are often performed by culturally or racially distinct
groups, each of which has its own distinct system of legal rights'.[13]
Although more complex than Horvath's tripartite division, Rex's
system of colonial stratification also privileges ethnicity and race as
major factors which 'serve to bind together groups with quite different
social functions and relations to the means of production'.[14]

A third distinguishing feature of colonial cities is the concentration of
social, economic, and political power in the hands of the colonizers,
often a racially distinct group. David Simon, following Horvath, con-

siders the stark asymmetry of power between the colonized and the colonizers to be by far the most powerful independent variable influencing both social processes and urban spatial structure.[15] Dominance-dependence, as a description of the ensemble of colonial power, is impressed on and perpetuated through the dualistic structure of the colonial urban landscape, characterized by segregated European and indigenous quarters with their own distinct type of economic activities, landuse patterns, and architectural styles.[16]

This position has led some scholars to assume further, whether implicitly or explicitly, that the colonial town or city is either wholly or largely a creation of its colonial masters. For example, Simon's 'schema of colonial urban development'[17] envisages that once 'subjugation' of the colonized peoples is achieved, metropolitan colonial power, exercised by 'soldiers, administrators, traders and settlers' who set up imported government machinery and other institutions based essentially on metropolitan norms buttressed by its supreme command of scarce factors of production, is the most profound, and practically the singular, influence on the making of the colonial city. Such a scheme ignores the impact of indigenous or immigrant agency and resources, whether separately or in interaction with the colonial power, in shaping colonial urban society and space. Similarly, McGee portrays turn-of-the-century Singapore as a planned colonial city which, despite a predominantly Chinese population, 'remained a city planned by Europeans, and inhabited by non-Europeans whose residential distribution continued to reflect the intentions of the European rulers to an amazing degree'.[18] Again, the focus is on the uncontested supremacy of the European colonizers in undeviatingly carrying through their own plans and intentions to fashion a city after their own image.

The above view of the colonial city is simply a confirmation of contemporary conviction. Windows which give insight into contemporary popular perceptions of the urban landscape such as postcards and travellers' guides to colonial Singapore confirm the standard impression of the city as one replete with grand edifices of colonial munificence: the Town Hall, municipal and commercial offices, European hotels, the Botanic Gardens, colonial bungalows, and the like. Non-European buildings and urban structures which occasionally find their way into these guides are severely selected (picturesque temples and mosques being highly favoured) and usually portrayed as piquant curiosities, cultural embellishments, or romanticized objects of the tourist gaze rather than symbols of a deeper and more integral component of the total urban landscape. The overwhelming numerical dominance of the Chinese in the town of Singapore seldom caused her British administrators to falter in their assumption that, despite concessions which had to be made as a result of the frustrating peculiarities of the city's Asian plebeian classes, Singapore was a British creation to be governed and moulded according to British principles. In order to examine the extent to which these contemporary perceptions of the colonial city can be

sustained, it is first necessary to examine some of the approaches in the
current literature to the shaping of the colonial built environment.

Approaches to the Shaping of the
Urban Built Environment in Colonial Cities

Spatial forms in Third World colonial cities of the nineteenth and
twentieth centuries have usually been approached from one of three
perspectives, in each of which the physical form of the city is seen
merely as a reflection of *dominant* cultural, social, or economic structures. These approaches are elaborated below.

*The Modernization Paradigm: The Colonial City as Transition
between the 'Traditional' and the 'Modern'*

In a scheme which posits that all cities inexorably pass through a linear
progression of stages until the form and function of a Western-style
modern city is attained, the colonial city is often considered to be
poised between traditionalism and modernization, and to articulate the
transition from one to the other. As such, space in the colonial city is
partitioned into two types: first, 'modernized' space such as the
Western commercial centre, the port, and the European suburbs; and
second, space which exists only as remnants of the pre-colonial era
such as the native bazaar or native sacred places. As modernization
inexorably progresses, the former type of space expands and the latter
is gradually obliterated. Such an approach is exemplified by John
Friedmann's model of urban growth and national development, based
on a linear model of economic development which culminates in the
attainment of Rostowian take-off into sustained growth and a Western-
style urban revolution.[19] Another variant of the modernization paradigm is offered by McGee's model of the South-East Asian city which
purports to trace morphological developments within the city through
'a cycle from colonial imposition to western replication'.[20] According to
this model, the South-East Asian colonial city comprised a dual economic structure—an upper circuit or 'firm-centred' economy characterized by Western capitalistic forms such as banks and trading firms,
and a lower circuit or 'bazaar' economy with pre-industrial and semi-
capitalistic forms of economic organization such as Chinese loan-
associations and mobile street markets.[21] The duality of the economic
structure is a 'basic determinant'[22] of urban landuse patterns. The
'firm-centred' economy is generative of morphological elements which
are imitative of the West such as the 'Western commercial zone' and
the port with its complex of warehouses and wharves. The employers
and operators in the 'firm-centred' sector live in residential suburbs
well separated from their place of work in the central business district
and adjacent government offices. In contrast, the bulk of the urban
population associated with the 'bazaar' sector live in inner city areas of
mixed, high-density land use in which industrial and retailing functions

are combined with the residential (as epitomized in the shophouse). With the accelerated expansion of the firm-centred sector, 'bazaar' activities decline and the city is gradually transformed by 'assuming patterns very similar to those of the western city'.[23]

The developmentalist assumption implicit in the modernization paradigm has come under attack on several fronts, not least because it is grounded on categories invented in response to Western experience and as such, is 'unable to account for the historical specificity of social development in Third World regions'.[24] The reality of cities in developing countries (a large proportion of which are former colonial cities) provides little empirical evidence to support the view of the city as a dynamic generator of economic and social development. Empirical research on Third World urbanization has shown it to be a different process from the experience of industrialized countries in several ways.[25] The demographic transition in Third World cities, for example, assumed a vastly different form, with relatively high birth rates and low death rates, in comparison to European urbanization at equivalent stages. Rural–urban migration preceded rather than followed or accompanied industrialization, at a rate beyond the absorptive capacity of cities, resulting in burgeoning squatter settlements, large-scale unemployment and underemployment, heightened socio-economic inequalities, and increasing levels of urban primacy. Third World colonial cities hence cannot be easily subsumed within a modernization trajectory, for various elements of the colonial, urban built environment such as the informal 'bazaar' sector, shophouse development, and squatter zones seem far more persistent than the modernization paradigm suggests. The empirical infidelity of this paradigm can be traced to its neglect of the historical and structural causes of spatial inequality at the local, national, and international levels embodied first in the process of colonialism itself and subsequently in international dependency relations.[26]

The Cultural Explanation: The Colonial City as a Product of 'Culture-Contact'

A second approach to understanding the built form of colonial cities is to view their morphological elements as spatial expressions of the imposition of an alien culture on an existing one. Among pioneering work in this tradition is the classic article by anthropologists Redfield and Singer on 'the culture role of cities'.[27] Ranging over large periods of civilization, the main objective of this paper is to analyse the role cities play in the development, decline, or transformation of culture. According to their hypothesis, cities are either 'orthogenetic' (the city of the moral order; the city of culture carried forward) or 'heterogenetic' (the city of the technical order), where 'local cultures are disintegrated and new integrations of mind and society are developed'.[28] Among the latter category are 'colonial cities ... the mixed cities on the periphery of an empire which carried the core culture to other peoples'.[29]

With the development and extension of the modern industrial world economy, eastern cities converge towards Western urban forms as 'variants of a single cultural and historical process' and as such, 'the problem of urbanization in Asia' can be viewed simply as 'a problem of westernization'.[30] Later formulations in this vein refer to the colonial city as representing the introduction of 'western' urban forms into 'non-western' countries or as a 'cultural hybrid' or 'dual city' subsuming both 'traditional' and 'modern' elements. Janet Abu-Lughod, for example, refers to the colonial city as exhibiting a 'physical duality' which is 'but a manifestation of cultural cleavage'.[31]

Continuing in the tradition of privileging 'culture' as an explanatory tool of the built form of cities, scholars such as Norton S. Ginsburg, Anthony D. King, Amos Rapoport, and Robert Reed have argued that the colonial city provides an unrivalled laboratory for cross-cultural research.[32] King's theory of colonial urban development proposes to examine 'cultural responses to the environment ... within a particular distribution of power' and emphasizes the importance of the transformed and transplanted culture of the European community in shaping urban spaces such as the cantonment, the bungalow-compound, and the hill-station, which respectively epitomized the essence of military, residential, and social space in colonial societies.[33] While his theory acknowledges the roles played by a number of variables in shaping colonial urban space, it is to *culture*[34] that he gives pride of place as 'the first and overriding variable' out of which other 'components of colonialism such as the *economic–technological order* and the *power structure of colonialism* arise' (italics in the original).[35] Despite claims that the colonial city is a product of a 'culture-contact' situation between metropolitan and host societies, the tendency is to emphasize the influence of the colonizing culture in space-making processes at the expense of the colonized, who usually remain as locally abundant but faceless units of labour ministering to the colonizers' needs.[36] Accepting at face value that contact with the colonized occurred on conditions entirely constructed and dictated by the dominant, there is little conceptual space for interrogating the nature of interaction between colonizer and colonized. By focusing exclusively on the separate impress of either European or indigenous culture on the colonial landscape, questions of negotiation and conflict between two or more cultures in shaping, representing, and using the colonial built environment are inevitably ignored.

In a similar vein, the studies of Robert Reed and Preeti Chopra investigate the cultural and religious influence of the colonizing country on the colonial city.[37] In discussing the translation of the Spanish model of grid cities in the Philippines, Reed illustrates the colonizers' concern with imprinting Ibero-American culture and civilization on the urban landscape. In this respect, the study is significant in its emphasis on the cultural influences on colonial city form but given its principal concern with the imperial designs of Spanish colonialism, the 'extramural peoples'—Filipino, Chinese, Japanese, and *mestizo*—were accorded much less attention. Chopra's work which focuses on two elements of French colonialism—visionary individuals and the policy of assimilation—in transforming the urban built environment of Pondicherry, a French enclave in India, also gives pride of place to the grand edifices of colonial domination without making space for the colonized.

The 'culture-contact'/'dominance-dependence' metaphor has also been employed in understanding colonial urban experiences in another way: by focusing on the policies and discourses of the colonial power in 'the constant framing and creation of the natives as the "other" in order to facilitate subordination'.[38] As with the case of overemphasizing the active dominance of the colonizing culture in shaping the built environment and dispossessing other cultures, this reworking of the metaphor, one which draws inspiration from Edward Said's powerful treatise that the 'Orient' is itself a creation of European colonialism,[39] places emphasis on the plans and projects of the colonial power in justifying and maintaining cultural domination and control over the indigenous populations. Two examples should suffice to illustrate the nature of scholarship in this vein: Hosagrahar Jyoti's interpretation of British New Delhi as 'a theater of colonial discourse' where India was constructed as a disorderly and degenerate 'other' in need of British authority, protection, and redemption; and Shirine Hamadeh's contention that the notion of the 'traditional city' was an ideological construct born out of colonial discourse and used to sustain French domination in North African colonies.[40] While the theoretical and methodological merits of such work should not be belittled, the pitfalls are best summed up in King's own reflections on 'the study of colonial domination and dispossession' he had helped develop: 'The focus on European colonialism, by occupying the historical space of those urban places and people of which it speaks, has marginalized and silenced two other sets of voices: the voices of resistance and the voices of, for want of a better term, "the vernacular".'[41]

The Political Economy Approach: The Colonial City as a 'Function of Dependent Peripheral Capitalism'

A third approach to the built environment of colonial cities is constructed on the foundations of the political economy approach.[42] From this perspective, the colonial city cannot be treated in its own terms but is seen to be enmeshed in a wider set of productive forces and social relations pertaining to the capitalist world economy. Colonial cities should be viewed as 'integral cogs in a broader, predominantly capitalist system generating inequality, poverty and often dependency' and it is these 'global structures and processes' which constitute the 'key determinants' and provide the 'problematic' for analysis.[43] This approach involves the historical correlation of particular phases of global capitalist expansion with the internal dynamics of urban development at the periphery. Thus, the expansionary phase involving mercantile capital was associated with the establishment of 'stabilizing points' for the concentration and collection of commodities from the hinterland. This form of 'pin-prick' or 'enclave' colonialism had limited impact on existing urban hierarchies and morphologies as the European presence tended to be restricted to specific areas of the city where trade concessions were held.[44] In contrast, the advent of industrial capitalism and its demands for raw material sources and captive markets required the full incorporation of the periphery into the capitalist world economy. The Third World was to be held in a 'distorted, externally oriented and dependent relationship of unequal exchange' *vis-à-vis* metropolitan countries and as such, urbanization was not an autonomous process but purely 'a function of dependent peripheral capitalism'.[45] Colonial cities were hence planted as 'headlinks' and designed to facilitate European capitalist penetration. The penetration of capitalism within a traditionally subsistence-based economy catalysed a clash of two incompatible modes of production, with the consequent destruction and restructuring of the pre-capitalist economy to form an externally oriented 'colonial space-economy'.[46] A new grid of productive forces, embedded in new social relations, reworked the spatial organization of the society in such a way that it became subservient to the demands of capital. The built environment of colonial cities incorporated into the capitalist world system hence bore the impress of capital in a number of ways. Colonial investment and expenditure tended to favour the construction of dendritic communication systems linking the hinterland to colonial cities to expedite the funnelling of commodities to the metropolitan core; financial, banking, insurance, and warehousing complexes proliferated to facilitate entrepôt trade, whilst the persistence of slums, squatters, and tenements testified to the neglect of housing and welfare facilities in the colonial city. Although work on the articulations between the social processes of urbanization and the spatial built form of the city using the urban political economy approach has largely focused on the European and North American contexts, detailed analyses of Third World colonial cities using this approach are now emerging.[47]

The world system/dependency approach has spawned a wide-ranging array of criticisms of which one particular set of arguments advanced by 'urban praxis theorists' such as Michael P. Smith, Manuel Castells, and R. Brenner is directly pertinent to the question of the shaping of the colonial, urban built environment. These critics assert that the approach does not adequately conceptualize class conflict and instead implies that capitalism emerges and reproduces itself according to the motivations and interests of capitalists.[48] Consciousness, politics, and culture are reduced to the logic of capitalist accumulation and there is little concern with how local communities interact with global capitalist processes, whether in the form of resistance, accommodation, or in other creative ways. Eric Wolf offers a similar critique of André G. Frank and Immanuel Wallerstein's work on the capitalist world system and the underdevelopment of the periphery:

For both (Frank and Wallerstein), the principal aim was to understand how the core subjugated the periphery, and not to study the reactions of the micro-populations.... Their choice of focus thus leads them to omit consideration of the range and variety of such populations, of their modes of existence before European expansion and the advent of capitalism, and of the manner in which these modes were penetrated, subordinated, destroyed, or absorbed, first by the growing market and subsequently by industrial capitalism. Without such an examination, however, the concept of the 'periphery' remains as much of a cover term as 'traditional society'.... [T]his examination still lies before us if we wish to understand how [the local populations] were drawn into the larger system to suffer its impact and to become its agents.[49]

Without interrogating the reaction and strategies of the colonized, the built environment is reduced to being the product of a capitalist logic rather than an arena of conflict between social groups which have differing vested interests in the city.[50] The approach has hence been criticized for being 'economistic', for 'precluding the scope of local human agency and non-materialistic motivation', for being character-ized by a 'materialist orientation to the urban', and for not adequately addressing the question of social, cultural, or symbolic uses of space within the colonial city.[51]

The Colonial City as Contested Terrain

The combined outcome of these perspectives is that all too often the colonial city is not treated in its own terms, and the contemporary significance of the spatial configurations of the city for the actual inhabitants who live out their habitual lives within its confines remains uninterrogated. In their review of urban theory in general, Smith and Tardanico observed that whilst '[t]he impact of capital accumulation strategies of businesses and state policies of governments have been frequently viewed as the central elements in the making, unmaking and remaking of cities', what is 'equally important but more neglected has been the impact of common people—their consciousness, intentional-ity, everyday practices, and collective action—on the planning and implementation of state and business policies as well as their conse-quences for the social production of cities'.[52] There is little attempt to study the actual use of physical space within the colonial city, let alone investigate how different elements of the urban landscape become invested with different meanings and purposes and how they are repre-sented. The emphasis is often on the dominant forces at work, be it the dominant culture or mode of production, without similar attention being given to the underside, where the colonized engage in daily rou-tines perceiving, utilizing, contesting, and reconstituting the urban landscape on their own terms. As such, there is little conceptual room for the insertion of conflict and collision, negotiation and dialogue, between colonizer and the colonized in shaping the urban landscape.

The task set out here is to move beyond looking at the colonial city as a product of inexorably dominant forces to examining the contemporary

social meanings of its built environment as transformed by social, political, and economic conflicts between different groups with different
claims on the city. One way of approaching this is to grapple with how
the urban built environment is differentially perceived and utilized by
the colonial authorities and other social groups, and to examine why
conflicts over the use and definition of space arise, and how they are
resolved. This focuses explanation on the practical nature of everyday
life as lived within the colonial city rather than the abstracted nature of
economic organization or superorganic culture. It views the socially
constructed environment of colonial cities as 'not simply an unpeopled
"landscape" acquiring an "imperial imprint" but [as] a disciplinary
terrain, a mechanism for inducing new practices, an arena [in] which
new discourses are created, a resource for some, a weapon for others,
with which to harass, reclassify, categorize and control'.[53] In this conception, the colonial landscape as contested terrain not only 'articulates the ideological intent of the powerful who plan and shape the
landscape in particular ways' but also 'reflects the everyday meanings
implicit in the daily routines of ordinary people associated with the
landscape'.[54]

The colonial urban landscape is hence not simply a palimpsest
reflecting the impress of asymmetrical power relations undergirding
colonial society, but also a terrain of discipline and resistance, a
resource drawn upon by different groups and the contended object of
everyday discourse in conflicts and negotiations involving *both* colonialists and colonized groups. It embodies the negotiation of power
between the dominant and the subordinated in society, each with their
own versions of reality and practice.[55] The next section addresses the
nature of power relations and the way power is exercised through
specific strategies in a colonial context while a subsequent section
returns to the theme of how this expression of power is mediated
through the form and fabric of the colonial urban landscape. The final
section discusses how these questions will be pursued in the specific
context of colonial Singapore and provides a synopsis of the overall
structure of the book.

The Exercise of Power in Colonial Societies

While the indelible imprint of asymmetrical power relations on the
colonial urban landscape epitomized by the contrast between
magnificent European buildings and squalid Asian tenements, is un-
deniable, it has encouraged too summary a dismissal of the 'power' of
subordinated society to control and effect changes in the urban built
environment. The lack of formal authority of colonized groups is often
equated with minimal contributions to the shaping of urban society
and space under colonial rule. This stems from a rather narrow conception of what 'power' in a societal context is, and also a lack of
serious interrogation into the nature of power relations within colonial
urban society.

Power, in a generalized sense, means 'transformative capacity', that is, 'the capability to intervene in a given set of events so as in some ways to alter them'.[56] Any assertion of power draws upon two forms of resources: the allocative, or dominion over material facilities; and the authoritative, or dominion over the activities of human beings themselves. Colonial social systems, like all social systems of any duration which incorporate and exhibit forms of domination and dependence, regulate the distribution of these resources so as to perpetuate particular configurations of power advantageous to the colonialists.

Although the application of overt force is necessary during critical flashpoints of conflict, it is what Anthony Giddens describes as 'the institutional mediation of power', that is, the application of power as a constant if undramatic force 'running silently through the repetition of institutionalized practices',[57] which is crucial in entrenching the supremacy of the colonialists and perpetuating the subjugation of the indigenous and immigrant masses. While the process of colonial domination presupposes a superior, organized force, it is in vital respects a cultural project in which legitimate modes of control are consolidated through 'rituals and routines of rule'.[58] This was particularly so in a multiracial commercial emporium like Singapore where overt military force as a means of control would have been unsuited to the free-trading ethos. It was hence mainly through various institutional means of control that the colonialists established the necessary structural apparatus to provide ramifying channels through which their power penetrated to encounter and control the basic continuities of indigenous and immigrant social life. Each of these institutions represented a means of power with the capacity to influence the lives of the colonized peoples and the shape of the urban landscape in which they lived. By the last quarter of the nineteenth century, the proliferation and strengthening of institutions of control such as the police, the Chinese Protectorate, the municipal authorities, and various advisory boards had extended the scope of colonial control to encompass several aspects of quotidian activity, including the daily management of the urban environment. These extensions of the state apparatus exercised a form of power which Michel Foucault calls 'pastoral power' to emphasize its focus on salvation (in terms of the reform of a people's health or habits) and its use of individualizing techniques.[59] Although each institution had its own 'field of operation', they shared the use of common disciplinary forms to mould the colonized body and the space it inhabited so as to facilitate social order and economic advancement, the twin imperatives which supplied the rationale for action and which came to dominate colonial policy.

In elucidating the process of colonial domination and its specific strategies of control, it is worth reiterating at this point the observation that colonialism does not simply involve political and economic coercion but also ideological and cultural impositions which are often inseparable from the material. The colonial encounter often takes on a ritualized form whose maintenance is dependent on the export of

notions, systems, and practices which displace indigenous forms or re-
create them in the image of the colonial power. Such a form of power,
variously called 'cultural hegemony' or 'conceptual domination', is
rooted in the control over 'the *definition* of people and places',[60] that is,
in the capacity and legitimacy accruing to colonial power to define,
prescribe, and impose upon society a whole range of operative concepts and discourses which are assumed, on an a priori basis, to be
superior, correct, or beneficial. By laying down the categories through
which reality is to be perceived and conversely, by 'deny[ing] the existence of alternative categories, to assign them to the realm of disorder
and chaos, to render them socially and symbolically invisible',[61] the
powerful attempt to impose their image of order on colonial society.
This form of power is not only clearly visible in the colonial authorities'
claim to the right to construct social and racial categories (for example,
to ascribe racial identities, polarize racial distinctions, or demarcate and
apportion racial space), but also in their attempt to define what consti-
tutes health as opposed to disease, science as opposed to 'quackery',
order as opposed to disorder, or public 'good' as opposed to public
'nuisances'. In other words, colonial power is exercised through the
construction and imposition of definitional categories about social and
spatial relations which permeate both the private and public domains of
colonial society. This form of power is akin to what Foucault analysed,
in a wider context, as the 'objectivizing of the subject' through 'dividing practices': the confinement of the poor, the insane and vagabonds
in the seventeenth-century Hôpital Général, the new classification of
disease and associated practices of clinical medicine in early nineteenth-
century France, and the medicalization, stigmatization, and normaliza-
tion of sexual deviance in modern Europe.[62] In colonial societies,
conceptual domination is occasionally inscribed in the legal code, but
more often than not, mediated through institutional practices. Once
the categories of ideological discourse has been constructed, sanctions
must be used to defend them against possible challenges, and often
this defence is made through 'the continuous repetition, in diverse
instrumental domains, of the same basic propositions regarding the
nature of constructed reality'.[63]

It has also been argued, by Foucault in particular, that power should
be neither arrogated to the state or political system nor identified with
particular individuals.[64] Instead, Foucault has argued that power
emerges from 'local arenas of action' and should be viewed as 'a
"microprocess" of social life or pervasive feature of concrete, local
transaction'.[65] Its effects cannot simply be defined as repressive but are
instead productive and enabling, 'run[ning] through the whole social
body' rather than confined to the state or political system.[66] Power is
not conceived as a property but expressed in specific strategies: its
effects of domination are attributed not to 'appropriation', but to dispositions, manoeuvres, tactics, techniques, functionings; that one
should decipher in it a network of relations, constantly in tension, in
activity, rather than a privilege that one might possess; that one should

take as its model a perpetual model rather than a contract regulating a transaction or the conquest of a territory. In short, this power is exercised rather than possessed.[67]

These 'specific techniques of power' or strategies of control include what Foucault has called 'generalized surveillance', that is, certain types of practices or disciplines which order society through spatial and temporal regulation, for example, the enclosing and partitioning of space, the organization of the distribution of individuals in space, and the co-ordination and allocation of activities in time. In particular, disciplinary power is fundamentally concerned with the organization of space:

Disciplinary space tends to be divided into as many sections as there are bodies or elements to be distributed.... Its aim was to establish presences and absences, to know where and how to locate individuals, to set up useful communications, to interrupt operations, to be able at each moment to supervise the conduct of each individual, to assess it, to judge it, to calculate its qualities or merits. It was a procedure, therefore, aimed at knowing, mastery and using. Discipline organizes an analytical space.[68]

In short, 'discipline fixes; it arrests or regulates movements, it clears up confusion, ... it establishes calculated distributions'.[69] By organizing 'calculated distributions' of individuals in 'analytical space',[70] it is then possible to subject individuals to an efficient and continual observation or what Foucault calls a 'normalizing gaze, a surveillance that makes it possible to qualify, to classify and to punish'.[71] As Timothy Mitchell has observed, the tendency of this form of disciplinary power is to 'colonize' in the sense of being able to 'infiltrate' and 're-order' in 'a manner which is continuous, meticulous and uniform'.[72]

The growing scope and legitimized intensity of power accruing to colonial institutions and their strategies of conceptual domination, spatial regulation, and surveillance, however, does not imply a concomitant degeneration of the colonized peoples into a state of subjugation and powerlessness. Alisdair Rogers, following S. Lukes's analysis of different levels of power, observes that if all power resides ultimately in the properties of inviolate institutional structures, then it is not power at all, since everything is structurally determined with no possibility of effective action on the part of individuals and collectivities.[73] To attribute an absolute *omnipotence* to the 'apparatus' of disciplinary power analysed by Foucault would, as Colin Gordon has argued, confuse the domain of *discourse* with those of *practices* and *effects* for what is intended and articulated by the 'powerful' within the domain of discourse (such as those of sanitary science or urban planning) may fail to materialize in its entirety when transposed to the domain of actual practices and techniques or produce unintended consequences and effects.[74] It would also amount to denying that those which disciplinary power seeks to control are capable of counter-strategies which can challenge disciplinary power and modify its effects. Within the colonial city, the 'power' of the ordinary masses is not necessarily inversely related to that of the colonialists. As Giddens has contended, 'no matter how great the scope and intensity of control superordinates possess, since their power presumes the active compliance of others, those others can bring to bear strategies of their own, and apply specific types of sanctions'.[75] In other words, even 'the most dependent, weak and the most oppressed ... have the ability to carve out spheres of autonomy of their own'.[76] The disciplinary techniques of colonizing power proposed by Foucault do not automatically preclude the exercise of counter-strategies on the part of those who are subject to such powers.[77] Foucault himself admits that there were possibilities for 'revolts to the gaze' and contends that resistances to power 'are all the more real and effective because they are formed right at the point where relations of power are exercised [and that] like power, such resistance is multiple and can be integrated in global strategies'.[78]

The configuration of power which assumes shape in the colonial city is hence not only dependent on the nature of rule, the form of institutions, or the scope and intensity of disciplinary power imposed by the colonialists, but also on the 'strategic conduct'[79] of the colonized peoples who must be seen as knowledgeable and skilled agents with some awareness of the struggle for control, not just passive recipients of colonial rule or,[80] as the colonialists often labelled them, 'ignorant and prejudiced' people whose obstructionist habits were the manifestations of a lack of civilization. 'Struggle' and 'resistance' which lie at the core of relations of power cannot simply be explained by recourse to abstract concepts like 'the logic of contradiction' but should be analysed in the concrete, 'in tactical and strategic terms, positing that each offensive from one side serves as a leverage for a counter-offensive from the other'.[81] Strategies are not the preserve of dominant groups but also used by the dominated to secure their own objectives and that, in fact, 'some of the most sophisticated strategies are those developed in response to the strategies of others'.[82]

Organized into cohesive ethnic communities and subgroups which were in part reinforced by colonial racial policies, the Asian communities which inhabited the city of Singapore could draw on a range of resources to resist attempts at hegemonic control by the dominant colonizing culture. Their major resource was membership of an ethnic culture which bestowed a systematic set of values and ways of perceiving, group reinforcement of these beliefs, attitudes and social mores, and a network of institutional support. The Chinese who came to the Nanyang,[83] for example, brought with them an entire array of organizations such as clan and dialect associations, trade guilds, temples dedicated to a panoply of Chinese deities, and secret societies which provided the institutional structures within which social, cultural, religious, and recreational activities were performed.[84] Through these institutions, Chinese groups had access to a certain range of services which supported immigrant life such as the provision of medical care, job protection, education, entertainment, and facilities which catered to the observance of the rites of passage without recourse to the colonial host society. These organizations also provided convenient institutional focal points for the consolidation of power and the organization of counter-strategies to confront whatever means of control imposed by the colonialists. These counter-strategies included 'active' forms such as rioting, holding demonstrations, or going on strikes as means of expressing grievances. However, more common than 'active' and 'heroic' forms of protest, and in the long run less costly in terms of effort and sacrifice, were the 'passive', rather unspectacular means of countering and inflecting colonial control.[85] The community could adopt an outward attitude of apparent acquiescence (or at least non-protest), but in reality disregard or even thwart the measures imposed by the colonial power.[86] They could drag their feet over or dig their heels into requests for co-operation. Even in aspects of daily life controlled by legislation, the Chinese still held some power over the conduct of their lives and the pursuit of their livelihoods according to their own habits and priorities simply because the administrative machinery could not run faultlessly. 'Openings' in the system could be exploited to thwart the execution of particular policies and even to influence the decisions and activities of those who apparently held power over the masses. These forms of resistance against colonial authority were more commonplace than the organized revolt because 'they require little or no coordination or planning; they often represent a form of individual self-help; and they typically avoid any direct symbolic confrontation with authority or with élite norms'.[87] Compliance was withdrawn unobtrusively, without calling attention to the act itself or upsetting the larger symbolic order of dominance and dependence prescribed for the colonial world. Even when such forms of resistance became widespread enough to awaken the colonialists to the inefficacy of their policies at grassroots level, they were often too dispersed, too anonymous, and far too commonplace to allow immediate effective action against the actual culprits. It is often in these ways, through the politics and practices of everyday life, that the Chinese labouring classes participated in the production and consumption of social spaces in colonial Singapore. The 'concrete space of everyday life', to borrow Henri Lefebvre's term, is not only enframed, constrained, and colonized by the disciplinary technologies of power, but also 'the site of resistance and active struggle'.[88]

In recent years, there has been greater advocacy for 'history from below' or 'people's history' which, in the Thompsonian tradition of 'rescuing the poor stockinger ... from the enormous condescension of posterity',[89] vindicates the importance of the coolie's contributions to the making of society and the necessity of preventing ordinary people from becoming 'truants of their own past'.[90] In the context of turn-of-the-century Singapore, whilst the scope for greater attention to be devoted to the 'subject people' of the colonial city cannot be denied, presenting colonial Singapore as a coolie town, a 'veritable Capital of Cooliedom'[91] swarming with immigrant labourers free from the structures imposed by colonial power is as partial as concentrating exclusively on the achievements of Residents, Governors, and their councils in shaping urban society and space. In the words of Edward Said,

> no matter how one tried to extricate subaltern from élite histories, they are different but overlapping and curiously interdependent territories.... For if subaltern history is constructed to be only a separate enterprise—much as early feminist writing was based on the notion that women had a voice or a room of their own, entirely separate from the masculine domain—then it runs the risks of just being a mirror opposite the writing whose tyranny it disputes. It is also likely to be exclusivist, as limited, as provincial and discriminating in its suppressions and repressions as the master discourses of colonialism and élitism.[92]

The localized lifeworlds of the Chinese labouring classes thus cannot be understood apart from the larger political economy of colonialism.

If a balanced perspective which does not privilege one or the other is to be achieved, it is necessary to focus attention at the interfaces where colonialist and coolie meet, and where social relations and the negotiation of power between different levels of colonial society in the form of specific strategies can be examined. It is in daily encounter that power can be studied 'in its external visage, at the point where it is in direct and immediate relationship with ... its object, its target, its field of application, ... where it installs itself and produces its real effects'.[93] It is at this level too that subjugation is achieved, challenged, or inflected. The dominant–subordinated interfaces in society hence 'faithfully mirror the complex weave of competition, struggle and cooperation within the shifting physical and social landscapes'[94] and it is at these interfaces that an historical geography incorporating both colonizers and colonized could be reclaimed. In the words of Eric Wolf, 'we can no longer be content with writing only the history of victorious élites, or with detailing the subjugation of dominated ethnic groups ... (but should) strive ... to abrogate the boundaries between Western and non-Western history'.[95]

Power Relations and the Colonial Urban Built Environment

Colonial authorities, through local institutions of urban governance such as the municipal authorities, attempted to structure the urban built environment in such a way as to facilitate colonial rule and express colonial aspirations and ideals. The urban built environment refers to what David Harvey defines as the 'totality of physical structures—houses, roads, factories, offices, sewage systems, parks, cultural institutions, educational facilities and so on'.[96] Colonial landscapes ideally reflected the power and prestige of the colonialists, were ordered, sanitized, and amenable to regulation, and structured to enhance the flow of economic activities such as trade and communications which were crucial to the entire colonial economy. The 'norms and forms' which shaped the built environment of the colonial city, however, served not only to reflect colonial aspirations but were also used 'both consciously and unconsciously, as social technologies, as strategies of power to incorporate, categorize, discipline, control, and reform' the inhabitants of the city.[97] It was often through manipulating and controlling various spatial aspects of the built environment—through 'implantations, distributions, demarcations, control of territories and organizations of domains'[98]—that colonial power operated. Towards these ends, an institutional framework putting into effect a social technology for planning, orchestrating, and controlling the organization and shaping of urban space had to be created. The components of such a framework varied between the cities of different colonial powers, and included the church (particularly in the Latin American colonies of Spain and Portugal), trading companies (such as the British and Dutch East India Companies), the military (particularly in the laying out of army cantonments) and the local state or municipality.

Within a colonial city, however, various elements of the built environment were differentially perceived and interpreted by different communities based on their own values, priorities, and resources. As Jon Goss has argued, there is no one privileged discourse which can fully penetrate the meaning of the built environment and instead, it should be viewed as a complex 'multicoded space' which is continually reinterpreted in 'everyday usage ... by everyday people who may be "reading" or "writing" different languages in the built environment'.[99] Conflict over the definition, meaning, and use of the built environment resulted as local communities resisted the imposition of an all-encompassing, colonial structure on the landscape. The process of conflict and the resolution of conflict were articulated in and through the context of the colonial landscape. Landscape served to mediate between the everyday lives of the individuals, on the one hand, and the institutional structures which constrained and enabled those lives, on the other. Whilst it must be recognized that there were specific structural inequalities within colonial society which affected the means and capacity to shape and modify the urban built environment, it must also be acknowledged that both dominant and subordinated groups were capable of human action and strategies which were motivated and intended as well as those habitually carried out in practical consciousness. Landscape is hence 'synergistic': 'it is created and it creates';[100] it is constructed, destroyed, and transformed by individuals, communities, and institutions within specific cultural and socio-economic contexts. It exhibits what Stephen Daniels calls a 'duplicity' in embodying the tension between the 'visual ideology' of those in authority on the one hand, and the more plebeian imprint of daily practice on the other.[101]

From the perspective of those in authority, the colonial city of Singapore was to be constructed with a legible system of clearly signposted streets and walkways, water supply and sewers, parks and open spaces. Institutional structures of control over the built environment were established in order to facilitate the realization of colonial economic, political, and ideological aspirations. As the commercial capital of the British Far Eastern Empire and a clearing-house for both

commodities and labour, the built form of the city had to be constructed and regulated in a manner which facilitated trade, communications, movement, and a high turnover of people. The built
environment was not only functional, but as with colonies elsewhere
which 'constituted a laboratory of experimentation for new arts of government capable of bringing a modern and healthy society into
being',[102] the colonial city of Singapore, the 'Queen of the East' and
symbol of British pride and progress, was to manifest the advancement
of science and civilization in modern techniques of town planning and
sanitary engineering. As a spatial system which reflected colonial consciousness, the built environment was segmented in such a way as to
manifest and re-create the separation of the inhabitants of the colonial
city into racial containers, and the division of their activities along a
multiplicity of lines such as public/private, sacred/profane, and progres-
sive/offensive. Underlying colonial discourse was the scheme that the
city of Singapore expressed and facilitated the aims of Empire in functional, symbolic, and spatial terms. Such a representation of the colonial landscape was, however, not necessarily accepted by those who
inhabited the city but was instead challenged, both at the level of discourse and in everyday social practice. Through their own institutional
structures and individual strategies, social groups within the city
attempted to appropriate the built environment for their own purposes
and make it effective according to their own economic and cultural
aspirations. The colonial urban built environment was hence not separately shaped by either colonial control or the agency of those who
inhabited its terrain, but embodied and expressed the tensions and
negotiations, conflicts and compromises between different groups.

Rhetorical Form and Recoverable Reality:
The Plan of the Book and the Use of Sources

The main aim here is to examine how the urban built environment of
colonial Singapore was shaped between 1880 and 1930 by conflict
and negotiation between a colonial institution of control—the
Municipal Authority of Singapore—and the Asian communities who
lived and worked in the city. Chapter 2 examines the development of
municipal government as an institution of control over the urban
built environment charged with responsibilities in two main spheres:
the maintenance of a sanitary environment and the promotion of
public order in the city's spaces. It was no mere coincidence that
order and sanitation were most closely identified with municipal functions for European colonial society found it expedient to debate the
social consequences of rapid urban growth and the large influx of
Asian labour migrants using the imagery of a public health menace
and a threat to orderly society.

Parts I (Chapters 3–5) and II (Chapters 6–8) examine the dialec-
tics of power between the municipal authority and the Asian communities as negotiated within specific areas of municipal responsibility

concerned with planning, regulating, and representing the built environment of the city. Part I is largely concerned with tensions between, on the one hand, municipal efforts in controlling and improving sanitary conditions in the city through a variety of strategies including the surveillance of Asian practices (Chapter 3), the modification of spatial built form (Chapter 4), and the replacement of Asian 'utilities' systems with municipal facilities (Chapter 5), and, on the other, Asian attempts to resist, evade, and inflect control on the basis of their own socio-economic organizations and perception and management of city life. Part II focuses on the municipal project of producing an urban public landscape which was orderly, disciplined, and amenable to the demands of urban development and efficiency. This involved ordering the city's public spaces through naming (Chapter 6), defining and assigning 'proper' uses to particular spaces (Chapter 7), and relegating 'traditional' uses of urban space to the periphery in order to accommodate the demands of urban expansion (Chapter 8). In each of these areas, differences in economic purpose and socio-cultural values expressed by the municipal authorities and the Asian communities brought different perceptions to bear on the urban built environment, resulting in different and occasionally conflicting social and spatial strategies in conceiving of and using it. The concluding chapter (Chapter 9) reflects on some of the general conclusions which can be drawn from the preceding chapters.

The attempt to elucidate the nature of quotidian conflict and the negotiation of power between the municipal authorities and Asian communities in shaping the urban built environment without privileging one or the other is complicated by the inherently inegalitarian nature of available source materials. The 'annals of the labouring poor' in colonial Singapore, as elsewhere, seldom match official colonial and municipal records either in volume or detail. In reconstructing the aims and strategies of the common people in representing and using the urban built environment, the fragmentary records of Chinese organizations such as clan and dialect associations and medical institutions, the occasional autobiography left by Chinese immigrants, and contemporary descriptions of urban life in colonial Singapore by Chinese visitors and settlers are of some assistance in lifting the veil of silence from the historical record.[103] In general, however, given the fragmentary nature of purely Chinese records, much of the evidence is drawn from official reports and documents produced by the various departments of the colonial government and the municipality. Since these records were created for the purposes of government such as monitoring questions of the demand and supply of labour, urban facilities and commodities, administering justice, collecting taxes, and providing data for devising improved means of social control, they are inherently biased when used as a source to illumine the daily practices and strategies of the governed. In particular, the aims and aspirations of the common people are not immediately recoverable from these records as they are interpreted through the filters of colonial discourse and hence refracted by the assumptions of those in power. These deficiencies cannot be totally overcome, and in any case, 'the records of the past never speaks for itself'[104] but has to be interpreted. In an attempt to recover the motives and attitudes of the ordinary people as credibly as possible, two guidelines have informed the use of official sources.

First, official documents used in the research are largely those which to some degree recorded or represented the 'encounters' between the municipal authorities and the Asian communities. Municipal minutes, the reports of commissions of enquiry set up to investigate Asian housing, living, or medical conditions, and the local English press occasionally contained the petitions, requests, and opinions of local communities filtered through to the authorities by community leaders and representatives. Court records provide another source which represent the 'negotiation' between colonial might as represented by the police magistrates and district judges, and the common people, albeit under conditions of unequal power. Despite constraints built into these records such as the limited space given to Asian opinion and problems of translation from Chinese and other Asian languages, they provide some indication of the 'voice' of those who traced daily trajectories within the urban fabric.

Second, even where Asian views were not verbally articulated and recorded in official sources, these sources often provide a substantial record of the actions and non-actions of the common people which can be used as a key to their attitudes and strategies.[105] Actions ranging from riots and strikes to everyday acts of non-compliance, evasion, and the refusal to change 'customary' practices or adopt innovations introduced by the authorities which were recorded in colonial and municipal annals as irrational behaviour, apathy, ignorance, or superstition could be reread from a different perspective, avoiding the racist assumptions which often underlay colonial discourse. Whilst not all everyday behaviour of the common people was equally 'rational', much of it was certainly comprehensible if interpreted within the contemporary socio-economic and cultural context, and some of it at least was knowledgeable and rational in the sense that it generated actions which were effective means of achieving definable (if somewhat limited) practical ends.

The colonial and municipal records hence can be 'reread' for the opinions of the plebeian classes expressed through petitions, community leaders, and court evidence as well as for their attitudes and strategies inferred from recorded action. Morever, within the conceptualization used here, silence in the official sources can be instructive: if the collation and organization of information is the first step to a successful system of surveillance as argued by Foucault and Giddens, the failure of colonial sources to record the actions of the colonized is a testimony to the invisibility of the common people to the 'normalizing gaze' of the authorities and their occasional success in evading municipal control of their daily practices.

1. David Simon, 'Third World Colonial Cities in Context: Conceptual and Theoretical Approaches with Particular Reference to Africa', *Progress in Human Geography*, 8 (1984): 493.

2. Gideon Sjoberg, *The Pre-Industrial City*, New York: Free Press, 1960.

3. Gideon Sjoberg, 'Cities in Developing and in Industrial Societies: A Cross-Cultural Analysis', in Philip M. Hauser and Leo F. Schnore (eds.), *The Study of Urbanization*, New York: John Wiley and Sons, 1965, p. 220.

4. Norton S. Ginsburg, 'Urban Geography and "Non-Western" Areas', in Hauser and Schnore, *The Study of Urbanization*, p. 315; Ronald J. Horvath, 'In Search of a Theory of Urbanization: Notes on the Colonial City', *East Lakes Geographer*, 5 (1969): 70–2; M. E. P. Bellam, 'The Colonial City: Honiara, A Pacific Islands' Case Study', *Pacific Viewpoint*, 11, 1 (1970): 66; Simon, 'Third World Colonial Cities', pp. 497–8; Anthony D. King, *Urbanism, Colonialism and the World Economy: Cultural and Spatial Foundations of the World Urban System*, London: Routledge, 1990.

5. George Balandier, 'The Colonial Situation: A Theoretical Approach', in Immanuel Wallerstein (ed.), *Social Change: The Colonial Situation*, New York: John Wiley and Sons, 1966, pp. 34–61; Horvath, 'In Search of a Theory', pp. 69–82; Simon, 'Third World Colonial Cities', pp. 493–514; Anthony D. King, 'Colonial Cities: Global Pivots of Change', in Ronald Ross and Gerard J. Telkamp (eds.), *Colonial Cities: Essays on Urbanism in a Colonial Context*, Dordrecht: Martinus Nijhoff, 1985, pp. 7–32.

6. Pluralistic characteristics are also evident in non-colonial cities but it is contended that what distinguishes the colonial city is the degree or scale to which these characteristics are manifest.

7. J. S. Furnivall, *Colonial Policy and Practice: A Comparative Study of Burma and the Netherlands Indies*, Cambridge: Cambridge University Press, 1948, pp. 304–5.

8. Horvath, 'In Search of Theory', pp. 76–8.

9. Simon, 'Third World Colonial Cities', p. 500.

10. D. A. Washbrook, 'Ethnicity and Racialism in Colonial Indian Society', in Ronald Ross (ed.), *Racism and Colonialism: Essays on Ideology and Social Structure*, The Hague: Martinus Nijhoff, 1982, pp. 156–7.

11. Ibid., p. 157.

12. John Rex, 'Racism and the Structure of Colonial Societies', in Ross, *Racism and Colonialism*, pp. 199–218.

13. Ibid., p. 207.

14. Ibid., p. 209.

15. Simon, 'Third World Colonial Cities', p. 499. As Nezar AlSayyad observed, 'dominance is not exclusive to colonial cities, but the use and manifestation of dominance in the colonial context is particularly blunt' (Nezar AlSayyad, 'Urbanism and the Dominance Equation: Reflections on Colonialism and National Identity', in Nezar AlSayyad (ed.), *Forms of Dominance: On the Architecture and Urbanism of the Colonial Enterprise*, Aldershot: Avebury, 1992, p. 5).

16. Bellam, 'The Colonial City', p. 68; Mariam Dossal, 'Limits of Colonial Urban Planning: A Study of Mid-nineteenth Century Bombay', *International Journal of Urban and Regional Research*, 13 (1989): 21.

17. Simon, 'Third World Colonial Cities', pp. 506–8.

18. Terence G. McGee, *The Southeast Asian City: A Social Geography of the Primate Cities of Southeast Asia*, London: G. Bell and Sons, 1967, p. 72.

19. John Friedmann, *Regional Development Policy: A Case Study of Venezuela*, Cambridge, Massachusetts: MIT Press, 1966.

20. Terence G. McGee, 'The Changing Cities', in R. D. Hill (ed.) *South-East Asia: A Systematic Geography*, Kuala Lumpur: Oxford University Press, 1979, p. 191.

21. According to Clifford Geertz, the 'firm-centred' sector is one 'where trade and industry occur through a set of impersonally defined social institutions which organise a variety of specialised occupations with respect to some particular productive or distributive end', whilst the 'bazaar' sector is founded on 'the independent activities of a set of highly competitive commodity traders who relate to one another mainly by means of an incredible volume of *ad hoc* acts of exchange' (Clifford Geertz, *Pedlars and Princes: Social*

Change and Economic Modernization in Two Indonesian Towns, Chicago: Chicago University Press, 1963, p. 28).

22. McGee, 'The Changing Cities', p. 188.

23. Ibid., p. 191. McGee's later work (with Warwick Armstrong) is more critical of the unalloyed modernization stance and seeks instead to place emphasis on understanding Third World cities as part of the framework of capitalist accumulation. As 'theatres of accumulation' as well as 'centres of diffusion', Third World cities are conceived as crucial elements in the accumulation at regional, national, and international levels of financial, commercial, and industrial capital, and at the same time centres from which are diffused the culture and values of Westernization (Warwick Armstrong and Terence G. McGee, *Theatres of Accumulation: Studies in Asian and Latin American Urbanization,* London: Metheun, 1985, p. 41).

24. P. J. Rimmer and D. K. Forbes, 'Underdevelopment Theory: A Geographical Review', *Australian Geographer,* 15 (1982): 197–211; Stuart Corbridge, *Capitalist World Development: A Critique of Radical Development Geography,* Totowa, NJ: Rowman and Littlefield, 1986; David Slater, 'Peripheral Capitalism and the Regional Problematic', in Richard Peet and Nigel Thrift (ed.), *New Models in Geography: The Political Economy Perspective,* Vol. 2, London: Unwin Hyman, 1989, p. 269.

25. McGee, *The Southeast Asian City;* Terence G. McGee, *The Urbanization Process in the Third World: Explorations in Search of a Theory,* London: G. Bell & Sons, 1971; Armstrong and McGee, *Theatres of Accumulation.*

26. Michael Peter Smith and Richard Tardanico, 'Urban Theory Reconsidered: Production, Reproduction and Collective Action', in Michael Peter Smith and Joe R. Feagin (eds.), *The Capitalist City: Global Restructuring of Community Politics,* Oxford: Basil Blackwell, 1987, p. 88.

27. R. Redfield and M. S. Singer, 'The Culture Role of Cities', *Economic Development and Cultural Change,* 3 (1954): 53–73.

28. Ibid., p. 59.

29. Ibid., p. 62.

30. Ibid., p. 54.

31. Janet L. Abu-Lughod, 'Tale of Two Cities: The Origins of Modern Cairo', *Comparative Studies in Society and History,* 7 (1964/5): 429–30 and *Cairo: 1001 Years of the City Victorious,* Princeton: Princeton University Press, 1971, p. 98.

32. Ginsburg, 'Urban Geography and "Non-Western" Areas', p. 319; Robert R. Reed, *Hispanic Urbanism in the Philippines: A Study of the Impact of Church and State,* Manila: University of Manila, 1967; Amos Rapoport, *Human Aspects of Urban Form: Towards a Man-Environment Approach to Urban Form and Design,* Oxford: Pergamon Press, 1977, pp. 351–5, and 'Culture and the Urban Order', in John A. Agnew, John Mercer, and David E. Sopher (eds.), *The City in Cultural Context,* Boston: Allen and Unwin, 1984, pp. 53, 57; Anthony D. King, *Colonial Urban Development: Culture, Social Power and Environment,* London: Routledge and Kegan Paul, 1976, pp. 13–15.

33. King, *Colonial Urban Development,* pp. 11 and 64.

34. King follows M. G. Smith in defining the core of a culture as its 'institutional system' including kinship, religion, law, property, and economy, each comprising 'set forms of activity, groupings, rules, ideas and values' (King, *Colonial Urban Development,* pp. 42–3).

35. King, *Colonial Urban Development,* p. 25. This emphasis on a 'culturalogical' explanation of colonial urbanism is supplemented by 'some of the materialist questions prompted by urban political economy' in King's later works such as *Urbanism, Colonialism and the World Economy: Cultural and Spatial Foundations of the World Urban System,* London: Routledge, 1990; and *Global Cities: Post-Imperialism and the Internationalization of London,* London: Routledge, 1990.

36. While King himself states that it is misleading to ignore the impact of 'indigenous social-spatial categories' such as those inherent in the system of caste on the urban built environment, this forms an insignificant part of his analysis. Instead, he justifies concentrating on the European colonial community on grounds of 'familiarity with its language and culture' and argues that the investigation of indigenous urban forms should

be left to members of the indigenous culture. King borrows from Malinowski's analysis of culture-contact but inverts his priorities: Malinowski's primary interest was with the transformation of institutions of 'native culture' as a result of contact with incoming (European) culture but King is principally concerned with the European colonial culture which results from the transformation of metropolitan cultural institutions in a colonial setting (King, *Colonial Urban Development*, pp. 16, 40, and 58).

37. Robert R. Reed, *Colonial Manila: The Context of Hispanic Urbanism and the Process of Morphogenesis*, Berkeley: University of California Press, 1978; 'From Suprabarangay to Colonial Capital: Reflections on the Hispanic Foundation of Manila', in AlSayyad, *Forms of Dominance*, pp. 45–81; Preeti Chopra, 'Pondicherry: A French Enclave in India', in AlSayyad, *Forms of Dominance*, pp. 107–37.

38. AlSayyad, 'Urbanism and the Dominance Equation', p. 8.

39. Edward W. Said, *Orientalism*, New York: Vintage, 1979.

40. Hosagrahar Jyoti, 'City as Durbar: Theater and Power in Imperial Delhi', in AlSayyad, *Forms of Dominance*, pp. 83–105; Shirine Hamadeh, 'Creating the Traditional City: A French Project', in AlSayyad, *Forms of Dominance*, pp. 241–59.

41. Anthony D. King, 'Rethinking Colonialism: An Epilogue', in AlSayyad, *Forms of Dominance*, p. 343.

42. This approach draws its key concepts from the Wallersteinian world-system theory and Latin American *dependentistas* writing. It is a macro-historical approach aimed broadly at analysing the economic and political relationship between core and peripheral areas within a capitalist world-system. The approach has stimulated a wide range of research directions of which an important strand is the interpretation of Third World urbanization in terms of the global core-periphery division of labour. There is no attempt here to review the vast literature connected with the world-system or dependency theories (for such endeavours, see Michael Timberlake, 'World-System Theory and the Study of Comparative Urbanization', in Smith and Feagin, *The Capitalist City*, pp. 37–65; David Slater, 'Capitalism and Urbanisation at the Periphery: Problems of Interpretation and Analysis with Reference to Latin America', in David Drakakis-Smith (ed.), *Urbanisation in the Developing World*, London: Routledge, 1988, pp. 7–21), simply to point out some of the implications of the approach for the built form of colonial cities.

43. David Simon, 'Colonial Cities, Postcolonial Africa and the World Economy: A Reinterpretation', *International Journal of Urban and Regional Research*, 13 (1989): 71.

44. D. A. Smith and R. J. Nemeth, 'Urban Development in Southeast Asia: An Historical Structural Analysis', in Drakakis-Smith, *Urbanisation in the Developing World*, pp. 127–31.

45. Simon, 'Third World Colonial Cities', pp. 495–6.

46. Stephen G. Britton, 'The Evolution of a Colonial Space-Economy: The Case of Fiji', *Journal of Historical Geography*, 6 (1980): 251–74.

47. Recent scholarship includes Riberio's study of late nineteenth- and early twentieth-century Rio de Janeiro, focusing on the spatial transformations of the city—in particular, the development of a capitalist form of housing production—brought about by the urban accumulation of mercantile capital (Luiz Cesar de Queiroz Riberio, 'The Constitution of Real-estate Capital and Production of Built-up Space in Rio de Janeiro, 1870–1930', *International Journal of Urban and Regional Research*, 13 (1989): 47–67). See also King, *Urbanism, Colonialism, and the World-Economy* which attempts to draw up a general approach which focuses on colonialism as the historical link between the built form of cities and the world economy.

48. Smith and Tardanico, 'Urban Theory Reconsidered', pp. 89–91; Robert Brenner, 'The Origins of Capitalist Development: A Critique of Neo-Smithian Marxism', *New Left Review*, 104 (1977): 25–92.

49. Eric R. Wolf, *Europe and the People Without History*, Berkeley: University of California Press, 1982, p. 23.

50. According to David Slater, leaving social groups and social space out of accounts is symptomatic of the 'econocentric' tendencies which characterized much of Third World urban and regional research along Marxist lines. He argues that within this perspective, much theorization about space has been concerned with economic restructuring,

labour markets, and changes in the spatial division of labour rather than the spatial impact of the state, civil society, ideology, and discourse (Slater, 'Peripheral Capitalism', p. 277).

51. King, *Urbanism, Colonialism and the World Economy*, p. 10; Simon, 'Colonial Cities, Postcolonial Africa and the World Economy', p. 71.

52. Smith and Tardanico, 'Urban Theory Reconsidered', p. 105.

53. Anthony D. King, 'Colonialism, Urbanism and the Capitalist World Economy', *International Journal of Urban and Regional Research*, 13 (1989): 15.

54. Brenda S. A. Yeoh and Lily Kong, 'Reading Landscape Meanings: State Constructions and Lived Experiences in Singapore's Chinatown', *Habitat International*, 18, 4 (1994): 17.

55. Kay J. Anderson, *Vancouver's Chinatown: Racial Discourse in Canada, 1875–1980*, Montreal: McGill-Queen's University Press, 1992, p. 28.

56. Anthony Giddens, *The Nation-State and Violence, Volume Two of a Contemporary Critique of Historical Materialism*, Cambridge: Polity Press, 1987, p. 7.

57. Ibid., p. 9.

58. Notwithstanding calls to 'distinguish between the impact of colonialism/imperialism as a political and cultural system (i.e. British) and as a mode (or part of a mode) of production (i.e. industrial capitalism)', it is maintained here that the economic effects of the colonial project are inseparable from the cultural, practical, and symbolic. The focus here, however, is primarily on conflicts between groups inherent in the cultural and ideological representation and consumption of the colonial built environment rather than its production by economic forces. This is not to deny the salience of 'concrete' economic forces but to focus on the domains in which conflicts over the significance of the urban landscape between colonizer and colonized tended to become manifest (Jean Comaroff and John Comaroff, 'Through the Looking-Glass: Colonial Encounters of the First Kind', *Journal of Historical Sociology*, 1 (1988): 7; Anthony D. King, 'The Impress of Empire', *Journal of Historical Geography*, 15 (1989): 194).

59. Michel Foucault, 'The Subject and Power', in Hubert L. Dreyfus and Paul Rabinow (eds.), *Michel Foucault: Beyond Structuralism and Hermeneutics*, Brighton: Harvester Press, 1982, pp. 213–15.

60. Kay J. Anderson, 'Cultural Hegemony and the Race-Definition Process in Chinatown, Vancouver: 1880–1990', *Environment and Planning D: Society and Space*, 6 (1988): 128.

61. Wolf, *Europe and the People Without History*, p. 388.

62. Foucault, 'The Subject and Power', p. 208.

63. Wolf, *Europe and the People Without History*, p. 388.

64. Michel Foucault, 'The Eye of Power', in Colin Gordon (ed.), *Michel Foucault: Power/Knowledge, Selected Interviews and Other Writings, 1972–1977*, Brighton: Harvester Press, 1980, pp. 156–9.

65. John A. Agnew, *Place and Politics: The Geographical Mediation of State and Society*, Boston: Allen and Unwin, 1987, p. 23.

66. Michel Foucault, 'Truth and Power' in Paul Rabinow (ed.), *The Foucault Reader*, Harmondsworth: Penguin, 1984, pp. 60–1.

67. Michel Foucault, *Discipline and Punish: The Birth of the Prison*, trans. Alan Sheridan, London: Penguin Books, 1979, p. 26.

68. Ibid., p. 143.

69. Ibid., p. 219.

70. In *Discipline and Punish*, Foucault exemplifies the operation of disciplinary power through 'calculated distributions' of individuals in 'analytical space' by invoking Jeremy Bentham's panopticon or inspection house designed in the late eighteenth century as a paradigmatic example. The panopticon consisted of a circular building containing a courtyard with an observation tower at the centre of the courtyard from which every part of every cell on the periphery which contained inmates could be observed. As the observed could not see the observer, they were forced to behave as if surveillance was constant, unending, and total. The panopticon represents 'a segmented space-time, supervised continuously and at every point, in which power is exercised without division

and in which each individual is constantly distributed, located, and examined' (Felix Driver, 'Power, Space and the Body: A Critical Assessment of Foucault's *Discipline and Punish*', *Environment and Planning D: Society and Space*, 3 (1985): 428). The 'panoptic' techniques of surveillance and gaze were not confined to 'total' institutions such as the prison and the hospital, but could be used in a wide range of institutions and at different levels of society.

71. Foucault, *Discipline and Punish*, p. 184.

72. Timothy Mitchell notes that although Foucault's analyses of disciplinary power were focused on France and northern Europe, 'the panopticon, the model institution whose geometric order and generalised surveillance serve as a motif of this kind of power, was a colonial invention ... devised on Europe's frontier with the Ottoman Empire' (Timothy Mitchell, *Colonising Egypt*, Cairo: American University of Cairo Press, 1989, p. 35).

73. Alisdair Rogers, 'Gentrification, Power and Versions of Community: A Case Study of Los Angeles', *University of Oxford School of Geography Research Paper 43*, 1989, p. 24. This echoes Foucault's view that 'power is exercised only over free subjects, and insofar as they are free'. 'Free subjects' are those 'faced with a field of possibilities in which several ways of behaving, several reactions and diverse comportments may be realised' (Foucault, 'The Subject and Power', p. 221).

74. Colin Gordon, 'Afterword', in Gordon, *Michel Foucault*, pp. 246–8.

75. Giddens, *The Nation-State and Violence*, p. 11.

76. Ibid.

77. Giddens argues that 'by lumping together the surveillance of the prison with that involved in other contexts of capitalist society [such as in the work-place or in daily practice]', Foucault does not 'adequately acknowledge that those subject to the power of dominant groups themselves are knowledgeable agents who resist, blunt or actively alter the conditions of life that others seek to thrust upon them' (Anthony Giddens, *A Contemporary Critique of Historical Materialism, Volume One: Power, Property and the State*, Basingstoke: Macmillan, 1981, pp. 171–3). Felix Driver, however, shows that Foucault's approach assigns an irreducible role to the resistance of human agents although this is by no means clearly spelt out in *Discipline and Punish*. He argues that while Foucault's work serves to undermine received notions of struggle such as that based on 'the logic of contradiction' favoured in much of Marxist analysis, it also lends itself to illuminating 'the whole panoply of concrete and local struggles which surround our everyday lives' (Driver, 'Power, Space and the Body', p. 443).

78. Michel Foucault, 'Power and Strategies', in Gordon, *Michel Foucault*, p. 142. In another interview, Foucault suggested that power relations should be analysed 'through the antagonism of strategies' and by 'taking the forms of resistance against different forms of power as a starting point' (Foucault, 'The Subject and Power', pp. 210–11 and 225–6).

79. Giddens defines this as the 'strategies of control within defined contextual boundaries'. 'Strategic conduct' implies a degree of choice and intentionality and gives primacy to the practical and discursive consciousness of the human agent but this does not mean that strategies develop outside of an institutional context containing structural constraints. See Anthony Giddens, *The Constitution of Society: Outline of the Theory of Structuration*, Cambridge: Polity Press, 1986, p. 288.

80. This understanding is clear in the writings of postmodernists and those who recognize that there are multiple struggles between 'system' and 'lifeworld'. Derek Gregory, for example, reiterates that 'ordinary people often have a remarkable sophisticated awareness of the impingements and encroachments of abstract systems on their everyday lifeworlds'; that people 'are not dupes living in one-dimensional societies'; and that there is often more than 'a single vantage point from which to map social order' (Derek Gregory, *Geographical Imaginations*, Oxford: Blackwell, 1994, p. 306).

81. Foucault, 'The Eye of Power', pp. 163–4.

82. Graham Crow, 'The Use of the Concept of "Strategy" in Recent Sociological Literature', *Sociology*, 23 (1989): 4.

83. This means 'Southern Seas', a Chinese term for South-East Asia.

84. Yen Ching-hwang, *A Social History of the Chinese in Singapore and Malaya 1800–1911*, Singapore: Oxford University Press, 1986, p. 317.

85. While 'active' and 'passive' resistance are terms used in contemporary sources themselves, such a binary division is unrealistic as resistance does not fall into strict categories. Gordon goes further to argue that the distinction between 'resistance' and 'non-resistance' is an unreal dichotomy for even 'the existence of those who seem not to rebel is a warren of minute, individual, autonomous tactics and strategies which counter and inflect the visible facts of overall domination, and whose calculations, desires and choices resist any simple division into the political and the apolitical' (Gordon, 'Afterword', p. 257). Strategies of resistance hence follow a continuum rather than fall into neat dichotomies. The term 'passive' is used here simply to indicate that certain strategies of resistance are not necessarily violent, highly visible, or necessarily organized.

86. In the context of peasant resistance in Kedah, Malaysia, James C. Scott calls this form of passive protest the ordinary 'weapons of the weak'. They include tactics ranging from foot dragging, dissimulation, false compliance, pilfering, feigned ignorance, slander, arson, sabotage to concealed, publicly unacknowledged strikes, and combined action amongst the poor (James C. Scott, *Weapons of the Weak: Everyday Forms of Peasant Resistance*, New Haven: Yale University Press, 1985). See also Walter Rodney, *History of the Guyanese Working People, 1881–1905*, Kingston: Heinemann, 1981, pp. 151–73, which chronicles the strategies of resistance and accommodation ranging from accidental breakages to force the shut down of factories, strikes, minor acts of violence, the formation of friendly societies, to increasingly politicized weapons among Indian immigrant labourers and the Creole working people in British Guiana in the making of 'a history of active struggle'. Whilst these strategies often met with failure against the combined power of planters and the imperial parliament, George Lamming in his foreword to the book (pp. xix–xi) argues that 'it remains true that every struggle planted a seed of creative disruption and aided in the process that released new social forces in the continuing drama between capital and labo[u]r. . . . Each struggle would alert the planters to the need of a new strategy of control, and each strategy served to introduce a new stage of conflict between the work force and the planter class'.

87. Scott, *Weapons of the Weak*, p. 29.

88. Concrete space is the origin of what Henri Lefebvre described as 'representations of space' or counterspaces which 'arise from the clandestine or underground side of social life' in contradistinction to 'spaces of representations', conceptions of space in which 'the dominant social order is materially inscribed' (Gregory, *Geographical Imaginations*, pp. 403–5).

89. E. P. Thompson, *The Making of the English Working Class*, Harmondsworth: Penguin, 1968, p. 13.

90. Derek Gregory, *Regional Transformation and Industrial Revolution: A Geography of the Yorkshire Woollen Industry*, London: Macmillan Press, 1982, p. 1. The reconstruction and legitimation of 'the lives and actions of the common people … the very stuff of history' is a well-established tradition of French and British Marxist historiography typified by George Rudé, E. P. Thompson, E. J. Hobsbawm, and Gareth Stedman Jones (Frederick Krantz (ed.), *History from Below: Studies in Popular Protest and Popular Ideology*, Oxford: Basil Blackwell, 1988). Within Malaysian (and Singaporean) historiography, the 'everyday world of ordinary people' has been largely neglected by colonial history with its traditional emphasis on dominant individuals and major events. The bid to redress the gross imbalance and reinstate the 'unsung heroes' of history—the common people—has been spearheaded by a number of recent works, including James Francis Warren's trilogy on the Chinese labouring classes of which the first two volumes have been published as *Rickshaw Coolie: A People's History of Singapore (1880–1940)*, Singapore: Oxford University Press, 1986, and *Ah Ku and Karayuki-san: Prostitution in Singapore, 1870–1940*, Singapore: Oxford University Press, 1993; Peter J. Rimmer and Lisa M. Allen's collection of essays entitled *The Underside of Malaysian History: Pullers, Prostitutes, Plantation Workers …*, Singapore: Singapore University Press, 1990; and Katherine Lian Bee Yeo's thesis, 'Hawkers and the State in Colonial Singapore: Mid-Nineteenth Century to 1939', MA thesis, Monash University, 1989.

91. Peter J. Rimmer, Lenore Manderson, and Colin Barlow, 'The Underside of Malaysian History', in Rimmer and Allen, *The Underside of Malaysian History*, p. 16.

92. Edward W. Said, 'Forward to Subaltern Studies', in *Selected Subaltern Studies*, New York: Oxford University Press, 1988, p. viii.

93. Michel Foucault, 'Two Lectures', in Gordon, *Michel Foucault*, p. 97.

94. David Harvey, 'On the History and Present Condition of Geography: An Historical Materialist Manifesto', *Professional Geographer*, 36 (1984): 7.

95. Wolf, *Europe and the People Without History*, p. x.

96. David Harvey, 'Labour, Capital, and Class Struggle around the Built Environment in Advanced Capitalist Societies', in Anthony Giddens and David Held (eds.), *Classes, Power, and Conflict: Classical and Contemporary Debates*, Basingstoke: Macmillan, 1982, p. 545.

97. King, *Urbanism, Colonialism and the World Economy*, p. 9.

98. Michel Foucault, 'Questions on Geography', in Gordon, *Michel Foucault*, p. 77.

99. Jon Goss, 'The Built Environment and Social Theory: Towards an Architectural Geography', *Professional Geographer*, 40 (1988): 398.

100. David J. Robinson, 'The Language and Significance of Place in Latin America', in John A. Agnew and James S. Duncan, *The Power of Place*, Boston: Unwin Hyman, 1989, p. 157.

101. This point is made by Daniels in comparing Marxist geographer, Denis Cosgrove's approach to landscape with that of the Marxist sociologist, F. Inglis. In contrast to Cosgrove who sees 'landscape' as an illusory 'way of seeing' which tends to obscure plebeian experience, Inglis argues that it is a 'living process', which, whilst not devoid of patrician meanings and manipulations, is also 'the solid embodiment of ... popular culture'. Landscape, concludes Daniels, is 'a "dialectical image", an ambiguous synthesis whose redemptive and manipulative aspects cannot finally be disentangled, which can neither be completely reified as an authentic object in the world nor thoroughly dissolved as an ideological mirage' (Stephen Daniels, 'Marxism, Culture, and the Duplicity of Landscape', in Peet and Thrift, *New Models in Geography*, pp. 205–7). A similar view is expressed by David Ley in his review of work focusing on 'the landscape as text ... to be read for the ideas, practices, interests and contexts constituting the society which created it'. According to Ley, 'landscape as text' articulates the synthesis between 'agency' and 'structure', and as such, 'replace[s] the phenomenalism of traditional studies of urban morphology and the cultural landscape with ... the rehumanization of urban geography' (David Ley, 'From Urban Structure to Urban Landscape', *Urban Geography*, 9 (1988): 100–1).

102. Paul Rabinow, 'Governing Morocco: Modernity and Difference', *International Journal of Urban and Regional Research*, 13 (1989): 32.

103. The period of interest here—1880 to 1930—is unfortunately too distant to be within the reach of oral testimonies collated by the Archives and Oral History Department of the Singapore Government of Chinese immigrants who migrated to Singapore in the 1930s and after.

104. J. H. Hexter, quoted in J. B. Harley, 'Historical Geography and Its Evidence: Reflections on Modelling Sources', in Alan R. H. Baker and Mark Billinge (eds.), *Period and Place: Research Methods in Historical Geography*, Cambridge: Cambridge University Press, 1982, p. 263.

105. E. J. Hobsbawm argues that often 'the prettiest sources' for reconstructing subaltern history are those which 'simply record actions which *must imply* certain opinions'. Given that for most of the past the common people were often illiterate, 'voting with one's feet can be as effective a way of expressing one's opinion as voting in the ballot box' (E. J. Hobsbawm, 'History from Below: Some Reflections', in Krantz, *History from Below*, p. 19).

2
Establishing an Institution of Control over the Urban Built Environment: The Municipal Authority of Singapore, 1819–1930

I think that the capacity for governing is a characteristic of our race, and it is wonderful to see in a country like the Straits, a handful of Englishmen and Europeans, a large and rich Chinese community, tens of thousands of Chinese of the lower coolie class, Arab and Parsee merchants, Malays of all ranks, and a sprinkling of all nationalities, living together in wonderful peace and contentment. It always seems to me that the common Chinese feeling is that we—an eccentric race—were created to govern and look after them, as a groom looks after a horse, whilst they were created to get rich and enjoy the good things of the earth.[1]

The Institution of Municipal Government

THE conception and establishment of a separate municipal authority based on popular representation to run local urban affairs in Singapore was a distinctively British experiment. The municipal authority was an integral part of the colonial power structure (Figure 2.1) and served as an institution of control over the built environment of the colonial city. Its historical evolution from embryonic committees with specific municipal functions appointed on an *ad hoc* basis to a fully-fledged, popularly elected Municipal Commission administering its own municipal fund separate from government revenue in 1887 was both a response to the growing need for a more sophisticated machinery to run a burgeoning city as well as 'a sort of compromise'[2] to satisfy local demands, largely articulated by the resident European population, for more control over local affairs. In terms of its objectives, its legislative framework and institutional style, the Municipal Commission of Singapore, with certain significant departures, was patterned after the municipal authorities of British towns and cities. As such, despite the fact that Chinese commissioners were appointed from 1870,[3] the Municipal Commission as an institution of representative government

FIGURE 2.1
Colonial Power Structure

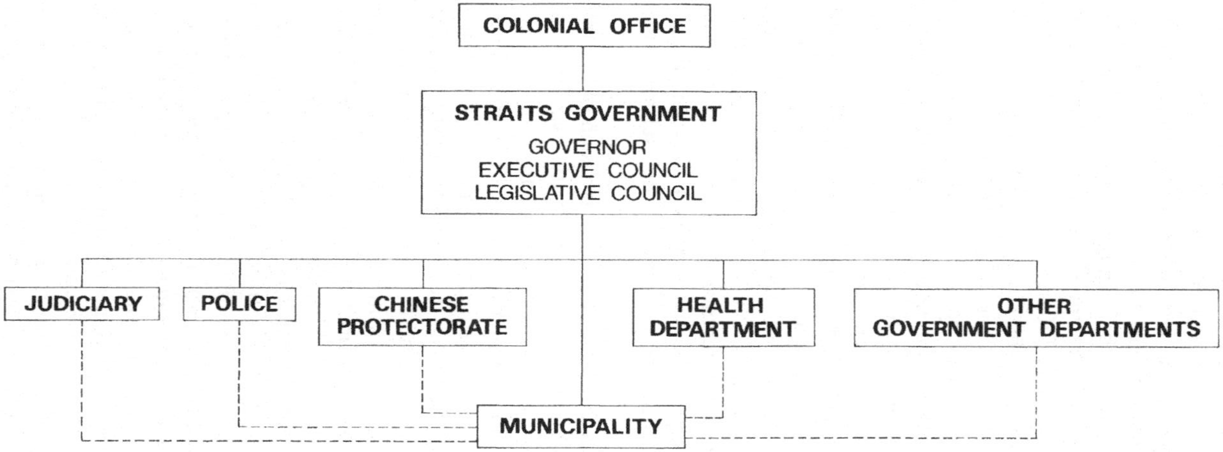

Source: Based aon Appendix Table A.2.

and control was alien to the majority of the Asian population of the town. For the Chinese, civic authority was vested in the myriad clan and *bang* organizations to which they belonged. The role of the Municipal Commission in improving civic life, the virtues of the elective principle, and the concept of the 'public good' were little appreciated if not totally misunderstood by the Asian people, as evident in the forlorn hopes of the municipal president, Alex Gentle, that proposed amendments to clarify the municipal ordinance would 'convince the ignorant people ... that sanitary rules [were] not the outcome of fitful zeal, or worse still, attempts at extortion on the part of a few proposed subordinate officers [of the Commission], but the settled purposes of the Government of the Colony, not to be evaded or resisted, but to be consistently enforced for the public good'.[4]

Conversely, it was equally evident that the Municipal Commission hardly addressed the needs and aspirations of the Asian plebeian classes on their own terms. Neither were they formally admitted as participants in the structure of local government. Institutional inequality in which only a minority of the city's residents were legally and socially incorporated into the urban power structure was basic to the hierarchical nature of colonial society. Municipal purposes were pursued according to both political and pragmatic considerations, but seldom with any reference to the views of the Asian plebeian masses, who were stereotypically disdained as ignorant and recalcitrant whenever they failed to conform to the 'settled purposes' of municipal authority.

The municipal authority, with its vision of urban life, its vested interests, and overriding local authority, was the chief social architect responsible for shaping the urban built environment of colonial Singapore. This chapter examines the historical evolution of municipal government and reformism as a means of simultaneously meeting the demands of European residents for autonomy and control of local affairs and solving increasingly severe social and environmental problems in a rapidly growing and overcrowded city inhabited largely by Asians. Following a brief overview of the early years of urban administration up' to the 1880s, the chapter turns to the demographic and spatial structure of the city which had crystallized at the end of the nineteenth century, a period of unprecedented urban growth and an intensification of congestion and other environmental problems. This provides the backcloth for a review of the widening municipal debate, the accumulation of greater powers, and the uneven history of municipal autonomy in the four decades after 1887, the date at which the Colony's first comprehensive municipal bill tailored to local conditions was passed. Two other concerns are also addressed: first, the racial composition of the Municipal Commission and the debate on the question of the 'fitness' of Asian participation in municipal government; and second, the tensions within the power structure of the municipal authority and the dialectical nature of its impact as an institution of control on Asian society.

The Administration of Urban Affairs, 1819–1887

Up to the fourth decade after the founding of Singapore by Stamford Raffles[5] in 1819 under the auspices of the British East India Company,[6] no municipal law existed in the town of Singapore. It was administered directly by the central government, which appointed *ad hoc* committees as the occasion demanded to attend to drains, street lighting, and the regulation of buildings.[7] One of the first of these committees was the town committee comprising three Europeans[8] appointed by Raffles himself in November 1822 to assume responsibility for 'an economical and proper allotment of the ground intended to form the site of the principal town'.[9] In 1827, another committee comprising five of the town's leading European residents was appointed by the Resident Councillor to supervise the making of drains outside of houses and the apportioning of cost among various proprietors.[10] None of these committees had any continuous life, nor were they in any way elected on a representative basis.

The campaign for public representation and participation in municipal government was led primarily by European residents anxious to foster local autonomy and control over urban affairs. In 1839, mounting grievances over the deteriorating sanitary conditions of the town and government indifference to local affairs led to much discontent and agitation among European residents. This resulted in the passing of Act XII of 1839 which allowed for the establishment of an assessment fund to be used for municipal purposes.[11] The fund was disbursed by executive officers of government and used mainly for the upkeep of a small police force to maintain order in the town.[12] It failed to quell European demands for representation in local affairs and in January 1845, a motion put forward by W. H. Read[13] at a public meeting proposing that the assessment funds be controlled by three persons—one appointed by the government and two by the ratepayers—was carried by a large majority.[14] Hopes for greater responsibility over municipal affairs through elected representation, however, were frustrated by the introduction of a new assessment bill in June 1846 providing for a municipal committee of five assessors—two officials and three non-officials—nominated by the Governor to administer the funds.[15] The bill was passed and came into effect as Act IX of 1848 but this so-called 'first milestone in the development of municipal government'[16] aroused little public enthusiasm. Instead, the question of public representation continued to feature prominently in Grand Jury presentments,[17] the local English language press,[18] and petitions to government.[19] It was also felt that the municipal committee, given its limited powers, was incapable of dealing with the sanitary ills and offensive nuisances which had long featured in the Grand Jury's sessional presentment, a document which the press praised as deserving the accolade of 'The Resident Councillor's and Municipal Committee's Manual'.[20] There was also considerable ill feeling that funds raised on an assessed tax on town property were used for the construction of

new roads and bridges which, by opening up virgin tracts and enhancing land values, only served to enrich central government through
lucrative land sales rather than municipal coffers.[21] Public protest[22]
against the centralization of power in the hands of the Executive was
not entirely futile for in 1854 steps were taken to extend the Indian
municipal laws to the Straits Settlements, although it was not until
1856 that municipal acts were passed.[23] Act XXVIII of 1856 provided
for the establishment of a Municipal Board comprising five members—
the Resident Councillor (and later the Colonial Secretary after the
1867 Transfer) as president, one other official and three representatives
chosen by ratepayers paying annual rates over 25 rupees.[24] Two other
complementary acts defined the powers and functions of the Municipal
Board: Act XXV provided for assessment and collection of municipal
rates and taxes; and Act XIV (commonly known as the Conservancy
Act) directed municipal commissioners to 'administer the funds applicable to the purpose of conservancy and improvement' (Plate 1) and
empowered them 'to make, cleanse, light and water streets, to remove
filth, to control erections of new huts, to pull down ruinous houses, ...
to drain the town' as well as to enforce penalties, supply water, regulate
slaughterhouses, and license dangerous trades and burial grounds.[25] A
fourth Act, XIII of 1856 (the so-called Police Act), provided for the
establishment and organization of a police force in each town. The
administration of the police was vested in a commissioner of police
appointed by the Governor. As such, although the police force was
maintained by municipal funds, the Municipal Board had no control
over it.[26]

Although this was the first time that the municipal committee
became partially open to the electoral process, the European public felt
that it did not go far enough in giving ratepayers power and a voice in
municipal affairs. In a series of resolutions passed at a public meeting
chaired by Read on 29 July 1856, the European public expressed their
dissatisfaction with the municipal acts. In their opinion, the number of
municipal commissioners should consist of seven elected and two government members for any fewer elected members would not adequately
represent the ratepayers.[27] They also objected to the prerogative vested
in the Governor to fill up vacancies in the municipal committee resulting from the failure of election and protested against the 'arbitrary and
unconstitutional measure' which placed the control of the police in the
hands of the central government.[28] In the agitation for representation
and municipal reform, the Asian communities had played little part,
but when the municipal and police acts became operative on 1 January
1857, the Chinese population closed their shops and went on strike
against the new legislation. The strike was attributed to the work of
secret societies and is often cited as an illustration of their power:

At the word of command given by the heads of those [secret] societies, within
a day—within an hour,—the town of Singapore was hermetically sealed to
trade, and there was not a door or window open or a man at work at his busi-

ness in the town.... It not only affected the Chinese, but also the Klings and others.... They did not take action at all, but they did not go about their usual avocations, because they were afraid to go against a power of which they saw only the effect, and not the origin of which they knew nowhere.[29]

The strike, however, passed without incident and business returned to normal within a few days when the government adopted conciliatory tactics by explaining the nature and objects of the acts to leading Chinese merchants and also assuring the population of the government's good intentions through an official proclamation.[30]

In the early years after the municipal acts came into effect, the Board attempted to abate some of the long-standing nuisances in the town, including raising some of the main streets to prevent flooding, clearing up the filth and stench in public places, and replacing dilapidated bridges with iron ones.[31] The town's first municipal engineer, surveyor, and architect, J. W. Reeve, was appointed in 1858.[32] However, despite these efforts, the compulsory levy demanded by government for the upkeep of the police force depleted the annual municipal revenue by more than half, and little was left over for carrying out functions contained in the Conservancy Act.[33]

On 1 April 1867, the Straits Settlements[34] were transferred from the Government of India to the Colonial Office in London but its new status as a Crown Colony did not effect any fundamental changes in the structure of the municipality and merely involved an adoption of the Indian acts. The new colony's first Governor, Colonel Harry St George Ord, however, objected to the employment of government officers as members of a Board which was not under government control.[35] He believed that the municipal system as it existed was defective for there was 'neither necessity nor inducement for the members to take an active part in municipal affairs', and 'no actual responsibility ... to any one' for the 'economical' disbursement of funds.[36] Ord favoured placing municipal affairs directly under a government institution comprising a mix of government officials and unofficial members representing the ratepayers but nominated by the Executive, with expenditure on public works controlled by legislature.[37] The Governor's views, tantamount to an erosion of public power and a denial of the elective principle, were, as might be expected, considered a retrograde step by the European public. An anonymous Straits merchant, annoyed by what was seen as the Governor's 'monkey-like restlessness which meddle[d] with everything', catalogued a list of complaints against Ord, including his attempt to 'menace [the municipality] with total abolition off the face of the earth' and went as far as to pronounce him 'unfit to rule a commercial colony like the Straits Settlements'.[38] In 1872, a subcommittee of the Legislative Council comprising three officials and two non-officials was appointed to enquire into the working of the municipality. Its report recommended that the municipality as an elective Board should not be abolished but reformed to admit a paid president to be appointed by the Governor.[39]

This proposal was stoutly resisted by the unofficial members of the Legislative Council and eventually withdrawn.[40]

Singapore's new status as the capital of a Crown Colony did little to improve local sanitary conditions. In June 1872, the municipal commissioners reiterated that the heavy contribution demanded by the government for police maintenance rendered it 'impossible' to effect urgently needed sanitary and improvement works.[41] They pointed out that sanitary conditions in the town, its drainage, and water supply were so 'eminently unsatisfactory' that 'the gravest results might be anticipated were an epidemic to break out', and complained of the 'perfunctory manner' in which the police had carried out conservancy and street obstruction laws.[42] The town, according to a senior army medical officer, was 'a nursery of disease', for 'not only [were] the drains choked with every description of filth, but the ordinary streets and thoroughfares [were] considered the proper receptacles for all kinds of accumulated impurities, and the atmosphere [was] filled with unwholesome emanations, in the highest degree dangerous to health'.[43]

The arrival in 1873 of a new Governor, Sir Andrew Clarke, augured well for the municipality for Clarke was in favour of the public taking a wider interest in the administration of local affairs. In his speech on the opening day of the Legislative Council in March 1874, he raised the question of 'extending the powers and increasing the means of [municipal] bodies'.[44] In 1875, Ordinance [II] of 1875 was passed relieving the Colonial Secretary of the post of president of the Municipal Board, as well as of the requirement of sitting as a member altogether. For the first (and also the last) time, the municipal commissioners were given the liberty of electing their own president. The promise of any further liberal reform, however, died with Clarke's departure from the Colony in May 1875 and for the rest of the decade, municipal reform stagnated at a low ebb.

In the early 1880s, under the governorship of Sir Frederick Aloysius Weld, there were again tentative indications that the government was considering taking steps to strengthen the power of ratepayers over municipal affairs. Weld remarked in 1883 that 'whereas the Government had been content to act the part of Police and tax-gatherers in the past, the time had come when efforts must be made by means of existing institutions to induce the people to take an intelligent interest and assist in the administration of their affairs'.[45] However, little was actually done to consolidate the powers of the municipalities in the Straits Settlements for Weld was more concerned with strengthening British control of the Protected Malay States than the 'gross details of administration' in the colony.[46] It was only towards the eve of his departure from the Straits that Weld set in motion a bill to amend and consolidate the law relating to the municipalities. In August 1886, he deputed the Colonial Secretary, the Attorney-General, and the Inspector-General of Police who was then the president of the municipality, to draft a bill for consolidating and amending the law relating to municipalities. This bill, the first strand of an increasingly ramifying web of legislation passed in the Crown Colony to control the built environment, was to precipitate a widening debate and unprecedented interest in the running of municipal affairs.

Urban Development and the Spatial Structure of the City in the Late Nineteenth Century

The last quarter of the nineteenth century saw a rapid expansion of Singapore's trade and the consolidation of her position as a premier entrepôt in the Far East. The opening of the Suez Canal in 1869 established the supremacy of her geographical position, for the Malacca Straits soon supplanted the Sunda Straits as the major waterway from Europe to the Far East and Singapore became an important transaction point for the East–West trade and a coaling station for steamers. The late nineteenth century also heralded the emergence of a world market and a system of trade in which the needs of capitalist interests in the industrializing West were met by a reorientation in the productive and accumulative activities of the 'periphery'.[47] British commercial capital abroad began responding to attractive possibilities for profitable investment in production in the colonies. In Malaya, with Singapore functioning as a bridgehead, the extension of British political control over the Malayan hinterland from 1874 went hand in hand with Western capitalist penetration of the interior. European capital investments in the Malayan tin industry and plantation agriculture generated large-scale bulk movement of primary products which were channelled for export through the port of Singapore. Singapore also acted as a base for European trading agencies and merchant houses which not only dominated the import–export trade but also handled the financial, commercial, shipping, insurance, and other related services connected with the rapidly expanding trade. Her runaway success and ascendancy as a premier port and trading centre were perceived to move inexorably in tandem with the progress of Queen Victoria's 'Glorious Reign'. As the municipal commissioners' Jubilee address in 1897 declared: 'Founded in the year of Your Majesty's birth, [Singapore] was but a small place of a few thousand inhabitants and with little trade at Your Majesty's Ascension in 1837. Now it is a city of 200,000 inhabitants, one of the largest sea ports in the world, visited in 1896 by ships whose combined tonnage exceeded 8½ million tons, and is the collecting and distributing centre for all the vast trade of Southern Asia and the Eastern Archipelago'.[48]

Singapore's rapidly expanding economy, coupled with a liberal open door policy on immigration, drew ever-increasing numbers of immigrants in the last few decades of the nineteenth century. Between 1871 and 1931, the total population of the Settlement increased by 24–43 per cent each decade (Figure 2.2),[49] from about 97,000 persons in 1871 to over 0.5 million in 1931 (Figure 2.3).[50] The dominant component of population growth was migrational surplus for natural increase was negative prior to 1921, partly because the gross imbalance

FIGURE 2.2
Decennial Percentage Change of the Population of
Singapore Settlement, 1871–1931

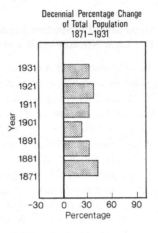

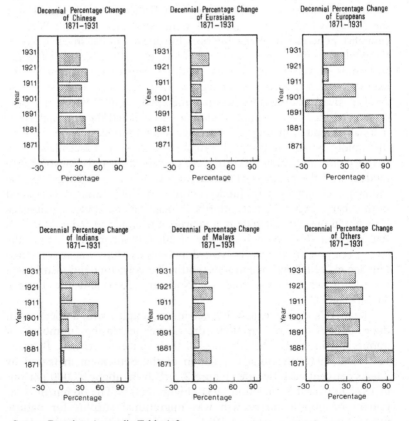

Source: Based on Appendix Table A.2.

FIGURE 2.3
Population Growth of Singapore Settlement, 1871–1931

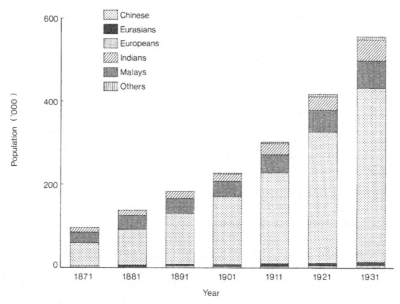

Source: Based on Appendix Table A.1.

in Singapore's sex ratio in the nineteenth and early twentieth centuries (Table 2.1) depressed crude birth rate to a low level and also because of the prevalence of high crude mortality rates (Table 2.2).

By 1881, 68.5 per cent of the total population resided within municipal limits, a proportion which was substantially increased to 83.2 per cent in 1891 partly as a result of the expansion of the municipal limits to include eight additional outer census divisions (Figure 2.4).[51] This redrawing of municipal boundaries was a reflection of the rapid physical expansion of the built-up area in the last quarter of the nineteenth century as demographic pressure on urban land and housing within the central area of the town reached unprecedented levels. Up to the 1870s, the town and its residential suburbs had remained circumscribed within a 2–2½ mile radius from the Singapore River. By the 1880s, however, it began to expand beyond these confines, initially as ribbon developments along the main trunk roads such as Tanjong Pagar Road, River Valley Road, Orchard Road, Bukit Timah Road, Thomson Road, Serangoon Road, and Kallang Road. New middle-class residential suburbs were established to the west of the town between River Valley and Thomson Roads to accommodate wealthier families fleeing the environmental malaise and congestion of the central area. The bulk of the labouring classes, however, remained confined to the highly congested inner districts where accessibility to work could be maximized and transport costs minimized. The degree of population

TABLE 2.1

Distribution of Population by Sex in the Settlement of Singapore,
1871–1931

Year	Males	Females	Sex Ratio[*]
1871	72,183	22,633	3,189
1881	104,031	33,691	3,089
1891	138,452	43,150	3,209
1901	169,243	57,599	2,951
1911	215,489	87,832	2,453
1921	280,918	137,440	2,044
1931	352,167	205,578	1,713

[*]Males per thousand females.

congestion in the inner districts as compared to the outer is reflected in
Figures 2.5 and 2.6, the former in terms of areal densities and the
latter in terms of house densities.

The population residing within municipal boundaries of which, at
the turn of the century, 74 per cent were Chinese belonging to various
dialect groups (the Hokkiens, Teochews, Cantonese, Hylams, and
Hakkas being the five main *bang*), reflected a highly cosmopolitan
flavour, including sizeable minorities of 'natives of the Malay
Archipelago' (14 per cent, made up chiefly by Peninsular Malays,
Javanese, Boyanese, and Bugis), 'natives of India' (8 per cent, comprising mainly Tamils, Bengalis, and Parsees), Eurasians (2 per cent),
Europeans (1 per cent), and 'other nationalities' (1 per cent, made up
principally by Arabs, Jews, Sinhalese, and Japanese).[52] Unlike in certain
colonial cities in British India and tropical Africa where a strict system
of racial segregation with 'the formation of a European quarter as
distant as possible from the native huts' was prescribed as being 'far

TABLE 2.2

Birth and Death Rates in the Settlement of Singapore, 1881–1930

Period	Crude Death Rates	Crude Birth Rates
1881–5	32.9	15.0
1886–90	37.3	16.4
1891–5	35.4	17.3
1896–1900	39.4	17.1
1901–5	47.1	21.2
1906–10	43.2	22.3
1911–15	37.6	23.9
1916–20	35.2	26.5
1921–25	29.1	29.5
1926–30	29.0	34.3

Source: Saw Swee Hock, 'Population Trends in Singapore, 1819–1967', *Journal of South
East Asian History*, 10 (1969): 46.

FIGURE 2.4
Municipality of Singapore, Census Districts

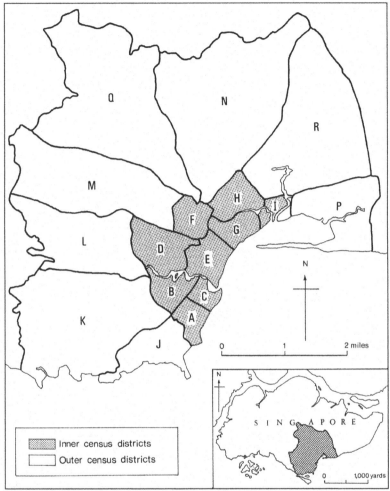

more effective than any other prophylactic measure' against diseases,[53] the government in Singapore was wary of any overt form of residential segregation for fear that it might alienate Chinese capital crucial to the production of the urban built environment. Walter Napier, a prominent solicitor and member of the Legislative Council, explained why the provision in the Municipal Ordinance Amendment Bill of 1903 to give the Municipal Commission power to prescribe the class of buildings to be erected in specified streets and areas and hence confer the power to control the degree of segregation of land use and races had to be struck out.

FIGURE 2.5
Population Density by Census Districts, 1901

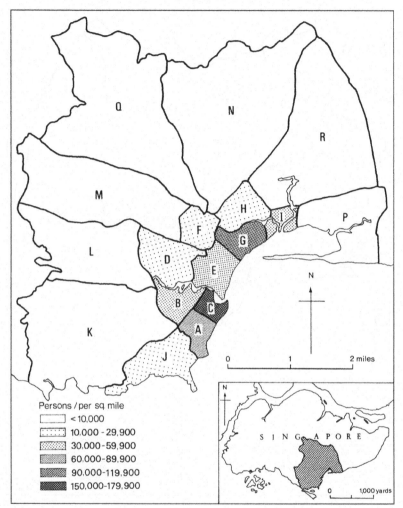

Source: Based on Appendix Table A.7.

It would probably be said that Orchard Road, out from Tanglin Kechil, is a residential quarter for Europeans, and ought to be kept as such, and it might be thought that a by-law should be made prohibiting shophouses ... [but] ... [w]e have to look at what effect it would have on the general prosperity of the place.... [It] would create a feeling of insecurity in property and the persons, Chinese and others, who had brought money into the place, in order to invest it in land, would feel that security which they always felt before, had gone, and they would never have the same inducement to buy property.[54]

There was, however, a tendency towards territorial concentration by race and dialect group partly facilitated by the original settlement

FIGURE 2.6
Average Number of Persons per House by Census Districts, 1901

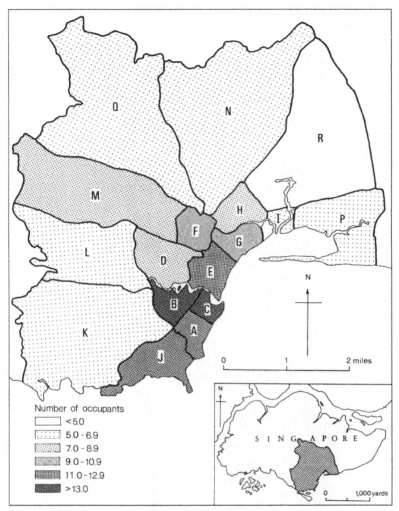

Source: Based on Appendix Table A.7.

pattern dictated by Raffles's 1822 plan to locate different races, dialect groups, and trades in different *kampung*[55] (Figure 2.7),[56] and partly because of the tendency of migrants to gravitate towards districts where the institutional support structures such as clan-based welfare and religious institutions or the control of particular occupational niches belonging to one's group had already been established.[57] As seen in Figures 2.8 and 2.9, whilst the Chinese tended to be fairly ubiquitous in all census districts, their highest concentration where they commanded over 85 per cent numerical supremacy was in Districts A, B, and C, where 'old' Chinatown was originally located. A

42

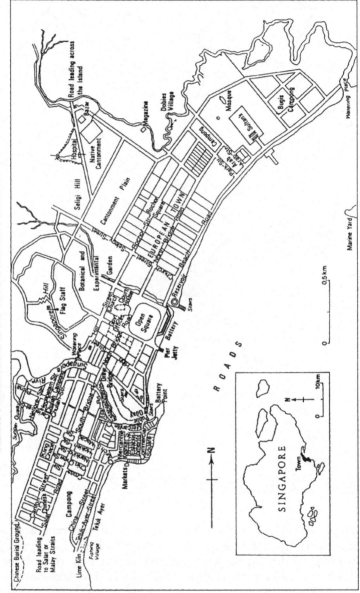

FIGURE 2.7

Plan of the Town of Singapore by Lieutenant P. Jackson, Drawn According to
Raffles's Instructions to the Town Committee of November 1822

FIGURE 2.8
Municipal Racial Composition by Census Districts, 1891 and 1901

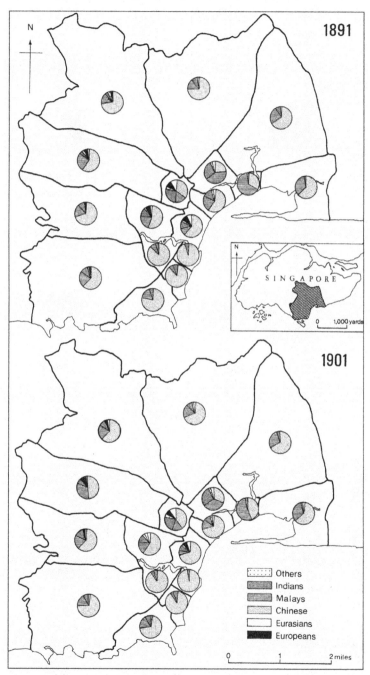

Source: Based on Appendix Table A.4 and A.5.

FIGURE 2.9
Percentage Distribution of Principal Races by Census Districts, 1901

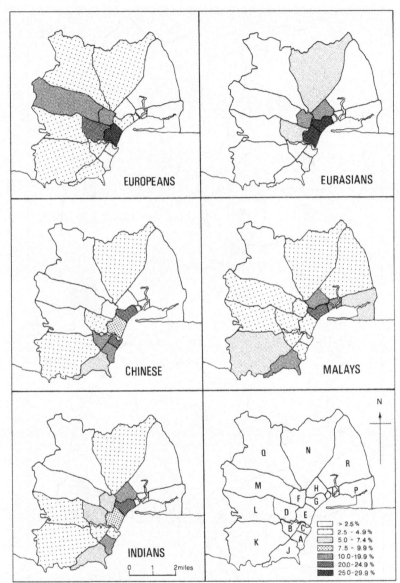

Source: Based on Appendix Table A.6.

second area where the Chinese overwhelmingly predominated was the area directly north of the Singapore River—Districts E (the old 'European town') and G (Kampong Glam, traditionally the domain of the Malay Sultan)—which formed an overspill area for Chinese immigrants across the river from their traditional core area. District J, the port area, also constituted another overspill area adjacent to 'old' Chinatown as well as a convenient location for Harbour Board coolies and other port workers. In contrast to the Chinese, the Malay/Indonesian population tended to concentrate in *kampung* on the northern edge of the city, on poorly drained land fringing the tidal swamps around the Kallang and Rochor Rivers (Districts G, H, I, and P in Figures 2.8 and 2.9). The location of the Sultan Mosque in this quarter (District G) also formed a powerful focus for the Malay Muslim population. The Indian population were mainly focused on the Serangoon Road–Kampong Kapor area (Districts F and H in Figures 2.8 and 2.9), although smaller pockets could also be found in various parts of the city such as concentrations of South Indian chettiars, Tamil Muslim traders, money-changers, petty shopkeepers, boatmen, and quayside workers on the fringe of the central business core in the Kling Street–Market Street area; Sindhi, Gujerati, and Sikh textile merchants in the High Street area; Gujerati and other Muslim textile and jewellery merchants in the Arab Street region; and Tamil, Telugu, and Malayali dock workers in the harbour area.[58] The Eurasian population were located to the north of the Singapore River in Districts E, F, G, and H (Figures 2.8 and 2.9), whilst the Europeans, originally allocated land in 'European Town' (Districts E and G) by Raffles, had increasingly moved to suburban locations, particularly the Tanglin–Claymore–Lower Bukit Timah area (Districts D, F, and M in Figures 2.8 and 2.9). Here were predominantly European enclaves where 'the white walls and red-tiled roofs of villas and bungalows, usually built with cool, arched alcoves and arcades around them' set off by the 'luxuriant verdure' of the suburbs stood sequestered in large gardens and compounds.[59] These residences were the homes of European merchants and government officials. Occasionally, some were turned into lodging houses for mercantile assistants as in the case of William Geoffrey, Edmund Hughes, and Robert Cameron, who found themselves established in Radcliffe Lodge in Tanglin, 'a comfortable and beautifully situated residence' with 'an excellent view of the country', 'neatly kept grounds of about three acres' and 'a large tennis lawn in front of the house'.[60]

One indication of the level of residential segregation between racial groups is given by indices of dissimilarity between the spatial distribution of different races.[61] Table 2.3 shows that the highest level of residential segregation occurred in comparing the distribution of the European population *vis-à-vis* those of the Asian communities (Chinese, Malays, Indians, and 'other nationalities') as opposed to comparisons between Asian communities. This is a reflection of the contrast between Europeans who favoured spacious suburban residential

TABLE 2.3
Interracial Residential Segregation as Measured by Indices of
Dissimilarity at Census District Level, 1901

Racial Group	Europeans	Eurasians	Chinese	Malays	Indians	Others
Europeans	–	34.6	56.6	53.3	48.1	48.5
Eurasians		–	50.9	39.5	31.2	25.7
Chinese			–	39.4	26.2	46.5
Malays				–	27.5	38.1
Indians					–	30.1
Others						–

Source: Calculated from Appendix Table A.6.

locations set apart from the place of work in the commercial area and the bulk of the Asian plebeian classes who were concentrated in the central areas in shophouses which combined the functions of work and residence.

Not only was Asian immigrant society in Singapore divided into distinct 'races', but each 'race' was further fragmented into several sub-groups. The Chinese, for example, were divided into Hokkiens from the southern area of the Fujian province; Teochews from Chaozhou province; Cantonese from the central Guangdong province; Hakkas or Khehs who originated from a number of prefectures in the Fujian and Chaozhou provinces; Hylams from Hainan Island; and a number of smaller groups, including Hokchius from Fuzhou in eastern Fujian. Most of these dialect groups were concentrated in Chinatown (Districts A, B, and C) and Kampong Glam (Districts E and G) (Figure 2.10). The only significant exception was the Hylams whose distribution tended to follow that of the Europeans, largely because they were mainly employed as domestic servants in European households. Cutting across territorial and linguistic claims on their loyalty were family clan organizations or *kongsi* which were organized for mutual protection and benefits, triad societies organized according to secret rules, and industrial guilds which monopolized particular trades and occupations.[62]

By the turn of the century, the principal components of colonial urban morphology were clearly imprinted on the built form of the city. A Western-style central business district, Commercial Square, later renamed Raffles Place, with its assemblage of principal banks, trading agencies, merchant houses, the post office, and the shipping office thrived as the commercial heart of the city. Across the Singapore River from Commercial Square was the government administrative quarter with its collage of government offices, the Supreme Court, the Town Hall, the Anglican and Roman Catholic cathedrals, the chief European hotels, and English schools. On the seaward fringe of this quarter was the Esplanade, which formed 'a favourite resort of [European] residents'.[63] In distinct contrast to these imposing edifices symbolic of

FIGURE 2.10
Percentage Distribution of Principal Chinese Dialect Groups
by Census Districts, 1901

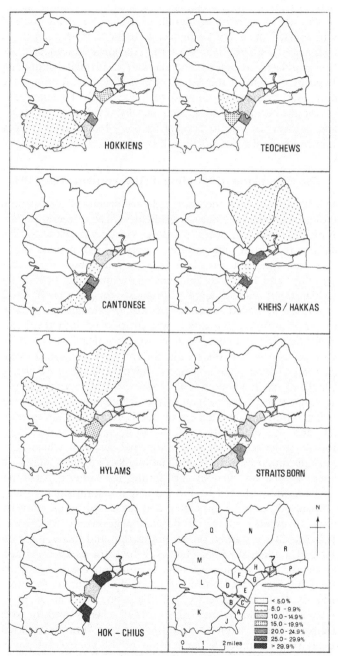

Source: J. R. Innes, *Report of the Census of the Straits Settlements Taken on the 1st March 1901,* Singapore: Singapore Government Printers, 1901, pp. 21–37.

British commercial capital and government, the Asian districts were complicated mosaics of specialized trade areas, bazaars, densely packed tenement housing, and concentrations of eating houses, theatres, and brothels 'as close together as the teeth of a comb'.[64] Whilst racial segregation was not as extreme or ingrained as in some of the colonial port cities in India,[65] there was recognizable separation between the suburban bungalow-type residences of the European colonialists and the densely populated mixed commercial and residential landuse districts inhabited by the bulk of the Asian communities. The framing of a form of municipal government tailored to the control of such a commercially vibrant, rapidly growing, and racially divided city with its contrasting landscapes was to emerge as one of the major concerns of those in power in the late nineteenth century.

The Expansion of Municipal Government, 1887–1930

In the 1880s, faced with the unprecedented strains of urban and demographic growth, the colonial government sought to increase the scope of municipal powers in order to enforce stronger measures of environmental control. From time to time, the local English press echoed the opinion that '[t]he great need here [was] to bring about big conceptions of civic duty and final abandonment of the infamous policy of *laissez faire* which ha[d] made Singapore the pigsty that it [was]'.[66] On 2 May 1887, a draft bill hailed as one which would 'inaugurate a new and ... better era in our Municipalities' was introduced into the Legislative Council to amend and consolidate existing municipal laws.[67] The principal object of the bill was to put an end to the anomalous state of affairs under which rates and taxes which were mainly expended for the benefit of the inhabitants of the town included levies on those living in the country districts.[68] Much of the debate centred on questions of municipal autonomy, the degree of public representation in municipal affairs, and the delimitation of the municipal franchise. Among a number of proposed changes in municipal law, that which raised the fiercest opposition from the unofficial members of the Council and the European public was that of substituting the hitherto part-time (and unpaid) president elected by the municipal commissioners themselves with a paid officer to be appointed and dismissed by the Governor-in-Council. In defence of the proposal, the government explained that given the rapidly expanding town population and increasingly onerous pressure of municipal responsibilities, a full-time executive officer who could devote his whole attention to municipal work was urgently required. Government control of the appointment of the president, claimed the Colonial Secretary, J. F. Dickson, was 'consistent with its duty to secure equally the rights of all classes and to keep the Municipality free from corruption'.[69] The unofficial members, however, saw the move as an erosion of the control ratepayers had over municipal affairs, as 'anti-reform'.[70] Thomas Shelford, the leading unofficial member, strenuously argued that 'the Board represent[ed] the ratepayers; they deal[t] with the funds of the ratepayers, and it seem[ed] only fair that the Principal Executive Officer should be selected by them.... To do otherwise was virtually to hand over the Municipality to the Government'.[71]

In addition, the unofficial members also objected to the Governor's powers to veto any item on the municipal budget proposed by the commissioners and to fix the maximum rate of taxation for they 'reveal[ed] the same spirit of distrust in the Board as [was] found throughout the Bill'.[72] They also contended that the bill was regressive for whilst it retained the principle of representation, the proportion of members nominated by the Governor was increased to at least half or more (in the case of an odd number) of the Municipal Board.[73] One of the unofficial members, J. Allan, described the bill as offering the municipal commissioners 'a shadow, while the substance [was] retained by the Governor' and went as far as to argue that he preferred 'to see the municipalities converted into Government Departments pure and simple' than to see such a repressive bill become law.[74] On its part, the government protested that it had no intention of interfering unnecesarily with municipal affairs but maintained that government control of the budget was essential to secure the confidence of capitalists in the matter of loans.[75] The Attorney-General, J. W. Bonser, argued that government interest had a right to be well represented as it contributed to 'the greater part of the support for the Police and the maintenance of Schools, Hospitals and Gaols—charges which in other towns [fell] on the Municipalities'.[76] The Colonial Secretary added that if there was any distrust to be deplored, it was that which the unofficials harboured against a government which was 'the embodiment of law and order', comprising men 'trained to administration; [and] who [had] no interests to serve, save those of good administration'.[77] Strong centralization of authority was crucial, he continued, for only the Governor, 'by the careful use of "the iron hand with the velvet glove" ... [could] hold together the different races and classes and creeds of which the community [was] composed'.[78]

Equally heated was the debate on the government's proposal to impose a fine of $500.00 on any commissioner who declined to serve on the Board without good reasons. Agreeing that it was no easy task to induce men of calibre to serve on the Board, Shelford reminded the government that its intention to 'cripple [the Board] in every direction ... by placing over them a Government officer' and 'by rendering well nigh every action, every duty they perform subject to the approval of the Governor' would further stifle whatever little public enthusiasm there was for municipal service.[79] The imposition of fines to force compliance was the crudest of devices, and Shelford stressed that the unofficial members '[would] have nothing to do with it'.[80] In his words, 'We will not join you [the official members] in your role of schoolmaster, whipping men through their pockets, oblivious apparently of the motto which tells us you may drag the horse to the brook but you cannot make him drink.'[81]

A proposed change originating from the municipal commissioners themselves was that of limiting the elective franchise by raising the qualifications of voters from the payment of the minimum rate of 25 rupees to $25.00 per annum, or the occupation of a house of the minimum annual value of $250.00. J. P. Joaquim, an unofficial member who was also a municipal commissioner, explained that the proposal was prompted by the difficulties involved in keeping track of changes within the body of small householders and in making reliable voters' lists which tended to become obsolete by the time they were published.[82] The Attorney-General also defended the proposal on the grounds that a smaller electorate might increase voters' commitment to vote as they might feel that their individual votes were of importance. The franchise could be extended to draw in others when this minority had been 'educated' to value their 'privileges'.[83] The official conviction echoed the commonly held view that electoral privileges should not be granted unless they were duly appreciated, and that 'population and property should go hand in hand in the power to exercise political rights; otherwise mere numbers, influenced by every transient passion, drown the voice of moderation and reason'.[84] Shelford, however, argued that not only was the raising of the franchise 'opposite to the spirit of legislation in every other part of the world', it further alienated the already 'apathetic and ignorant' poorer classes whose co-operation was necessary for successfully carrying out the provisions of the bill.[85]

Less controversial than the question of municipal autonomy, the degree of public representation, and the nature of the franchise was the overall objective of the bill to 'cloth[e] the Municipality with extensive powers as a sanitary authority'.[86] As with Victorian cities in Britain, municipal reform was closely identified with the improvement of public health and sanitation through a commitment to environmental control.[87] The editor of *The Straits Times* welcomed the 'ample provision' in the bill to enforce stringent sanitary rules and expressed hope that 'the reformed Municipality [would] with the proverbial thoroughness of new brooms sweep away the hindrances in the way of sanitary reform to the fullest extent'.[88] The elastic powers invested in the commissioners 'to act promptly and effectively in the cause of public health or in the interest of public morality' were redeeming features of the bill which 'alone contributed largely to counterbalance, in the public view, many of its objectionable clauses and modify the acrimony which they gave rise to'.[89] To press on with sanitary reform 'with unswerving tenacity', alleged the press, was a vital municipal task given the 'utter indifference [of the Asiatic population] to any conditions of sanitation, and the filth in which they [were] accustomed to revel'.[90]

So firmly and ably did the unofficial members present their case against some of the proposals contained in the draft bill that in the legislature of 14 July 1887, several concessions were made by the government including limiting the Governor's control of the municipal budget to only expenditure for works costing over $10,000; lowering the minimum qualifications of voters to the payment of rates of

$12.50 per annum, or the occupation of a house of annual value of $150.00; and the withdrawal of the proposal to fine unwilling commissioners.[91] The government was, however, adamant that the clause providing for the appointment and dismissal of the president by the Governor-in-Council was carried. As the Governor, Sir Frederick Weld, explained to the Secretary of State for the Colonies, 'looking to the possible preponderance of Chinese wealth and influence in these Settlements, [he] was not prepared to give way on this point'.[92] It has been speculated that the main reason why the government objected to a paid, elected president was its fear that this might lead to the domination of the Municipal Board by wealthy Chinese ratepayers who might utilize the Board for their own ends.[93]

Three and a half months after its introduction into the Legislative Council, the municipal bill was passed and scheduled to come into effect on 1 January 1888 as Ordinance IX of 1887.[94] The municipality was divided into five electoral wards—Tanjong Pagar, Central, Tanglin, Rochor, and Kallang (Figure 2.11)—and arrangements made for elections to be held in January 1888. The post of president of the Municipal Commission was offered to Dr Irvine Rowell[95] whom Weld described as 'the best man [he] [could] find for the post ... [given] his [Rowell's] aptitude for organisation and his energy in carrying out sanitary improvements'.[96]

The new municipal law provoked a flurry of petitions from residents and ratepayers of the Straits Settlements to the Secretary of State for the Colonies, Sir H. T. Holland, requesting certain amendments to the new law. The first to reach the Secretary of State was that from the Straits Settlements (Penang) Association which urged the suspension of the bill until it had been made intelligible to the people by translation into' the principal languages of the Colony.[97] Holland agreed to await the arrival of further representations from the inhabitants of the Straits Settlements and telegraphed the Governor to postpone the implementation of the municipal ordinance until 1' July 1888.[98] Sir Cecil Smith, the new Governor who succeeded Weld in October 1887, however, managed to persuade Holland that the implementation of the ordinance could not be postponed as this would cause great practical difficulties:

It would cause great inconvenience owing to the arrangements already made with regard to the elections and other details.... It cannot be expected that a moribund body like the existing Commissioners can do justice to the very important duties with which they are entrusted, nor with any prospect of public benefit could it be anticipated that a new set of Commissioners would be able to enter on their functions in the middle of the year with new machinery for working the Municipality and with Estimates partly used up.... It will [also] involve the postponement of pressing sanitary improvements and other important public works.[99]

The ordinance came into effect in March 1888 but this did not stop the stream of petitions requesting amendment. The Singapore memorial which reached the Secretary of State in February appeared to be a

FIGURE 2.11
Municipal Electoral Wards, 1887–1913

purely Asian initiative. In his comments on the memorial, Smith
observed that 'there was not the signature [out of a total of 490 col-
lected] of a single member of the European or Foreign Mercantile
Community attached to the memorial' and given the 'very little support
it ha[d] received from the European Community', he had 'known
nothing about it'.[100] Among the issues raised were that of the nominated president, the practice of assessing property based on gross
rather than nett annual value, the inclusion of water rates within a
maximum consolidated rate of 15 per cent rather than separate assessment for property and water, the rating of vacant lands and empty
houses, and the substitution of house-to-house collection of rates by
half-yearly payment at the municipal office. Whilst the bulk of the petition reflected the concern of Asian property owners against heavy or
escalating assessments, the last-mentioned issue also impinged upon
the poorer classes. The memorialists claimed that the change in the
system of payment of rates, apparently impelled by the frequent
embezzlement of monies by collectors employed by the municipality,
would 'inflict hardship upon a large number of poor people, ... com-

posed as they [were] of varied nationalities and possessing calendars differing from the English and from each other'.[101] As a result, the new system '[would] prove a great hardship to owners ignorant of the English language and thus unable to know when their assessment [was] due, and on women too poor to employ Agents, who [would] be bound to break through their customs of seclusion and present themselves at the Municipal Office, or suffer the consequences'.[102]

Memoranda were also received from the London-based Straits Settlements Association, largely accusing the Straits government of trying to get its hands on 'the Municipal Funds, their expenditure and the real administration of Municipal affairs' by means of a bill which 'bristle[d] throughout with references to the Governor'.[103] Concern was also expressed regarding the Asian response to an ordinance with extensive implications for the conduct of everyday life. T. Cuthbertson, a former unofficial member of the Straits Settlements Legislative Council, pointed out that as the bill was not translated and published in any of the 'native' languages, there were no means by which the 'native' community in the Straits could acquaint themselves with its provisions. Had this been done, claimed Cuthbertson, 'there would have been a very distinct expression of opinion on the part of the native community adverse to the Bill'; instead, 'a Bill which [was] of special interest to the non-English speaking inhabitants of the Colony ha[d] been passed without their having an opportunity of understanding the changes which were being introduced'.[104] Another former unofficial member of the Council, S. Gilfillan, criticized the bill for the 'petrify[ing] multiplicity and minuteness of its provisions' which restricted the 'free action of the Municipal Council', and added that this would deepen Asian suspicion that the Council was merely another 'new department of the Government' rather than a representative body.[105] In response to the petitions[106] received, however, the Secretary of State saw no reason to rescind the ordinance but agreed to review it after sufficient time had elapsed 'for experience of its merits and defects'.[107]

The 1887 Municipal Ordinance formed the nucleus of all subsequent amendments in the next half century and although it underwent considerable elaboration as municipal government steadily enlarged its spheres of influence, imposed new duties, and assumed new obligations, the basic pattern of municipal affairs set out by this ordinance continued to persist. In November 1889, an amending ordinance was introduced, making small concessions to the public such as giving the municipal commissioners an increase of power over their president and extending the municipal franchise by including on the electoral roll joint occupiers of premises, or those who jointly pay rates to an amount sufficient to qualify them all if equally divided between them.[108] The social purpose of municipal reform was, at this stage, still focused on sanitary control of the urban environment, rather than a wider concern for the cultural and recreational needs of the population or what Helen Meller called 'civilization'.[109] A motion introduced to extend the powers of the municipal commissioners to encompass the provision of public recreation and entertainment was strenuously debated, with half the Legislative Council opposing it on the grounds that municipal energy should be expended on the onerous task of improving the sanitary conditions of the city rather than 'extraneous matters' such as recreation and amusement.[110] The proposal was ultimately carried on the strength of the Governor's casting vote, on the understanding that only a limited portion of municipal funds should be spent on the provision of public recreation.[111] The municipal commissioners themselves, however, remained wary of taking on board additional responsibilities and in 1894 negated the government's proposal to add the charge of maintaining public libraries to their duties.[112]

It was not until 1895–6 that another bill was introduced to amend and consolidate municipal law (Plate 2). One of the principal alterations was the inclusion of a large number of provisions culled from English Public Health Acts relating to the removal of obnoxious substances, the pollution of rivers and streams, sewage schemes, common lodging houses, and mandatory orders for the removal of nuisances.[113] Again, the main debate revolved around the question of municipal autonomy. The clarion call was first taken up by the Singapore Branch of the Straits Association which petitioned the Governor to relinquish certain executive powers of the Legislative Council, allow the municipal president to be elected by the commissioners themselves, divest himself of veto control over minor municipal salaries, and refrain from interfering with any decisions of the commissioners unless he had reason to believe that the commissioners were acting from improper motives.[114] During Council debates, Shelford expressed his disfavour that there were no new concessions granted to the commissioners beyond the power proposed to allow the Board to fix rates for the use of shops, stalls, sheds, pens, and standings in public markets without reference to the Governor.[115] He argued that the increased conservancy powers proposed by the bill imposed on the commissioners greater and more difficult responsibilities, which warranted 'some increased liberty of action [to] be given to their administrative power'.[116] However, he conceded that nominated presidents had hitherto upheld 'Municipal rights and Municipal privileges in a free and independent spirit' and that given signs that 'the Asiatics [would] before long, be largely represented on the Municipal Board', he no longer supported the call for an elected president but urged that commissioners should be consulted before the Governor made an appointment.[117] Shelford's amendment was accepted although no other constitutional changes were made and the bill, passed in late October 1896,[118] remained on the statute books until the end of the first decade of the twentieth century.

The question of municipal constitution was reopened in 1909 during a session of the Legislative Council. Hugh Fort, an unofficial member of Council, expressed 'grave misgivings' over the conduct of municipal affairs in general and the success of the ward system of elections in particular, and advocated setting up a commission of inquiry to look into the administration of municipal affairs.[119] His proposal was agreed to and a municipal enquiry commission chaired by J. O. Anthonisz was appointed in September 1909. The report of the commission of inquiry, laid before the Legislative Council a year later, was highly critical of the system of ward elections: 'It can hardly be said that the experiment of carrying on the municipality on representative lines has proved to be a success. [This is evidenced by] the small interest taken in municipal matters by the general body of voters, the frequent failures to get the necessary number of votes even when nominations take place and the dependence on Government which is generally shewn in the matter of Municipal representation.'[120]

It also found that 'there [were] few men of leisure in the Colony competent for the work or men who [were] able to devote the time necessary for the discharge of the numerous duties imposed on the commissioners by the Ordinance' and recommended that the ward system should be abolished and replaced by a Board comprising entirely of members nominated by the Governor but representative of 'prominent communities in the Municipality'.[121] The Board was to play a purely advisory role: it was 'to be in a position to advise on general policy of the Municipality and to consider supplies' but 'should not be allowed to interfere in the discharge of the ordinary routine duties of the Municipal staff'.[122] In August the following year, a new municipal bill was introduced into the Council, mainly to give effect to some of the recommendations contained in the commission of inquiry report.[123] It was to unleash a storm of protest on a massive scale over the proposed attempt to abolish commissioners and ratepayers' right to elect representatives to control expenditure of public monies paid in for the purpose of municipal administration. On 1 September, a meeting of ratepayers held in the Chinese Chamber of Commerce with Manasseh Meyer[124] in the chair unanimously passed a resolution against the proposed bill 'which create[d] a dangerous and revolutionary change in the present system'.[125] A petition was presented to the Governor drawing his attention to the fact that the proposed advisory board would be 'no adequate safeguard for the interests of ratepayers' as it had few 'real powers', and pointing out that in an era when 'the principle of local self-government ha[d] been extended in the British Empire ... it [was] somewhat astonishing that a contrary and reactionary principle should be applied to one of the most important seaport towns in the British Dominions'.[126]

In February 1912, the 1911 bill was withdrawn 'in view of the general opposition to the clauses which altered the constitution of municipalities by providing for the appointment of advisory boards in the place of Municipal Commissioners' and a new one introduced nine months later.[127] This bill retained the municipal commissioners, to number between five and fifteen, as a corporate body but removed the elective principle altogether. Instead, the power of nomination of all members of the Commission was to be vested in the Governor, whom the Attorney-General described as the only authority who had at his disposal 'the counsel of both European and Native bodies' and 'the information as to who [were] fit and proper persons qualified to sit as Commissioners'.[128] This time the bill was surprisingly well received by most of the unofficial members who 'congratulated [the Government] on the methods adopted with regard to the constitution of the Municipalities' and agreed that the Governor was in the best position to appoint suitable men to carry out municipal functions.[129] While pressing the point that the Governor should consult leading ratepayers before making appointments, the Chinese unofficial member, Tan Jiak Kim, concurred that ratepayers had forfeited their 'privilege of election' by virtue of their 'apathy'.[130] Only A. Huttenbach, the unofficial member from Penang, rose to give a spirited defence of municipal elections. He countered the assertion of 'Asiatic apathy' by pointing to the example of municipalities in Indian townships, where there was ample proof that 'natives [would] take great interest [in municipal affairs] provided they ha[d] a real voice in the management'.[131] He stressed the educational role which elections played: 'the school of Municipal life', he claimed, would fill 'the blank which [was] felt after people left school'.[132] World events too seemed to be tending towards progressive liberalization with 'Russia given a Parliament, the Chinese [in Java] the same rights as Europeans, ... [and] the revolution in China bring[ing] the people more voice'.[133] Finally, he appealed to 'Imperial [pride] and Colonial utility':

The British Empire has the reputation for giving, and claims wherever it governs to give, a voice to the most humble races, thereby raising their status and gradually giving them citizen's rights and responsibilities. This reputation has become a most valuable asset, its moral force at times is most powerful. Due solely to the wire-pulling of a few Europeans, this great policy is now for the first time to be reversed.... As regards Colonial utility, what most attracted people in the past was the belief that they enjoyed privileges here which they could not get elsewhere.... [Taking away elective freedom] will affect the popularity, and thereby the future, of the Colony.... We are here for the sake of peace, order and good government. This road might lead to ... silence, peace and order of—the graveyard.[134]

When Huttenbach moved an amendment to reinstate all election clauses, he was supported by Dr D. J. Galloway, who felt that among the better educated Asians, 'whatever lack of interest there [might] be in municipal elections there was no lack of interest in municipal affairs' and that the failure of elections lay at the door of the 'machinery', that is, the system of voting by wards in a community where 'the mere fact of living in some propinquity could [never] over-ride the claims of religion and race'.[135] The ensuing debate on the question whether the elective principle was suited to or appreciated by the Asian population, however, quickly lost ground when Tan Jiak Kim interposed to add the backing of both the Chinese Advisory Board and Chinese Chamber of Commerce to the government's proposal to adopt a complete nomination system.[136] Huttenbach's amendment was hence lost and the bill embodying the abolition of municipal elections and its substitution by government nomination was passed on 18 April 1913 as Ordinance VIII of 1913.[137] With few significant alterations, it was to remain the frame of municipal law in Singapore until the outbreak of the Second World War.

In the 1920s, the most significant changes in the constitution of the Municipal Commission lay in the increase in the number of commissioners and the partial delegation of the power of nomination to certain public bodies. Under the amending Ordinance XXIX of 1920, the maximum number of commissioners was increased from fifteen to twenty-five in order to facilitate the division of the Municipal Board into committees to address specific administrative duties.[138] In January 1921, as a sequel to this amendment, the Governor, Sir Laurence Guillemard, proposed 'a purely tentative measure' of asking certain public bodies to nominate members to serve on the Commission, subject to the Governor's approval.[139] These included the Straits Settlements (Singapore) Association (three seats), the Singapore Chamber of Commerce (two seats), the Chinese Chamber of Commerce (two seats), the Straits Chinese British Association (one seat), the Eurasian Association (one seat), the Mohammedan Advisory Board (one seat), and the Hindu Advisory Board (one seat). This proposal found much favour and was applauded as a means of increasing the representativeness of the Commission without the pitfalls of an electoral system. It gave one commentator ground to assert that 'the history of the Singapore Municipal Commission ... [was] the history of a steady increase in the number of Commissioners, giving a greater measure of representation and bringing a wider viewpoint to bear on matters municipal'.[140]

The renunciation of the right of publicly electing representatives to municipal government was generally hailed as an abandonment of 'all pretence of popular representation [which] had never been more than a fiction and had led to abuses'.[141] The general populace had forfeited their right to a voice in municipal affairs by virtue of their apathy and ignorance. In general, 'the voters probably neither [knew] nor care[d] about the idea of sanitary improvement, which [was] the mainspring of municipal action. They show[ed] their small interest by the low number of votes, and their representatives [were] usually animated principally by the desire to vote against anything which costs money.'[142] On a number of occasions, candidates who stood for ward elections had failed to secure more than twenty votes, the requisite minimum required by the Municipal Ordinance to constitute a valid election.[143] As such, it had become 'the habit of the Governor, after such farces ... to step in, exercise his prerogative, and "appoint" the non-elected nominee to the vacant seat'.[144] According to *The Straits Times*, the experiment with representative government had failed because circumstances were not ripe.

It cannot be said that the British have been unwilling to adopt the principles of free representative Government in their Colonies and dependencies. They have erred, if at all, in making experiments with such Government too readily.... The existing [electoral] system stands condemned not only by reason of the deplorable corruption which has grown up under it, but by the waste of energy, the overlapping of powers, and the unsatisfactoriness of its results.[145]

It was felt that a system of nomination would redress some of these ills and work for the good of the municipality. With such a system where

the Commissioners [were] not ... dependent for their seats to what passe[d] for public opinion, the proceedings of the Board [would be] characterised especially by an absence of log-rolling influence.... The President, being the only whole-time member of the Board, [would be] the source of policy and since he ha[d] no supporters, it [was] necessary that his proposals be sound if they [were] to be accepted by a Board which [knew] no parties and work[ed] only for the good of the town.[146]

'Public opinion' in a city largely inhabited by Asians was hence considered highly suspect. There was to be no return to a system of election by ratepayers[147] and for the following decades, the Governor himself (and those he chose to consult), remained the final arbiter of who were 'fit and proper persons' to be recruited into the realm of city government.

The Composition of the Municipal Commission

As seen, the history of municipal constitutional reform largely reflected the tensions between granting greater local autonomy in the administration of urban affairs and the fears that this would lead to the domination of Asian interests in city government. Even those who favoured extending municipal government by placing its constitution on a more popular basis conceded that a policy of liberalization had to be kept within certain constraints 'in a Colony such as this, consisting largely of Eastern people'.[148] While it was under the governorship of Weld that the first moves towards a ward electoral system was made, Weld himself held reservations as to whether the 'Asiatic psyche' was receptive to a representative system of local government. In June 1884, he declared:

I have always advocated self-government and self-reliance in English-speaking communities, and have ever held that ... [a] united British Empire, was best obtainable through local self-government and local self-reliance, yet I am not one of those *doctrinaire* statesmen who believe that what is good for Englishmen in an English community is good for all races, at all times, and under all circumstances.... Personal government is ... a necessity for Asiatics; it is the outcome of their religious systems, of their habits of thought, and of long centuries of custom. In municipal elections it is inconceivable how little interest is taken by even rich Asiatics.[149]

The Governor's power of veto in order to 'retain a hold over the purse strings of the Municipalities, ... restrain unworthy contracts ... and purify [municipal] administration' was accepted as necessary given the presence of large 'Eastern and Asiatic populations' in the colony.[150] Whilst much of the debate regarding municipal government had pivoted on the question of public representation and autonomy, it was implicitly assumed that the question of representation applied solely to Europeans and the incorporation of a limited number of Asian leaders. Such an assumption was often couched in terms of the 'fitness' of particular races to lead in municipal affairs as exemplified by the following statement from the *Singapore Free Press*: 'It is to the Europeans that the introduction into this colony of civic representative government is due, and it is to them that the native inhabitants of all races have a right to look for that leadership in these matters that they should so well [be] fitted to afford.'[151]

Fears were often expressed that widening the municipal franchise or relaxing government control of membership of the Municipal Board would result in the Board being swamped by 'a crew of Chinese and Arabs ... who were already holders of a vast proportion of landed property in Singapore'.[152] During a debate organized by the Singapore Debating Society in February 1887, the proposition contending for the abolition of the Municipal Commission justified the motion by claiming that the system of franchise based on rateable value inevitably led to 'an intolerable nuisance':

It is right and proper [argued the proposition] that [the Asiatics] should have a voice in public matters, and they are already well represented both on the Legislative Council and the Municipal Council.... But to make them absolute masters of the situation and to entrust them with the control of public monies is sheer lunacy. We should profit by the miserable effects experienced in India by the introduction of natives into Municipal bodies; it is notorious that this unwise and ill-considered step has been attended with the most evil results.[153]

While arguing that a 'progressive Colony' like Singapore should be chary of giving up municipal self-government, 'one of the most precious of a Briton's privileges', the opposition was equally adamant that the European majority had to be maintained on the Commission.[154] Pursuing a similar theme, the editor of *The Straits Times* cautioned that dividing the municipality into wards and arousing interest in municipal elections among 'the native ratepayers' was in the long run detrimental 'unless fenced in with safeguards and restrictions to ensure its proper workings'.[155] He asserted that

when power is put more within the reach of native landholders and houseowners, and they become fully aware of their having a determining voice in the incidence of the rates levied on them, self-interest will assuredly come into play. Asiatics set no store by sanitation and town improvement which to Western ideas are indispensable. The election of commissioners pledged to carry out their narrow-minded views and lessen the burden on the rates will put an effective stop to the march of sanitary and hygienic improvement.[156]

To meet such a contingency, *The Straits Times* favoured strengthening the countervailing power of government nominee members as 'an effective check upon any vagaries on the part of elected members'.[157]

Given the high ratepaying qualifications governing the eligibility to vote, only a small sector of Singapore society was able to participate in the arena of urban government.[158] Besides the primary conditions of being male and of adult age, those qualified for the municipal vote must either have paid rates amounting to a minimum of 12.5 per cent[159] for property situated in the ward, or occupied for six months a house within the ward of a minimum annual rateable value of $150.00. The electoral list drawn up in 1896, ten years after ward elections were introduced, contained approximately 1,440 names,[160] constituting less than 1 per cent of the total municipal population.[161] To qualify for election as a commissioner, the conditions were even more stringent: the candidate must be a male, British subject; have completed his twenty-fifth year; be able to speak and read the English language; reside within the municipality or within 2 miles of its limits; and either have paid rates for the half year amounting to $20.00 and upwards, or occupied a house of annual rateable value of not less than $480.00[162] In 1896, the number of 'persons duly qualified to be elected municipal commissioners' tallied up to less than 200.[163] However, despite strong disqualifiers, the Chinese remained highly influential at municipal elections by sheer force of numerical superiority. The results of the first elections held when the Municipal Ordinance of 1887 came into effect bore ample testimony to this:

In every case the gentlemen who were selected as fit and proper persons to represent the various wards of Singapore at the meeting of Chinese ratepayers which took place a week before the election, were precisely those names which stood at the top of the poll in the different ward lists.... The majorities in each case were so substantial as to completely destroy the idea that there was in any single instance even an approach to a genuine contest. The result of the election has prove[d] that all five seats are at the disposal of the Chinese whenever they choose to occupy them. Their numerical superiority in every ward makes this certain beyond doubt and it is therefore quite hopeless for any candidate to contest any ward in future if he have [sic] not already secured a considerable influence with the majority of the Chinese voters resident therein.[164]

The editor of *The Straits Times* warned that the ward system of elections would 'strengthen the Asiatic element on the Board', an outcome which he found 'disturbing' because 'with Asiatic commissioners, we may expect their conservative instincts to work in keeping to the old paths, they [having] not the faintest idea of the value of sanitation and hygienic reform'.[165]

The fear of being swamped by Asians in elections was counterbalanced by the acknowledgement that access to the body of people could only be effectively gained through the mediation of Asian leaders. The importance of securing the collaboration of leaders of the Asian communities in governance had been well attested to from the earliest days of the colony as exemplified by Raffles's appointment of *kapitan* or headmen who were responsible to the Resident for policing their respective *kampung*.[166] The success of riot control in the past had depended to a large extent on the presence of Chinese leaders to quell disturbances and to interpret government policies to the masses.[167] Co-option of Asian leaders into the Municipal Commission would also give its policies the stamp of legitimacy and strengthen the image of the Commission as the fount of social authority. While the 'civic gospel' which had carried much influence in British cities in making municipal councils the 'embodiment of civitas for all citizens'[168] was much more muted in colonial Singapore, there was concern that there should be a community of interests between the Municipal Commission and the inhabitants of the city. The local English press expressed this concern with typical condescension: 'We desire that even in this polyglot community there should be at least some approximation to the wholesome interest with which Englishmen are accustomed to regard all matters of self-government.'[169]

From the perspective of the Asian leaders themselves, the Council provided a means to realize or consolidate hegemonic aspirations other than through clan or *bang* leadership. Unlike Chinese-speaking, China-oriented merchants and businessmen who secured their power base and attained community leadership either through the control of key communal organizations or by capturing the leadership of the Singapore Chinese Chamber of Commerce,[170] avenues to power for emerging leaders of the Straits-born, English-educated community largely depended on official recognition of their role as social brokers mediating between the colonial state and the Chinese community.[171] As in British cities, municipal government provided a means of legitimizing social and political leadership in the city.[172] An analysis of the membership composition of the Municipal Commission provides further insight into the racial distribution of power and the section of Asian leadership recruited as city fathers. Between 1880 and 1887, Europeans overwhelmingly dominated the Municipal Board, taking up 90 per cent of available seats (Figure 2.12).[173] The Asian population had either one representative on the Board or none each year. In the twenty-six and a half years between 1888 and the first half of 1913 during which a ward system of elections operated alongside a system of government nomination, Europeans continued to dominate, constituting 79 per cent of the total number of all seats on the Municipal Commission, whilst the Chinese took up the remaining 21 per cent (Figure 2.12; Plate 3). Of those nominated by the Governor, the Europeans enjoyed a near monopoly with 95 per cent of all nominated seats, while the other 5 per cent was filled by two leading members of the Chinese community, Seah Liang Seah and Choa Giang Thye (Figure 2.13). Both were prominent Straits Chinese who were fluent in English and active in public affairs.[174] Of the elected seats, the Chinese won over a third of all seats, particularly those for Central and Rochor Wards (and occasionally for Kallang Ward), while the Europeans took up the remaining two-thirds (Figure 2.12). The Chinese who represented

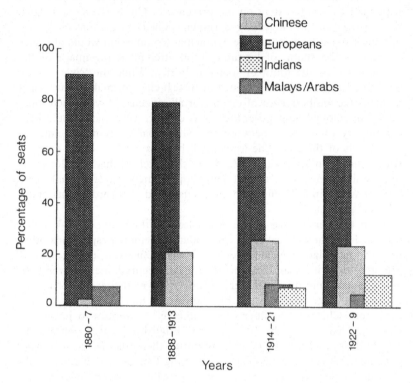

FIGURE 2.12
Racial Composition of the Municipal Commission, 1880–1929

their communities on the Board were largely Straits-born, English-educated Chinese loyal to the British Crown and already influential men in
various capacities. Tan Beng Wan, who defeated Lee Keng Kiat in the
contest for Central Ward in December 1888, was described as a gentleman 'with a good deal of English influence' and 'a Chinaman who
by adoption and training, [was] a Singaporean, in close accord with
English mercantile interests, and well-disposed to English methods'.[175]
On his retirement from the Legislative Council in June 1915, Tan Jiak
Kim, who also served on the Municipal Commission for many years in
the 1880s and 1890s, was accorded the accolade of being an 'enlightened' leader of the Chinese community, 'broad-minded enough to
grasp the western point of view and to weld it easily and smoothly
with the eastern point of view'.[176] Tan Kheam Hock, one of the last
commissioners to win a seat before elections were abolished, was
described as 'a man of progressive ideas [who had] identified himself
with all movements for the welfare of the Chinese community'.[177]
Many of the Chinese commissioners were active members of British-
oriented organizations such as the Straits Chinese Recreational Club[178]
and the Straits Chinese British Association[179] as opposed to traditional
Chinese *bang* or clan organizations.[180]

FIGURE 2.13
Racial Composition of Nominated and Elected Members of the
Municipal Commission, 1888–1913

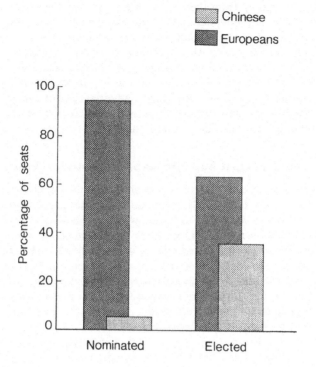

After the amending Municipal Ordinance of 1913 came into effect, abolishing elections, the composition of the Municipal Board showed greater variation in terms of its racial make-up. European/Eurasian domination was reduced to 58 per cent of the total number of seats between 1913 and 1921, while Chinese, Malay/Arab, and Indian commissioners constituted 26, 9, and 7 per cent respectively (Figure 2.12). In the eight years from 1922 following the partial delegation of the Governor's prerogative to appoint commissioners to public bodies, the ratio amongst the races remained fairly stable except for a significant increase in the number of Indian commissioners at the expense of Malay/Arab members.[181] It would appear that having abolished the 'vagaries' of elections and secured absolute control of the recruitment of commissioners as a safeguard, the colonial government was more prepared to admit a larger proportion of Asians into the Commission. Among the Asian commissioners themselves, while an electoral system had favoured the domination of the Chinese as the numerically superior community, the nomination system allowed a diversification of the racial composition of the Commission, with members nominated to represent different sectional and communal interests. Despite these changes, however, there was a strong degree of continuity in the class of Asians considered 'fit and proper persons' meriting nomination. By and large, those nominated were English-speaking, British-oriented men, drawn mainly from the merchant and property-owning class but there was also a preponderance of professionals.[182] With few exceptions, leaders of Chinese clan and *bang* organizations were seldom nominated; in many cases, they were automatically disqualified by their lack of command of the English language.[183] Hence, the representativeness of the Municipal Commission was a strictly limited one: it remained the preserve of Europeans and privileged members of the Asian English-educated élite who served to mediate between European policy-makers and the Asian communities.

The Municipal Authority as an Institution of Power

Possessing a significant, if by no means absolute, measure of discretionary powers in purely local municipal affairs, the Municipal Authority of Singapore serves as a lens through which to view the institutional mediation of power at a local level. As an institution of power, whilst its policies often carried vested interest and systematic bias in favour of particular groups within colonial society, the Authority was by no means a self-contained and monolithic power structure. Instead, policy-making within the Authority was subjected to tensions and negotiations between constituent parts of the power structure. At least three potential sites of tension could be identified.

First, although it was asserted that the proceedings of the Municipal Board were 'generally speaking, free from the dead hand of Government interference',[184] unfettered autonomy and independence from the supervision and pressure of the Straits government in the realm of policy-making were more apparent than real. Not only did the government control the appointment of the municipal president and nominated members of the Board, it also influenced the scope of municipal powers and responsibilities through controlling the formulation and ratification of municipal legislation. Amendments and additions to municipal legislation were often preceded by government-appointed commissions of inquiry or investigative surveys of different aspects of municipal administration by external experts and although these investigations possessed no power of enforcement in themselves, they effectively applied pressure in shaping municipal policies. The powers of the Municipal Commission were hence, to an extent, circumscribed by the larger powers of the Straits government and conflict between the Board and the government was inevitable as the former oscillated between dependence on and deviation from the latter. One of the main sources of tension lay in the question of delimiting the respective spheres of responsibilities pertaining to the government and the Municipal Authority. For example, in the 1890s, the colonial government attempted on several occasions to enlarge municipal responsibilities to include the registration of deaths and the notification and detection of infectious diseases.[185] These government initiatives were ambivalently received by the Municipal Board who, whilst welcoming increased powers which accompanied enlarged responsibilities, were at the same time concerned about the inadequacies of municipal financial and manpower resources to cope with each new responsibility. Tensions were also manifested in the negotiations over the collaboration of the municipal authority with government institutions such as the police, the Chinese protectorate, and the Chinese Advisory Board with which certain powers and functions overlapped. For example, the roles of the Municipal Authority and the police in the spheres of sanitary regulation and the maintenance of public order were closely intertwined, a situation which led frequently to accusations of neglect of duty on the part of the other party.[186]

A second locus of potential conflict lay in the relationship between the Municipal Council and the salaried officials employed to carry out municipal policies. Theoretically, as in the case of British municipalities, the Council was responsible for policy-making and the officials for day-to-day administration.[187] In practice, however, suggestions for policy changes often originated with the chief officers of the municipality such as the municipal health officer and the municipal engineer whose representations carried weight with the Council as they were given credence as the expert opinion of professional men. The principal officers heading various municipal departments were exclusively European men (Table 2.4; Plate 4) who had gained their professional qualifications and working experience in Britain or elsewhere in the British Empire, particularly India.

Third, the Municipal Board itself was by no means a unitary policy-making body. The strengthening of government control over Council membership after 1913 was accompanied by increasing ethnic diversification of the Council and the incorporation of a larger proportion

TABLE 2.4

Chief Salaried Officers of Two Main Municipal Departments, 1880–1929

Municipal Health Officer		Municipal Engineer	
1887–9	T. I. Rowell	1880–3	T. C. Cargill
1889–92	W. Gilmore Ellis (Acting)	1883–95	J. MacRitchie
1892–3	E. C. Dumbleton	1895–1900	S. Tomlinson
1893–1920	W. R. C. Middleton	1900–16[a]	R. Peirce
1920–3	J. A. R. Glennie		
1923–9	P. S. Hunter		

[a]In 1916, as a result of the great increase in the work of the Municipal Engineer's Department, duties were divided into separate subdepartments. B. Ball was appointed engineer for roads, sewers, and general engineering works; S. Williams was put in charge of the waterworks; J. P. Hallaway was appointed gas engineer; and J. H. Mackail electrical engineer.

of Asian commissioners. Although Asians who gained entry into the
Council tended to be pro-British and English-educated, this did not
necessarily prevent them from representing the views and interests of
the wider Asian communities. Complaints and protests against intrusive
municipal policies and policing affecting the Asian population as
diverse as eviction resulting from slum clearance and areal reconstruction, the cutting off of municipal water supply to coolie lodging houses,
the 'running in' of itinerant hawkers, and municipal interference with
burial plots occasionally found sympathy with the Asian commissioners
who advanced some of these claims at Board meetings. These representations served to check the excesses of municipal policy and allowed
some of the tensions and grievances rooted in Asian society to be
expressed and negotiated within the Commission itself. On the one
hand, the involvement of Asian leaders in municipal policy-making
served to help gauge Asian opinion and to provide relatively innocuous
channels for the expression of discontent. On the other hand, it must
be acknowledged that Asian commissioners carried a certain, if limited,
degree of influence on municipal policies. Compromises and concessions were occasionally made in deference to Asian opinion, although
these seldom caused the Municipal Board to swerve away from its continuing and conscious efforts to 'improve' the Asian environment and
way of life.

These tensions within the municipal institution, however, did not
prevent the systematic elaboration of an increasingly pervasive and
penetrating bureaucracy which sought to shape the urban environment,
control the Asian population, and improve their habits and customs.
The emergent civic institution was charged with duties ranging from
the collection of taxes, the regulation of sanitation, the provision of
public facilities, and the maintenance of public order. Increasingly
complex and detailed municipal regulations attempted to control and
dictate the everyday aspects of urban life, including the location of
markets, burial grounds, and offensive trades, the design and dimensions of cubicles, rooms, latrines, building blocks, footways, and streets,
the venue and duration of *wayang* (theatrical shows), *sembahyang* or
religious celebrations and processions, and the manner in which the
Asians removed and used night-soil, treated the sick, attended the birth
of children, and disposed of the dead. A *sinkheh* (newly arrived
Chinese immigrant) who arrived in colonial Singapore in the early
twentieth century was likely to encounter municipal might as wielded
by its minions in many guises, from the sanitary inspector who ordered
the pulling down of his cubicle in a coolie lodging house, the municipal
peon who served a limewash notice on the chief tenant of his house,
the registrar of vehicles who issued his licence to pull a *jinrikisha* (small,
lightweight, two-wheeled vehicle), the police officer (employed within
the municipality) who arrested him for obstruction of public places, to
the municipal apothecary who inspected his corpse for signs of infectious disease and the burial grounds inspector who ensured his burial
in a municipally sanctified plot.

It was thus through what Veena Oldenburg calls 'tireless, small, practical decisions made in the field'[188] that municipal policies ultimately impinged upon the lives and livelihood of the Asian population. It was, however, much easier to establish a municipal bureaucracy, to legislate for municipal powers, or even engage in public work or provide municipal facilities, than to instil civic consciousness and allegiance or to create civic order among a transient urban population riven by potentially divisive forces of race, language, religion, dialect, clan, and neighbourhood. In particular, the conception of a municipal authority working towards 'public good' was alien to Chinese society which revolved around a clan-centred culture visibly demonstrated in the urban landscape by the proliferation of *kongsi*, clan-owned temples, burial grounds, sick-receiving houses, schools, and lodging houses catering exclusively to the needs of members and affiliates. The innate prejudice among the immigrant Chinese against constituted authorities was already strong, for their experience of law as administered by the Mandarins in China had long been associated with injustice and extortion.[189] An institution with power and participation centralized in the hands of the few, given to the furthering of colonial rather than subordinate interests by means of an intrusive bureaucracy was not likely to dispel such prejudice.

The interventionist thrust of municipal regulatory systems, often legitimized by the language of enlightened reform and improvement, provoked a range of responses amongst Asian communities which sometimes culminated in violent backlash, but were more commonly articulated through strategies of evasion, non-compliance, and adjustment, or channelled through Asian leaders (including municipal commissioners, unofficial members of the Legislative Council, and members of the Chinese Advisory Board and other government organizations) with some degree of influence on the formulation of municipal policy. As David Arnold has argued, it is only through an awareness of the dialectical nature of the encounters between the ordinary people and the colonial state that it is possible to avoid assumptions of mass 'passivity' and 'fatalism'.[190] It is precisely this dialectic of power inherent in the encounter between, on the one hand, municipal attempts at imposing social and spatial control to create a city after the colonial image – orderly, sanitized, racially divided, hierarchical—and on the other, Asian agency in wresting concessions and asserting its own view of urban life which form the focus of the six chapters which follow.

1. Frederick A. Weld, 'The Straits Settlements and British Malaya [speech delivered on 10 June 1884]', in Paul H. Kratoska (ed.), *Honourable Intentions: Talks on the British Empire in South-East Asia delivered at the Royal Colonial Institute 1874–1928*, Singapore: Oxford University Press, 1983, pp. 46–7.

2. 'Municipal Government in the Straits Settlements', *Colonial Office Journal*, 4, 3 (1911): 220.

3. The first Chinese commissioner, appointed in 1870, was Tan Seng Poh, a wealthy spirit merchant and opium farmer and businessman in Singapore (Song Ong Siang, *One Hundred Years of the Chinese in Singapore*, Singapore: Oxford University Press, 1984, pp. 131–2).

4. ARSM, 1896, p. 11.

5. Thomas Stamford Bingley Raffles (1781–1826) began his career as a junior clerk in the office of the Secretary of the East India Company in Leadenhall Street. After various appointments on the Prince of Wales Island and in Java, he arrived in Bencoolen as Lieutenant-Governor in March 1818. In his capacity as agent and representative of the Governor-General of India, he was responsible for establishing the settlement and port of Singapore.

6. Raffles first landed on Singapore island on 28 January 1819. A preliminary agreement between Raffles and the Temenggong of Johore allowing the establishment of a factory at Singapore was reached two days later. This agreement was ratified by the treaty of 6 February 1819, concluded between Raffles as agent of the East India Company on the one hand, and Hussein Mohamed Shah, the Sultan of Johore, and Abdu'r-Rahman, the Temenggong of Johore, on the other. According to the treaty, in recompense for maintaining a factory at Singapore, the East India Company agreed to pay the Sultan 5,000 Spanish dollars annually, and the Temenggong, 3,000. Both the agreement and the treaty are reproduced in Ronald St. John Braddell, *The Law of the Straits Settlements: A Commentary*, Kuala Lumpur: Oxford University Press, 1982 (first published in 1915), pp. 142–7.

7. F. J. Hallifax, 'Municipal Government' in Walter Makepeace, Gilbert E. Brooke, and Ronald St. John Braddell (eds.), *One Hundred Years of Singapore*, London: John Murray, 1921, Vol. 1, pp. 316–7.

8. The three men selected for the task were Captain Charles Edward Davis of the Bengal Native Infantry, George Bonham of the Bencoolen Civil Service, and Alexander L. Johnston, head of one of the pioneer mercantile firms.

9. 'Raffles to Town Committee, 4 November 1822' (reproduced in Charles B. Buckley, *An Anecdotal History of Old Times in Singapore*, Singapore: Oxford University Press, 1984 (first published in 1902), p. 81).

10. Buckley, *An Anecdotal History*, p. 196.

11. ARSM, 1923, p. 11.

12. In 1843, for example, $12,000 was spent for the upkeep of the police, $1,900 on roads, and $18.62 on enclosing the Esplanade. The police force was, however, 'avowedly inefficient' (Buckley, *An Anecdotal History*, pp. 385 and 441).

13. William Henry Macleod Read (1819–1909) came out to Singapore in 1841 to succeed his father as a partner of A. L. Johnston & Co. He played a prominent role in local politics and was a strong advocate of free trade and communal liberties. He was later to become a municipal commissioner and also the first non-official member of the first Legislative Council when the Straits Settlements became a Crown Colony in 1867 (Buckley, *An Anecdotal History*, pp. 365–9; William Henry Macleod Read, *Play and Politics, Recollections of Malaya by an Old Resident*, London: Wells Gardner, Darton & Co., 1901).

14. Buckley, *An Anecdotal History*, p. 423.

15. C. Mary Turnbull, *The Straits Settlements, 1826–67: Indian Presidency to Crown Colony*, Singapore: Oxford University Press, 1972, pp. 320–1.

16. C. Mary Turnbull, *A History of Singapore, 1819–1975*, Kuala Lumpur: Oxford University Press, 1977, p. 67.

17. On 25 November 1854, for example, after drawing attention to a number of long-standing nuisances such as the disgraceful state of the jail, the 'ruinous' church, and the 'sickening malaria' arising from inefficient and filthy drains in town, the Grand Jury presented that the municipal committee had not tended to public welfare and should either be reformed or abolished entirely for 'as might be expected from its constitution and as its action has proved, it is a mere Government bureau, whose object and effect is to render the raising of money from the public more easy, and its

extraction from Government for public use more difficult. . . . It serves only as a shield to receive and break the shocks which would otherwise fall directly upon the Executive authorities' (Lee Yong Kiat, 'The Grand 'Jury in Early Singapore (1819–1873)', *Journal of the Malayan Branch of the Royal Asiatic Society*, 46, 2 (1973): 104).

18. For example, see *Straits Times*, 17 June 1846. Also see extracts from the *Singapore Free Press* in Buckley, *An Anecdotal History*, pp. 474–6.

19. Turnbull notes that a petition was drawn up at a public meeting asking the Calcutta government to withdraw the Assessment Act when it came into force in 1848 and that most ratepayers declined to serve on the new municipal committee (Turnbull, *The Straits Settlements*, p. 321). In 1847, a petition was also presented to British Parliament signed by 215 persons and calling for the assessment funds to be administered by 'a Committee of the ratepayers or other popularly elected body' (Buckley, *An Anecdotal History*, pp. 465–6). John Crawfurd, a former Resident of Singapore through whom the petition was presented to the members of Parliament, forcefully argued that the Indian Government had failed to take cognizance of conditions prevailing in the Straits Settlements, and that 'the administration of mere local affairs must, from its very nature, be best conducted by those who [were] in a position to understand it best, and who ha[d] the most immediate interest in conducting it efficiently and economically' and that 'these [were] . . . the inhabitants of each locality, and not the Executive Government' (Buckley, *An Anecdotal History*, pp. 472–3).

20. *Singapore Free Press*, 8 December 1854.

21. G. F. Davidson, *Trade and Travel in the Far East or Recollections of 21 Years Passed in Java, Singapore, Australia and China*, London: Madden & Malcolm, 1846, pp. 71–2.

22. In mid-nineteenth-century Singapore, despite the lack of representative institutions, the European mercantile community had several channels through which they could agitate against official policy and make their views known. These included the press, the Grand Jury, and the public meeting which acted as rallying points for growing opposition to the government (Turnbull, *A History of Singapore*, p. 70).

23. Hallifax, 'Municipal Government', p. 317.

24. One rupee was the equivalent of 0.444 Straits dollar.

25. ARSM, 1923, pp. 11–12.

26. All four acts were passed by the Legislative Council in India and were applicable to Calcutta, Madras, Bombay, and the Straits Settlements.

27. Buckley, *An Anecdotal History*, p. 632.

28. Ibid., pp. 632–3.

29. PLCSS, 20 June 1870, p. 45. Turnbull suggests that the strike was a result of the failure of communication and misunderstanding between the government and the Chinese population. In the words of the Governor, the Straits government found it 'utterly hopeless' to 'translate into hieroglyphic Chinese the technicalities of an English legal enactment'. As a result, exaggerated rumours regarding the magnitude of fines, the number of petty offences punishable, and the arbitrary powers of the police spread among the Chinese, causing fear and discontent (C. Mary Turnbull, 'Communal Disturbances in the Straits Settlements in 1857', *Journal of the Malayan Branch of the Royal Asiatic Society*, 31, 1 (1958): 96).

30. Turnbull, 'Communal Disturbances', p. 97.

31. Turnbull, *The Straits Settlements*, pp. 41–2.

32. Buckley, *An Anecdotal History*, p. 668.

33. Turnbull, *The Straits Settlements*, p. 88.

34. The Colony of the Straits Settlements comprised Singapore, Penang, and Malacca with additional entities attached from time to time for administrative convenience: the Dindings area of Perak was part of the colony from 1874 to 1934, Labuan from 1906 to 1946, the Cocos Keeling islands from 1886 to 1955, and Christmas Island from 1900 to 1958.

35. PLCSS, 1868, 'Appendix B: Copies of Correspondence Relative to the Reconstitution of the Municipal Councils: Circular Memo to be Laid before the Municipal Bodies at an Early Meeting by the Lieut. Governors and Colonial Secretary', p. iii.

36. PLCSS, 1868, 'Appendix B: Copies of Correspondence Relative to the Reconstitution of the Municipal Councils: Minute of His Excellency the Governor', p. vi.

37. Ibid., p. vii.

38. A Singapore Merchant, *How to Govern a Colony*, London: A. H. Bailey & Co., 1869, pp. 7–8.

39. PLCSS, 1873, 'Appendix 19: Report of the Sub-Committee of the Legislative Council Appointed to Examine into the Working of the Municipal Acts of 1856 and Their Applicability to the Present State of the Straits Settlements'.

40. PLCSS, 30 June 1887, p. B24.

41. PLCSS, 1872, 'Appendix 10: Correspondence with Reference to the Sanitary Condition of the Town of Singapore', p. lxvii.

42. This led to an acrimonious war of words between the municipal commissioners and the Inspector-General of Police as to the extent to which each body was responsible for implementing the conservancy and street obstruction laws under Acts XIV of 1856, XIII of 1856, and XLVIII of 1860 ('Correspondence with Reference to the Sanitary Condition of the Town', pp. lxvii-lxxii).

43. 'Correspondence with Reference to the Sanitary Condition of the Town', p. lxxiii.

44. PLCSS, 21 March 1874, p. 1.

45. PLCSS, 30 June 1887, p. B21.

46. Eunice Thio, *British Policy in the Malay Peninsula 1880–1910*, Vol. 1, Singapore: University of Malaya Press, 1969, p. 12. The extension of British political control in Malaya, the so-called 'forward policy', began in 1874 under the governorship of Sir Andrew Clarke when the 'Residential System' was introduced into three west coast Malay States—Perak, Selangor, and Sungai Ujong (one of the Negri Sembilan group of states). In 1888, a British Resident was established in Pahang and in 1895, the protected states of Selangor, Perak, Pahang, and Negri Sembilan were formed into a federation under a Resident-General (Emily Sadka, *The Protected Malay States 1874–1895*, Kuala Lumpur: University of Malaya Press, 1968, pp. 375–6).

47. Jomo Kwame Sundaram, *A Question of Class: Capital, the State and Uneven Development in Malaya*, Singapore: Oxford University Press, 1986, p. 138.

48. MPMCOM, 9 June 1897.

49. Based on Appendix Table A.2.

50. Based on Appendix Table A.1.

51. During the next four decades, municipal boundaries did not change substantially except for minor additions in 1906, the inclusion of a portion of the Tanjong Katong district in 1918, and the inclusion of Siglap and parts of Bedok, Ulu Bedok, and Paya Lebar in 1928 (Hallifax, 'Municipal Government', p. 339; ARSM, 1928, Assessment and Estate Department, p. 8A). The proportion of the population residing within municipal limits stabilized at 80–85 per cent.

52. See Appendix Table A.3.

53. *Papers Relating to the Investigation of Malaria and Other Tropical Diseases and the Establishment of Schools of Tropical Medicine*, London: HMSO, 1903, p. 21. Spatial segregation between the 'white' town and the 'native' areas was most marked in areas with a substantial indigenous population and a long tradition of urbanization (Robert J. Ross and Gerard J. Telkamp, 'Introduction', in Robert J. Ross and Gerard J. Telkamp (eds.), *Colonial Cities: Essays on Urbanism in a Colonial Context*, Dordrecht: Martinus Nijhoff, 1985, p. 5; Susan J. Lewandowski, 'Urban Growth and Municipal Development in the Colonial City of Madras, 1860–1900', *Journal of Asian Studies*, 34 (1975): 346–7; Anthony D. King, *Colonial Urban Development: Culture, Social Power and Environment*, London: Routledge & Kegan Paul, 1976, pp. 108–9).

54. PLCSS, 6 March 1903, p. B10.

55. The word *kampung* (*kampong*, *campong*), meaning quarter or subdivision of a town, is possibly a corruption of the Portuguese word *campo*, and was used in Portuguese Malacca at the beginning of the seventeenth century by Godinho de Eredia with reference to 'Campon China', a ward in Malacca (Henry Yule and A. C. Burnell,

Hobson-Jobson: A Glossary of Anglo-Indian Colloquial Words and Phrases and of Kindred Terms, London: John Murray, 1903, p. 242).

56. In November 1822, Raffles appointed a town committee to oversee the laying out of the town according to his instructions which stipulated, *inter alia*, that separate quarters or *kampung* be demarcated for specific racial and occupational groups. The European town was alloted the expanse to the east of the government reserved area whilst the Asian races were to be settled in well demarcated *kampung*. The Chinese *kampung* was to take up the whole of the south bank of the Singapore River, except the riverside locations already alloted to European merchants. Raffles also contemplated a separate division for traders from Amoy (Hokkiens) to the east of the European Town and the Sultan's area in recognition of their importance as principal Chinese traders. The Sultan's residence was allocated 56 acres east of the European Town, lying between Rochor River and the sea. The Arab *kampung* was located between the European Town and the Sultan's residence, whilst the Bugis were removed beyond the Sultan's residence. The Chulias were alloted a division up the Singapore River on the south bank. This plan was committed to cartographic representation by Lieutenant Philip Jackson, the assistant engineer attached to the detachment of Bengal artillery, either in December 1822 or January 1823. According to H. F. Pearson, the plan represented an idealized project rather than an actual ground plan and there is a certain amount of doubt as to the degree to which the plan was carried out (Raffles to Town Committee, 4 November 1822', reproduced in Buckley, *An Anecdotal History*, pp. 81–6; H. F. Pearson, 'Lt. Jackson's Plan of Singapore', in Mubin Sheppard (ed.), *Singapore: 150 Years*, Singapore: Times Books International, 1982, pp. 150–4). Jackson's plan was first published in John Crawfurd, *Journal of an Embassy to the Courts of Siam and Cochin-China, etc.*, London: Henry Colburn & Co., 1828, facing p. 529). B. W. Hodder notes that whilst John Cameron writing in 1865 said of the many *kampung* that 'though [they] ... were probably first occupied by the races whose names they bear, no such distinction appear[ed] now to exist', this is contradicted by H. Norman's claim in 1895 that each race had its own quarters, citing Kampongs Malacca, Kling, Siam, and China. What could be deduced from these statements is the increasing complexity and flux in the racial pattern of the city as the population continued to increase rapidly. Whilst many of the older *kampung* had survived only in name, new racial concentrations continued to emerge throughout the nineteenth and early twentieth centuries (B. W. Hodder, 'Racial Groupings in Singapore', *Malayan Journal of Tropical Geography*, 1 (1953): 27).

57. Cheng Lim-Keak, *Social Change and the Chinese in Singapore*, Singapore: Singapore University Press, 1985, pp. 28–9.

58. Kernial Singh Sandhu, 'Some Aspects of Indian Settlement in Singapore, 1819–1969', *Journal of Southeast Asian History*, 10 (1970): 197.

59. Weld, 'The Straits Settlements', pp. 48–9.

60. J. T. L., 'William Geoffrey's Profit', reprinted from the *Straits Times Annual*, December 1906, Singapore: Straits Times Press, 1906, p. 5.

61. The index of dissimilarity measures the 'spatial distance' or more precisely, the difference between the areal distribution of two variables. It may be interpreted as a measure of displacement, that is, the degree (in percentage terms) of one group which has to move to a different area in order to make their distribution identical with that of a second group. To compute the index, the percentage distribution of each variable in each areal unit is first calculated. The index between two variables is half the sum of the absolute values of the differences between the respective distributions, taken area by area. It ranges from zero (for two absolutely similar percentage distributions) to 100 (for two totally dissimilar percentage distributions) (Otis Dudley Duncan and Beverly Duncan, 'Residential Distribution and Occupational Stratification', in Ceri Peach, *Urban Social Segregation*, London: Longman, 1975, pp. 52–3).

62. Michael R. Godley, *The Mandarin-Capitalist from Nanyang: Overseas Chinese Enterprise in the Modernisation of China, 1893–1911*, Cambridge: Cambridge University Press, pp. 46–7; Cheng, *Social Change and the Chinese*, pp. 13–23.

63. G. M. Reith, *1907 Handbook to Singapore*, revised by Walter Makepeace, Singapore: Oxford University Press, 1986 (first published in 1892), p. 32.

64. Li Chung Chu, 'A Description of Singapore in 1887', trans. by Chang Chin Chiang, *China Society of Singapore 25th Anniversary Journal* (first published in Chinese in 1895), 1975: 25.

65. For a study of the urban morphology of colonial Madras, Bombay, and Calcutta, see Meera Kosambi and John E. Brush, 'Three Colonial Port Cities in India', *Geographical Review*, 78 (1988): 32–47.

66. *Straits Times*, 21 November 1919.

67. PLCSS, 30 June 1887, p. B29.

68. PLCSS, 6 May 1887, p. 752.

69. PLCSS, 2 July 1887, p. B40.

70. PLCSS, 30 June 1887, p. B25.

71. Ibid., p. B26.

72. Ibid., p. B27.

73. Ibid., p. B26. Since the elective principle was introduced in 1856, elected members had comprised the majority by a ratio of three to two.

74. PLCSS, 2 July 1887, p. B38.

75. PLCSS, 30 June 1887, p. B27.

76. Ibid., p. B32.

77. PLCSS, 2 July 1887, p. B41.

78. Ibid.

79. PLCSS, 30 June 1887, p. B26.

80. Ibid.

81. Ibid.

82. Ibid., p. B36.

83. Ibid., p. B33.

84. 'Powers of Colonial Governments', *Colonial Magazine*, 3 (1840): 178.

85. PLCSS, 30 June 1887, p. B28–9.

86. *Straits Times*, 7 June 1887. The specific powers the municipality possessed in sanitizing the urban environment are discussed in Part I (Chapters 3–5).

87. Derek Fraser, *Power and Authority in the Victorian City*, Oxford: Basil Blackwell, 1979, pp. 167–9.

88. *Straits Times*, 7 June 1887.

89. *Straits Times*, 4 May 1888.

90. *Straits Times*, 7 June 1887; 4 May 1888.

91. PLCSS, 4 July 1887, p. B47; 7 July 1887, p. B50; 14 July 1887, p. B67.

92. CO273/148/22039: Weld to Holland, 19 September 1887.

93. Khoo Kay Kim, 'The Municipal Government of Singapore 1887–1940', Academic Exercise, University of Malaya, Singapore, 1960, pp. 11–12.

94. PLCSS, 18 August 1887, p. B116.

95. Dr T. Irvine Rowell received his medical training in Aberdeen and was at the time of his appointment to the presidentship the principal civil medical officer and port health officer of Singapore (G. E. Brooke, 'Medical Work and Institutions', in Makepeace, Brooke, and Braddell, *One Hundred Years of Singapore*, p. 518).

96. CO273/148/22039: Weld to Holland, 19 September 1887. In order 'to give [the municipality] the means to satisfactorily direct sanitary improvements', the government also offered Rowell the opportunity of continuing as health officer to Singapore for which he would receive the Bill of Health fees on the condition that he also discharged the duties of health officer to the municipality. The president was hence also the municipality's first health officer (CO273/148/24539: Smith to Holland, 5 November 1887: 'Dickson to Rowell, 6 October 1887'.

97. NL955, Despatches from the Secretary of State to the Straits Settlements, 8 December 1887, No. 291: 'Correspondence between the S.S. (Penang) Association to Colonial Office, 13 August 1887'.

98. CO273/148/25140: Smith to Holland, 14 November 1887.

99. Ibid. The Governor's statements largely reflected the views of the municipal commissioners themselves as expressed in their letter to the Governor protesting against

any postponement of the commencement of the ordinance (MPMCSM, 12 November 1887).

100. CO273/151/4511: Smith to Holland, 31 January 1888.

101. 'The Humble Memorial of the Undersigned Inhabitants of Singapore, in the Colony of the Straits Settlements', reprinted in *Straits Times*, 24 December 1887.

102. Ibid.

103. NL955, Despatches from the Secretary of State to the Straits Settlements, 13 April 1888, No. 106: 'Memorandum on the Straits Municipal Bill by T. Cuthbertson, formerly an unofficial member of Council'; 'Memorandum on the Straits Municipal Bill by C. R. Rigg, for many years Secretary to the Municipal Commissioners, Singapore'.

104. 'Memorandum on the Straits Municipal Bill by T. Cuthbertson'.

105. NL955, Despatches from the Secretary of State to the Straits Settlements, 13 April 1888, No. 106: 'Memorandum on the Straits Municipal Bill by S. Gilfillan, formerly an unofficial member of Council'.

106. Petitions were also received from the Singapore Wharf proprietors, the residents of Penang, and the residents of Malacca (NL955, Despatches from the Secretary of State to the Straits Settlements, 13 April 1888, No. 106: Herbert to Smith).

107. NL955, Despatches from the Secretary of State to the Straits Settlements, 13 April 1888, No. 106: Herbert to Smith.

108. PLCSS, 7 November 1889, p. B125; 8 November 1889, p. B131; 5 December 1889, p. B150.

109. Quoted in Fraser, *Power and Authority*, p. 168.

110. PLCSS, 14 November 1889, pp. B133–B138.

111. Ibid., pp. B138–B140.

112. MPMCOM, 29 August 1894.

113. PLCSS, 21 October 1895, p. B112.

114. *Straits Times*, 11 May 1896.

115. PLCSS, 14 May 1896, p. B79.

116. Ibid., p. B80.

117. PLCSS, 25 June 1896, p. B136.

118. PLCSS, 29 October 1896, p. B280.

119. PLCSS, 6 August 1909, p. B102.

120. PLCSS, 23 September 1910, 'Report of the Municipal Enquiry Commission', pp. C194–C195. The Report also found fault with several other aspects of municipal administration such as deviations from the procedure laid down in the Municipal Ordinance with regard to the preparation and use of assessment lists, the lack of financial control on the part of the municipal president, and the want of initiative on the part of the Health Department in implementing sanitary schemes.

121. Ibid.

122. Ibid., p. C195.

123. PLCSS, 4 August 1911, pp. B99–B100.

124. A well-known Jewish businessman and property owner, Manasseh Meyer had served as a municipal commissioner in the 1890s. Meyer's 'great knowledge of property and local circumstances' made him 'exceptionally useful as a City Father' (Walter Makepeace, 'Concerning Known Persons', in Makepeace, Brooke, and Braddell, *One Hundred Years of Singapore*, p. 463).

125. Song, *One Hundred Years' History of the Chinese*, p. 468.

126. Ibid., pp. 469–71.

127. PLCSS, 15 November 1912, p. B159.

128. Ibid.

129. PLCSS, 13 December 1912, pp. B183–B184.

130. Ibid., p. B185.

131. Ibid., p. B188.

132. Ibid., p. B189.

133. Ibid.

134. PLCSS, 14 February 1913, p. B49.

135. Ibid., p. B50.

136. Ibid., p. B51.

137. PLCSS, 18 April 1913, p. B121.

138. PLCSS, 25 January 1921, p. B2. Initially, three subcommittees—Health, Fire and Vehicles; Public Works and Conservancy; and Water, Electricity, and Gas—were approved (MGCM, 4 June 1920). By 1929, there were seven standing committees, each having five or more commissioners as members, with work apportioned as follows: Secretariat, Assessment, Vehicles, and Dog Registration; Health, Markets, Slaughter Houses, Town Cleansing, Sewerage, and Street Watering; Municipal Engineer's Department, Roads, Canals, Piers, Bridges, Drainage, and Stores; Building, Architect's Department, and Fire Brigade; Water, Gas, and Electrical Departments; Parks and Open Spaces; and Finance and General Purposes (J. W. Harries, 'The Singapore Municipal Commission', *British Malaya*, 4, 4 (1929): 115).

139. PLCSS, 25 January 1921, p. B2.

140. Harries, 'The Singapore Municipal Commission', p. 115.

141. *Singapore Free Press Centenary Number*, 8 October 1935.

142. 'Municipal Government in the Straits Settlements', p. 220.

143. MPMCOM, 11 March 1904; 31 January 1908.

144. Edwin A. Brown, *Indiscreet Memories*, London: Kelly and Walsh, 1936, p. 218.

145. 'Municipal Government in the Straits Settlements, p. 223.

146. *A Handbook of Information Presented by the Rotary Club and the Municipal Commissioners of the Town of Singapore*, Singapore: Publicity Committee of the Rotary Club of Singapore, 1933, p. 29.

147. On one of the rare occasions when the question of restoring to ratepayers the right of electing a proportion of the municipal commissioners was raised, the issue was summarily dismissed on the grounds that it had 'proved unsatisfactory in the past, and would not be an improvement' (PLCSS, 16 March 1925, p. B22).

148. PLCSS, 30 June 1887, p. B24.

149. Weld, 'The Straits Settlements', p. 46.

150. PLCSS, 14 May 1896, p. B80.

151. *Singapore Free Press*, 23 January 1888.

152. *Straits Times*, 10 February 1887.

153. Ibid.

154. Ibid.

155. *Straits Times*, 14 May 1887.

156. Ibid.

157. *Straits Times*, 14 May 1887, 17 February 1887, and 10 August 1887.

158. Only 811 persons voted during the first municipal elections after the Municipal Ordinance of 1887 came into effect (*Singapore Free Press*, 30 January 1888). This represented only about 0.6 per cent of the total municipal population.

159. This was changed to payment of a minimum half-yearly rate of $6.00 by the amending ordinance of 1889.

160. SSGG, 18 September 1896, pp. 1575–600. How comprehensive this list was is highly suspect. Whilst the municipality had a sufficiently complete list of houses whose occupancy would carry a municipal vote, it possessed no information with regard to the persons occupying such premises or the duration of their occupancy. The only means by which the municipality came into actual contact with the tenant was through payment of the water rate. However, this rate was not levied against the tenant by name but against an anonymous occupier of the house. There could also be no reliance on the decennial town census to identify eligible tenants not only because it yielded no information on the duration of occupancy but also because it was too infrequently taken to keep pace with the rapid population turnover. Compilation of the list of voters was hence highly dependent on applications from the public to be included in the list and 'in many instances, ratepayers [did] not give the information sought' (*Straits Times*, 6 September 1887; MPMCOM, 20 November 1908).

161. This was minute compared to the municipal electorate of British towns which stood at about 18–20 per cent of the population at the end of the nineteenth century (E. P. Hennock, *Fit and Proper Persons: Ideal and Reality in Nineteenth-century Urban Government*, London: Edward Arnold, 1973, p. 12).

162. 'Ordinance [IX] of 1887: An Ordinance for Consolidating and Amending the Law Relating to Municipal Government', in *Straits Settlements Ordinance, 1887*, Singapore: Government Printing Office, 1888, p. 38; C. G. Garrad, *The Acts and Ordinances of the Legislative Council of the Straits Settlements from 1st April 1867 to 7th March 1898*, Vol. 2, London: Eyre & Spottiswoode, 1898, p. 1478.

163. SSGG, 18 September 1896, p. 1571.

164. The successful candidates were T. Scott (121 out of 123 votes) for Tanjong Pagar Ward; Tan Kim Cheng (287 out of 337 votes) for Central Ward; T. Shelford (144 out of 219 votes); Lim Eng Keng (62 out of 102 votes); and Tan Jiak Kim (20 out of 30 votes) (*Singapore Free Press*, 30 January 1888).

165. *Straits Times*, 2 February 1888. The mettle of hitherto untried municipal commissioners, the editor continued, would be tested as they confront the 'burning question' of the day—the clearance of obstruction on verandahs, the public footways of the city. The 'verandah question' is extensively discussed in Chapter 7.

166. 'Raffles to Farquhar, 25 June 1819', reproduced in Buckley, *An Anecdotal History*, pp. 56–8.

167. Yong Ching Fatt, 'Chinese Leadership in Nineteenth Century Singapore', *Journal of the Island Society*, 1, 1 (1967): 12–13. Mediation between the government and the Chinese population using community leaders as arbitrators was a basic government policy. Besides informal links forged through business connections, Chinese leaders were also more formally incorporated into the government structure through official appointments such as justices of the peace or as unofficial members of the Legislative Council. Before the suppression of secret societies in 1890, the government also consulted the headmen of these societies on matters affecting the Chinese and often resorted to apprehending them to do duty as special constables during riots. After secret societies were outlawed, a Chinese Advisory Board comprising leading members of different *bang* of the Chinese community selected by the government was established to 'enable the Government to ascertain the feelings of the Chinese community on any measures that it might raise while at the same time, giving the Chinese a medium through which to voice their opinion on matters affecting their interests'. The protector of Chinese was *ex-officio* member and acted as the chairman of the Board (Ng Siew Yoong, 'The Chinese Protectorate in Singapore, 1877–1900', *Journal of South East Asian History*, 2 (1961): 95; Lea E. Williams, 'Chinese Leadership in Early British Singapore', *Asian Studies*, 2 (1964): 178–9; Chu Tee Seng, 'The Singapore Chinese Protectorate, 1900–1941', *Journal of the South Seas Society*, 26, 1 (1971): 9–10).

168. Derek Fraser, 'Introduction: Municipal Reform in Historical Perspective', in Derek Fraser (ed.), *Municipal Reform and the Industrial City*, Leicester: Leicester University Press, 1982, p. 7.

169. *Straits Times*, 8 December 1888.

170. Founded in 1906, the Singapore Chinese Chamber of Commerce was the paramount organization aimed at promoting Chinese commercial interests in Singapore.

171. Yong Ching Fatt, 'Emergence of Chinese Community Leaders in Singapore, 1890–1941', *Journal of the South Seas Society*, 30, 2 (1975): 1–2.

172. For case-studies demonstrating how municipal government served the political and social aspirations of an emerging élite in British cities, see V. A. C. Gatrell, 'Incorporation and the Pursuit of Liberal Hegemony in Manchester 1790–1839', in Fraser, *Municipal Reform*, pp. 15–60; Adrian Elliot, 'Municipal Government in Bradford in the Mid-nineteenth Century', in Fraser, *Municipal Reform*, pp. 111–61; Hennock, *Fit and Proper Persons*.

173. A detailed list of municipal commissioners for 1880–1929 is found in Brenda S. A. Yeoh, 'Municipal Control, Asian Agency and the Urban Built Environment in Colonial Singapore', D. Phil. thesis, Oxford University, 1991, pp. 398–403.

174. Seah Liang Seah (1850–1925), described as 'a shrewd and successful business-man, landed proprietor and owner of gambier and pepper plantations' was also for many years during the 1880s a member of Legislative Council (Yong Ching Fatt, 'A Preliminary Study of Chinese Leadership in Singapore, 1900–1941', *Journal of South East Asian History*, 9 (1968): 265); Song, *One Hundred Years' History of the Chinese*, pp. 212–13). Choa Giang Thye (1865–1911) was a partner in a firm of commission agents and also a member of the Chinese Advisory Board and the Committee of the Po Leung Kuk (Song, *One Hundred Years' History of the Chinese*, p. 299; MPMCOM, 10 March 1911).

175. *Straits Times*, 8 December 1888.

176. Song, *One Hundred Years' History of the Chinese*, p. 197.

177. Ibid., pp. 257–9.

178. The Straits Chinese Recreational Club, 'the first club adopting English outdoor sports ever established by the Chinese', was started in the mid-1880s 'for the purpose of playing lawn tennis, cricket and practising English athletic sports'. Municipal Commissioners Chia Keng Chin and Ong Tek Lim numbered amongst its avid support-ers (Song, *One Hundred Years' History of the Chinese*, pp. 216–17).

179. Founded on 17 August 1900, the objectives of the Straits Chinese British Association were to promote loyalty to the British Crown and the welfare of Straits Chinese British subjects. The stronghold of the English-educated professional class, its membership included several municipal commissioners including Tan Jiak Kim, Seah Liang Seah, Tan Beng Wan, Lee Choon Guan, Ong Tek Lim, Lim Boon Keng, and Tan Kheam Hock. In contradistinction to transplanted clan, dialect, and territorial organ-izations which maintained strong ties with China, locally grown institutions like the Straits Chinese British Association and the Straits Chinese Recreational Club provided the bulwark of an ideology in which the tenets of British allegiance and social reform and emancipation from 'Manchu servitude' were central (Brenda Saw Ai Yeoh, 'The Decline of a Community—The Babas', BA dissertation, Cambridge University, 1985, pp. 96–7).

180. These tended to be the strongholds of the Chinese-speaking, Chinese-educated élite who in turn were seldom nominated by the colonial government to play a represen-tative role in local politics (Yong, 'A Preliminary Study of Chinese Leadership', p. 272).

181. Percentages of the total number of seats for this period were as follows: Europeans/Eurasians (58 per cent); Chinese (24 per cent); Malays/Arabs (5 per cent); and Indian (13 per cent).

182. Of the twenty-five to thirty Asians appointed to the Commission between 1913 and 1929, several were doctors (Dr Lim Han Hoe, Dr N. Veerasamy, Dr H. S. Moonshi, and Dr K. K. Pathy) and lawyers (Wong Siew Qui and Wee Swee Teow).

183. An exception was See Tiong Wah, who was both a municipal commissioner and president of the Hokkien Huay Kuan and the Chinese Chamber of Commerce in the 1920s. Unlike most traditional clan and *bang* leaders who were educated in Chinese, See Tiong Wah was educated at St Joseph's Institution in Singapore and spoke fluent English (Song, *One Hundred Years' History of the Chinese*, p. 104).

184. *A Handbook of Information*, p. 29. According to this handbook produced by the municipal commissioners and Rotary Club, the Board had free rein of municipal affairs except where the following were concerned: in the event of malfeasance the government maintained the right to step in and right matters; any loan-raising by the commissioners had to be sanctioned as to terms by the Legislative Council and, as to subjects, by the Governor; purchase of lands required the sanction of the Governor; and, by-laws made by the commissioners had to be confirmed by the Governor.

185. ARSM, 1895, 'Appendix Q: Registration of Deaths', pp. 130–45.

186. This is further elaborated on in Chapters 3 and 7.

187. Hennock, *Fit and Proper Persons*, pp. 7–8. In colonial Singapore, the municipal president was both a member of the the council as well as a salaried official and played an important role in mediating between the two bodies.

188. Veena Talwar Oldenburg, *The Making of Colonial Lucknow, 1856–1877*, Princeton: Princeton University Press, 1984, p. xix.

189. Eunice Thio, 'The Singapore Chinese Protectorate: Events and Conditions Leading to its Establishment, 1823–1877', *Journal of the South Seas Society*, 16 (1960): 41.

190. David Arnold, 'Touching the Body: Perspectives on the Indian Plague, 1896–1900', in Ranajit Guha (ed.), *Subaltern Studies V: Writings on South Asian History and Society*, Delhi: Oxford University Press, 1987, p. 56.

PART I

Sanitizing the Private Environment

AN emerging theme in current research into the health conditions of colonial societies has been the concern not so much with disease and medicine as purely epidemiological or medical phenomena, but with their instrumentality: that is, with their role in 'describing a relationship of power and authority between rulers and the ruled and between colonialism's constituent parts'.[1] It has been argued that the history of 'tropical medicine' and 'sanitary science' did not pursue an independent, progressive trajectory divorced from the economic, social, and political ambitions of colonialism, but instead posed as another form of domination.[2] The colonial medical and sanitary campaign not only served to legitimize imperial rule and to impart to it a gloss of munificence, an 'illusion of permanence',[3] but was in itself an 'exercise of disciplinary power'[4] which penetrated into the smallest details of everyday life. Its potency was reflected in the variety of metaphors used to describe colonial medicine: it was one of the 'tools of empire' that facilitated Western penetration,[5] a 'military campaign'[6] and a 'key weapon' in the subjugation of indigenous culture and the promotion of European allegiance.[7] Recently, studies of colonial medicine and sanitation have focused on the capacity of colonial power, through its complex of institutional and legislative machinery, to order, regulate, and sanitize the subordinated society. For example, Swanson traces the creation of urban apartheid in South Africa to the 'sanitation syndrome', and its associated societal metaphors of infection and disease,[8] whilst Thomas argues that in colonial Fiji, the extension of state power and its various political, moral, and cultural impositions were justified by their association or conflation with the programme of sanitation.[9]

While it is often acknowledged that the relationship of power inherent in colonial medical and sanitary projects was neither static nor uncontested, much less has been written about the attitudes and responses of the colonized body to these impositions of power. 'What we still need to discover', according to one commentator, 'is how Southeast Asians [and other colonized peoples] perceived the effects of imperial medicine, how they weighed its bungling insensitivity against occasional triumphs.'[10] There has been relatively little analysis of the specific strategies with which the colonized responded to co-operate with, inflect or counter the priorities and claims of imperial medical and sanitary schemes.[11]

Part I examines the project of sanitizing the colonial city of Singapore during the late nineteenth and early twentieth centuries as a dialectical encounter between the municipal authorities of the city and the Asian plebeian classes. By the late nineteenth century, the overwhelming concern for a sanitized environment in Singapore provided much of the ideological impetus which led to the passing of a ramifying web of municipal by-laws to regulate the daily routines and habits of the Asian plebeian classes. In short, sanitary control became the 'mainspring of municipal action'.[12] The first three sections of Chapter 3 examine contemporary colonial perceptions of death and disease in Singapore which were crucial in shaping the essentially environmentalist thrust of municipal schemes to improve public health. At the level of everyday practices, the sanitary project had to be translated into specific strategies of control. Chapters 3–5 explore the progression of municipal sanitary strategies from a primary focus on the surveillance of Asian daily practices in using the environment during the late nineteenth century (Chapter 3) to a greater diversification of strategies in the early twentieth century to include intervention with the built form of housing (Chapter 4) and the replacement of Asian systems of water supply and sewage disposal with municipal utilities (Chapter 5). The success of the sanitary project, however, did not depend solely on the capacity of colonial institutions to control and regulate the urban built environment, but also on the response and 'strategic conduct' of the colonized to the new regime of disciplinary techniques. Each chapter also explores Asian views with regard to health, disease, and the management of the urban environment, and specific counter-strategies to secure their own priorities. Through the interplay of strategies and counter-strategies, negotiation over the control of sanitary aspects of the urban environment played a key role in describing the relationship of power between rulers and the ruled.

1. David Arnold, 'Introduction: Disease, Medicine and Empire', in David Arnold (ed.), *Imperial Medicine and Indigenous Societies*, Manchester: Manchester University Press, 1988, p. 2.

2. Michael Worboys, 'British Colonial Medicine and Tropical Imperialism: A Comparative Perspective', in G. M. van Heteren, A. de Kneckt-van Eekelen, and M. J. D. Poulissen (eds.), *Dutch Medicine in the Malay Archipelago*, Amsterdam: Atlanta GA, 1989, p. 153.

3. Francis G. Hutchins, *The Illusion of Permanence: British Imperialism*, Princeton: Princeton University Press, 1967.

4. Michel Foucault, *Discipline and Punish: The Birth of the Prison*, trans. Alan Sheridan, London: Penguin Books, 1979, p. 198.

5. Daniel R. Headrick, *The Tools of Empire: Technology and European Imperialism in the Nineteenth Century*, Oxford: Oxford University Press, 1981.

6. Maryinez Lyons, 'Sleeping Sickness Epidemics and Public Health in the Belgian Congo', in Arnold, *Imperial Medicine*, p.107.

7. Malcolm Nicholson, 'Medicine and Racial Politics: Changing Images of the New Zealand Maori in the Nineteenth Century', in Arnold, *Imperial Medicine*, p. 79.

8. Maynard W. Swanson, 'The Sanitation Syndrome: Bubonic Plague and Urban Native Policy in the Cape Colony, 1900–1909', *Journal of African History*, 18 (1977): 387–410.

9. Nicholas Thomas, 'Sanitation and Seeing: The Creation of State Power in Early Colonial Fiji', *Comparative Studies in Society and History*, 32 (1990): 157.

10. Norman G. Owen, 'Towards a History of Health in Southeast Asia', in N. G. Owen (ed.), *Death and Disease in Southeast Asia: Explorations in Social, Medical and Demographic History*, Singapore: Oxford University Press, 1987, pp. 19–20.

11. The essays of David Arnold and Terence Ranger are significant exceptions. Focusing on specific infectious diseases in the Indian context, Arnold examines the management of epidemics and disease outbreaks to illumine the complex and shifting relationship among the colonial state, the indigenous élite, and the subaltern classes, illustrating the interventionist capacity of the colonial state on the one hand, and the potent though divergent responses of the colonized on the other (David Arnold, 'Cholera and Colonialism in British India', *Past and Present*, 113 (1986): 118–51; 'Touching the Body: Perspective on the Indian Plague, 1896–1900', in Ranajit Guha (ed.), *Subaltern Studies V: Writings on South Asian History and Society*, Delhi: Oxford University Press, 1987, pp. 55–90; 'Smallpox and Colonial Medicine in Nineteenth Century India', in Arnold, *Imperial Medicine*, pp. 45–65). Ranger pays particular attention to the African perception of the 1918 influenza pandemic and the emergence of African anti-medicine movements (Terence Ranger, 'The Influenza Pandemic in Southern Rhodesia: A Crisis of Comprehension', in Arnold, *Imperial Medicine*, pp. 72–188). See also the essays collected in Owen, *Death and Disease*.

12. 'Municipal Government in the Straits Settlements', *The Colonial Office Journal*, 4, 3 (1911): 220.

3
Municipal Sanitary Surveillance, Asian Resistance, and the Control of the Urban Environment

The science of sanitation ... is not hard to understand, and amounts to nothing more than carrying out cleanliness by scientific method.... Its laws, easily observed as they are, do not admit of neglect. Disregard of them swiftly brings down the penalty on the offending parties. The public health should override every other consideration when the prevention of disease is in question in a town like Singapore.[1]

Such is oft the course of deeds that move the wheels of the world: small hands do them because they must, while the eyes of the great are elsewhere.[2]

The Campaign for a Sanitized Environment in Colonial Singapore

AT the heart of the colonialist's concern with sanitation lay the fundamental issues of demography: that is, population, morbidity, and mortality. From the perspective of the colonial administrator, this had two aspects. First, it involved the question of 'the continued vigour and vitality of the [European] race when transplanted from temperate to torrid zones'.[3] At the height of Empire-building at the turn of the century, increasing numbers of British administrators, merchants, traders, and soldiers spent large parts of their lives in the colonies exposed to the ravages exacted by tropical conditions. The medical fraternity and the Colonial Office found it imperative to bring what were perceived to be largely climatic or environmental hazards prevalent in the tropical dependencies under some form of control. The setting up of Schools of Tropical Medicine in Liverpool and London in 1898 and 1899, respectively, to investigate 'tropical diseases' on 'systematic and scientific lines' was, according to Sir William McGregor, 'but one of the many means devised or fostered by the Rt. Hon. Joseph Chamberlain, then Secretary of State for the Colonies, for curtailing the toll of our fellow citizens in those insalubrious, over-sea territories of the empire'.[4]

Second, in port cities like Singapore, where prosperity was largely dependent on entrepôt trade and the energies of a continuous stream of Asian immigrants, the colonial state had also to look beyond the question of European demography and address the more intractable problems of Asian morbidity and mortality. Rampant spread of infectious diseases and high mortality rates would not only debilitate the local population, but cripple the colony's trade. That the latter was the colonial government's main preoccupation is clear from the reason given by the Colonial Secretary, J. A. Swettenham, for his injunction to the municipal commissioners to devise effective means of disease control within the municipality:

The prosperity of Singapore so entirely depends upon its use as a commercial emporium, and that use is so gravely jeopardised by the occurrence here of dangerous infectious disease which provokes other places to impose quarantine regulations against Singapore, that His Excellency [the Governor] cannot too earnestly recommend the Commissioners so to strengthen their Health Department so as to enable them to grapple with cases of disease.[5]

It was intolerable for a community whose main source of wealth depended on entrepôt trade to dwell under the shadow and suspicion of harbouring large numbers of cases of undetected infectious disease. As succinctly summarized by one of the local dailies, 'an epidemic mean[t] a great commercial loss to the place, apart from any more humane considerations'.[6]

In Britain, by the end of the nineteenth century, falling mortality rates seemed to indicate that the campaign for a sanitized environment had triumphed over the scourge of disease.[7] These sanitary reform movements and the crusade against epidemic smallpox and cholera in Britain created pressures for similar campaigns to rescue the colonies from the heavy toll exacted by disease. Advances made in the new sciences of bacteriology and parasitology imbued medical men working in colonial settings with a confidence that epidemic disease could be controlled and conquered through the application of western scientific knowledge and reason.

At the heart of the 'science' of sanitation lies the contention that by managing the environment and restructuring space using scientific principles, it was possible to banish disease and improve health. Sanitary science examined the urban geography of disease and in particular, 'its relationship with local environmental conditions and the location, distribution and migration of the population'.[8] As such, sanitary experts and medical men were, in Foucault's phrase, amongst the first 'specialists of space' who addressed four fundamental spatial concerns: 'that of local conditions such as climate and soil; that of co-existences between men themselves and between men and things, such as questions of density and proximity, water, sewage and ventilation; that of residences including environmental issues; and that of displacements such as the propagation of diseases'.[9] Through the sanitary discourse, the urban environment with its 'principal spatial variables' such as the 'disposition of various quarters, their humidity and exposure, the ventilation of the city as a whole, its sewage and drainage systems, the siting of abattoirs and cemeteries, [and] the density of population' became a 'medicalisable object' which demanded and justified 'scientific' intervention.[10] The confidence of sanitary experts in their 'science' was epitomized by the Malayan malariologist Dr Malcolm Watson's assurance that the day would come when, by precisely manipulating physical and chemical environments using scientific principles, one would be able to 'play with species of *Anopheles*, say to some "Go" and to others "Come", and to abolish malaria with great ease, perhaps at hardly any expense'.[11]

Sanitary science, and the campaigns to reform the environment which it inspired, began to gather impetus in the colonies from the last quarter of the nineteenth century onwards.[12] The preservation of health was no longer just a personal or domestic concern but an arena for public endeavour since individual health, even of hygiene-minded Europeans living in the tropics, was dependent on the sanitation of the shared environment. While the primary endeavour of sanitary science was to minister to the health of the colonizers, the living conditions of the colonized could not be ignored without certain perils: 'modern civilisation' is 'a condition of interdependence' and 'germs of diseases which obtain a footing among those who live in an unwholesome environment soon acquire the strength sufficient to enable them to overcome the superior resisting powers of those who themselves dwell in more healthy localities, and under better conditions'.[13] In a plural society like Singapore where the European minority lived in constant interaction with the Asian masses, the vulnerability of all social classes to diseases harboured by the Asian plebeian classes could not have been more evident. For example, Dr James Kirk, one of the principal medical practitioners in Singapore, expressed fears that the 'comparative immunity' enjoyed by the European class from disease would be nullified by 'the increasing invasion of the European residential quarters by native houses, especially those of the Chinese "shophouse" class'.[14]

Mortality rates in turn-of-the-century Singapore were unquestionably high among the immigrant and indigenous communities. Between 1893 and 1910, crude death rates ranged between 32.9 and 51.1 per thousand, with an average of 42.2 per thousand. Between 1911 and 1925, high death rates continued to prevail in 1911, 1912, and 1918 (over 42 per thousand), but the average had fallen to 32.3 per thousand, with the death rate dipping below 30 per thousand by the mid-1920s (Figure 3.1). The average death rates for the period between 1893 and 1929 varied dramatically among different races, particularly between the Europeans (14.0 per thousand) on the one hand, and the Malays, Chinese, and Indians (40.1, 39.9, and 32.7 per thousand respectively) on the other (Figure 3.2). The disparity is even greater if the infant mortality rates among different races are compared (Figure 3.3). Between 1900 and 1929, the

FIGURE 3.1
Annual Mortality Rates of the Total Municipal Population, 1893–1929

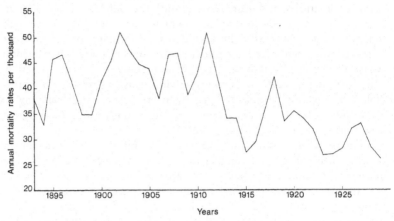

Source: ARSM, Health Officer's Department/Health Officer's Report/Municipal Health Office, 1893–1929.

average infant mortality rate for Europeans was 65.7 per thousand, far below those of the other racial groups—Eurasians (218.9), Indians (257.1), Chinese (305.4), and Malays (343.7).

These mortality figures were deeply embarrassing if not worrying to the Straits government as there were few extenuating circumstances which could explain their magnitude. Demographically, the colony's age structure comprised a large proportion of economically active adults and small proportions of the very young and old, a structure

FIGURE 3.2
Annual Mortality Rates of Principal Races in the Municipality, 1893–1929

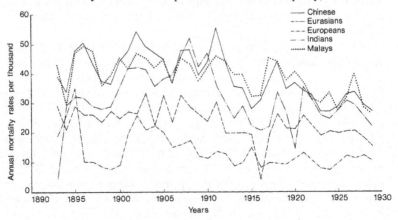

Source: ARSM, Health Officer's Department/Health Officer's Report/Municipal Health Office, 1893–1929.

FIGURE 3.3
Infant Mortality Rates of Principal Races in the Municipality, 1900–1929

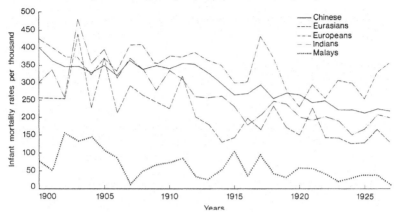

Source: ARSM, Health Officer's Department/Health Officer's Report/Municipal Health Office, 1900–29.

which should have produced low mortality rates. Epidemiologically, Singapore was relatively free from devastating epidemics such as small-pox, plague, or even cholera and enteric fever. Dr Charles Dumbleton, Singapore's first municipal health officer in the early 1890s, also argued that it was invalid to attribute high mortality rates to the insalubrity of the tropical climate or the weakness of the Asian constitution, because Singapore's death rates appeared unfavourably high even in compari-son with Indian and other eastern cities.[15] He concluded that the dis-parities between mortality rates in Singapore and the other cities could be traced to a difference in municipal purpose, principle, and commit-ment. He claimed that in other cities, 'the authorities ... have not only, by a careful study of vital statistics, discovered the cause of their mor-tality, but that they have also been alive to sanitary requirements, that the grants laid out for sanitary purposes have been the means of saving myriads of lives'.[16] In contrast, in Singapore, 'it cannot be said that any serious attention ha[d] been given to this important branch of Public Health by the Municipality'.[17] The experience of other cities embodied the ideal formula for successful sanitary reform: death and disease first had to be statistically collated and reduced to quantifiable variables amenable to scientific investigation and analysis. This must be coupled with sanitary zeal and knowledge of the principles of sanitary science on the part of the authorities. It was incumbent on municipal author-ities to seek the qualified and hence authoritative opinion of Western experts in the fields of public health and tropical engineering. Finally, science had to be backed by financial commitment to embark on sani-tary projects designed to ameliorate or even eliminate diseases. In reality, however, the ideal scheme was thwarted by a number of factors and could seldom be adhered to.

The Aetiology of 'Filth' Diseases and Colonial Perceptions of Asian Domestic Practices

The diseases which were of particular interest to the Municipal Health Department were those belonging to the 'zymotic' class.[18] According to the Registrar-General's returns for the municipality, zymotic diseases which consistently felled the greatest numbers throughout the 1890s, 1910s, and 1920s were tubercular diseases, beriberi, and malarial or remittent fevers (Figure 3.4).[19] Another group of zymotic diseases, far less dominant in numerical terms but of particular interest to the Municipal Health Department, were the 'dangerous infectious diseases' such as cholera, enteric fever, smallpox, and bubonic plague (Figure 3.5) which were capable of assuming epidemic proportions within a short space of time.

According to the aetiologies of zymotic diseases as established by contemporary epidemiological theory, diseases sprang from filthy habits and insanitary environments. The earlier preoccupation with the tropical climate and natural topography as the main sources of ill-health[20] had by the last quarter of the nineteenth century given way to theories which espoused human behaviour and the conditions of human habitation as the chief causes of disease. For example, in the early twentieth century, Revd J. A. Bethune Cook, a Presbyterian missionary who spent over a quarter of a century in Singapore, claimed that the climate of Singapore was 'most equable' and attributed the prevailing high mortality rates to 'the insanitary and immoral lives of the Asiatic races, ... Government and municipal neglect in the past, and the utterly unreasoning and unnecessary alcoholic habits of nearly all races'.[21] Medical opinion held that specific zymotic diseases such as

FIGURE 3.4

Mortality Rates from Selected Principal Diseases, 1893–1925

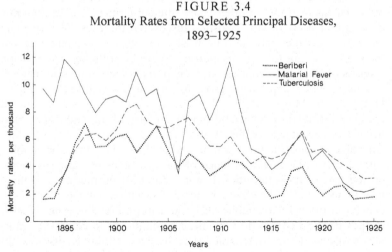

Source: ARSM, Health Officer's Department/Health Officer's Report/Municipal Health Office, 1893–1925.

FIGURE 3.5

Mortality Rates from Selected Epidemic Diseases, 1893–1925

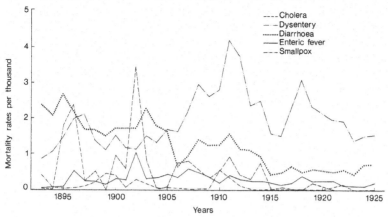

Source: ARSM, Health Officer's Department/Health Officer's Report/Municipal Health Office, 1893–1925.

cholera, enteric fever, dysentery, and diarrhoea were 'filth' diseases 'carried by dirty people to dirty places'.[22] Outbreaks of zymotic diseases such as the cholera scourge in 1895, which claimed 318 lives out of 430 reported cases within four months, were hence viewed as ominous warnings of further retribution for sanitary neglect and as an 'incitement to proceed energetically with such reform as [would] conduce to the removal of glaring sanitary defects which [had] been allowed too long to exist in the town'.[23] At the turn of the century, even a disease like beriberi, later proved to be a deficiency disease,[24] was perceived to be inextricably conflated with insanitary environments. Dr Max Simon, the principal civil medical officer, argued that 'the real cause of the disease [was] to be found in "*mal*-aria" in its broadest sense'.[25] He claimed that the prevalence of the disease was 'in proportion as the surroundings [were] insanitary'.[26] He categorized environments most conducive to the beriberi scourge into three groups: those which were naturally susceptible such as 'low-lying damp places, with badly drained alluvial soil'; environments rendered vulnerable by economic activities or despoiled by 'general overcrowding and Asiatic insanitation'; and entirely artificial environments such as to be found on ships 'long at sea with an Asiatic crew on board'.[27]

Typical of the nineteenth century division of causes of diseases into predisposing conditions and efficient causes, it was believed that suitable moisture and temperature conditions of the soil combined with organic contamination 'especially with excremental filth swarming with bacteria' provided conditions predisposed to an outbreak of zymotic diseases.[28] The 'latent poison' which developed as a result of the above predisposing conditions was either washed into surface wells or expelled into the atmosphere with the next rise of groundwater during the wet season. The 'poison' was then conveyed to human victims through a variety of means, the chief medium being contaminated well water used for drinking, bathing, rinsing, and cleaning household utensils and houses. However, it was claimed that the 'poison' could also be transmitted through the vehicle of food, shellfish collected from the rivers and foreshore,[29] unwashed hands, as well as emanations from evacuations voided by patients which could be inhaled through the mouth and swallowed.[30] 'Asiatic habits', including 'carelessness regarding the removal of excreta and household refuse' hence allowing filth to accumulate in the house, using polluted well water despite municipal

FIGURE 3.6
Percentage Distribution of Reported Cholera Cases, 1900–1904

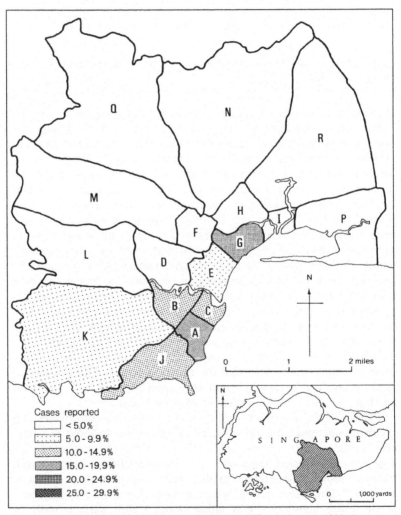

Source: ARSM, Health Officer's Department/Health Officer's Report, 1900–4.

attempts to close contaminated wells, ignorance and indifference to infectious diseases as exhibited by large numbers of men sharing occupancy of a room with an infected patient,[31] provided both the bacteria (contained in excrementitious filth) as well as the medium (polluted water and vitiated air) for disease to prevail.

The association of a 'filth' disease like cholera with the Asian populace, particularly the Chinese, was further strengthened by spatial statistics collated by the municipal health officer. Figure 3.6 shows the percentage distribution of reported cholera cases within the municipality for 1900–4, a period with serious outbreaks of cholera. The districts which tended to be most consistently and severely affected by cholera outbreaks were mainly district A (the southern portion of Chinatown), B (the area around Pearl's Hill), C (the northern portion of Chinatown), G (the area around Kampong Glam), and J (the Tanjong Pagar Dock area). Districts A, B, and C were traditionally Chinese-dominated areas, with the Chinese comprising 87.1 per cent, 89.6 per cent, and 96.4 per cent of the district population respectively, according to the 1901 census. District G, the traditional domain of the Malay Sultan and his retinue, had by the turn of the century been infiltrated by increasing numbers of Chinese, overspilling from the densely-packed traditional Chinese quarter across the Singapore River. High rates of cholera cases in district J were not unexpected as the port and harbour were considered the 'gateways to disease'[32] as many infectious diseases were thought to have been imported from ships calling at the port of Singapore. The impression that cholera tended to prevail in predominantly Chinese areas was further confirmed by the racial breakdown of cholera victims for the years with severe outbreaks of cholera: 84.5 per cent of all notified cases of cholera between 1895 and 1925 were Chinese,[33] mainly males of the coolie class in the 25–45 age range.

In sum, contamination, filth, and a dangerous disregard for dirt were from the municipal perspective symptomatic of Asian domestic practices. In turn, the latter were perceived as highly inimical to health and instrumental to the propagation of disease.

The Scourge of Tuberculosis and Asian Housing Conditions

While the fear of cholera, enteric fever, beriberi, and other so-called 'filth' diseases was inextricably conflated with repugnance against Asian domestic practices, increasing concern for the possible contagiousness of another principal disease—phthisis or tuberculosis—drew municipal attention to another aspect of their domestic arrangements, namely, Asian housing and sleeping conditions. By the mid-1890s, the bacterial mode of disease transmission for a disease like phthisis had gained currency among the medical fraternity in Singapore. Dr Lim Boon Keng and Dr W. R. C. Middleton described the disease as being spread by the inhalation of 'bacillus-laden air'.[34] The bacillus, according to the theory, was found in abundance in the sputa of infected persons, and in dried sputum the bacillus was capable of retaining its virulence for a very long period of time. The dangers from sputum arose 'not so much in the open air, as in the habitation or the room occupied by a patient'[35] where the air, breathed and rebreathed, became increasingly vitiated and foul. The principal local conditions which favoured the prevalence of the disease were first, overcrowding and ill-ventilation of dwellings, consequences of the Chinese 'talent' for 'huddling together',[36] and second, 'the habit of promiscuous spitting so universal among the Chinese'.[37] The association of tuberculosis with insanitary environments and filthy personal habits was often explained using the analogy of the parable of the sower: the 'seeds' (bacilli) were everywhere but it was the 'soil' (environment) which was important:

A Tuberculous soil is necessary before the Bacillus can get implanted and cause disease. All men ... are more or less prone to be Tuberculous, but only when a Tuberculous soil is made possible by such vicious conditions as Poverty, Insanitation, Over crowding, worry and anxiety, strenuous life, late and long hours, Sexual and Alcoholic excess, that a latent affection can be turned into an active disease. Not only is a favourable soil necessary for latency to become active, but also for the spread of infection from one person to another.[38]

Tuberculosis was hence considered a disease which could be prevented by cleanliness, good ventilation, and a moderate and temperate lifestyle. The reform of housing and the personal habits of the population were the chief aims of the sanitarian.

It was therefore not surprising that on encountering a death rate of eight per thousand from tuberculosis and eleven per thousand from respiratory and tubercular diseases in Singapore,[39] Professor W. J. Simpson, an expert with the reputation of being the 'stormy petrel of tropical hygiene'[40] who had been commissioned by the Governor to investigate the reason for high death rates in Singapore in 1906, immediately directed his attention to housing conditions in the city.[41] Professor Simpson's investigations led him to identify the incidence and recrudescence of tubercular diseases with a particular type of house form commonly found throughout the city: the Chinese built shophouse. Figure 3.7 records the *in situ* number of deaths from tuberculosis between 1901 and 1905 in two blocks of Chinese shophouses (taking up the whole of Upper Chin Chew Street and one side each of Upper Nankin Street and Upper Cross Street) typical of those found in the heart of Chinatown. It illustrated the high prevalence of the disease and its tendency to recur, sometimes assuming epidemic form in this type of housing.[42] Professor Simpson observed that while the city was spared the evils of narrow streets and winding lanes and instead inherited the legacy of orderly arrangement, wide streets, and open spaces as a result of Raffles's foresight in town planning, these potential advantages to the health of the city were increasingly nullified by the peculiar traits of Chinese house design and construction.[43] The Chinese, according to Professor Simpson, preferred to build horizontally rather than vertically: 'enlargement of the [Chinese] house is not

FIGURE 3.7

Incidence of Deaths from Tuberculosis in Two Shophouse Blocks
in Chinatown

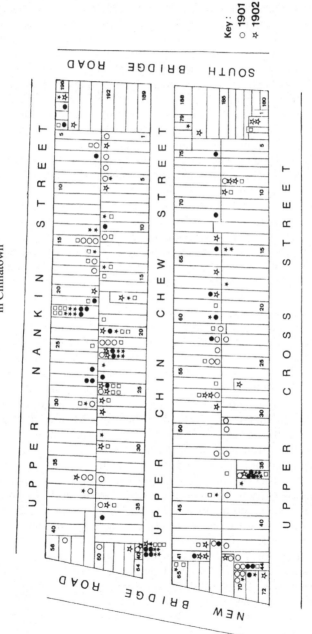

Source: W.J. Simpson, *Report on the Sanitary Condition of Singapore* [hereafter cited as Report on the Sanitary Condition],
London: Waterlow and Sons, 1907, pp. 10–11.

effected by adding another storey to heighten it, but by erecting another building behind, ... separated by a courtyard. There may be a series of buildings each with courtyards approached by passages leading one from the other.[44]

As the process of horizontal accretion was confined within the rectangular blocks into which urban land had been partitioned, it eventually encountered obstructions at the rear by coming into contact with similar additions made to the house immediately at its back but facing the next street of the block. At this point, the house-owner might either 'build vertically and contract the courtyards', or 'acquire the property behind and get a communication right through to the other street'.[45] Figure 3.8 illustrates a case where both options had been pursued. Originally two separate houses placed back to back in a rather narrow block, the processes of horizontal fusion and vertical additions had culminated in a four-storey tenement house which ran through from Sago Street to Sago Lane. This state of affairs, typical of many other blocks in the Chinese quarters, was condemned as 'destructive to health, lighting and ventilating of houses, and the efficient scavenging and drainage of the houses'[46] for several reasons.

First, such a style of house construction excessively covered the house site of a building and 'shut out from the interior an adequate and necessary supply of sunlight and fresh air'.[47] Tenements normally had only a single point of access through the front door on the ground floor, and in construction little heed had been paid to light and air planes. In Figure 3.8, the existing angles for open space (77–78°) fell far short of the recommended angles (45°, or at most 56°).

Second, in the case of tenement houses, defective house design was further aggravated by the subdivision of each floor into a large number of cubicle rooms (Plate 5). Since the houses had no lateral windows, these cubicles were 'windowless rooms, dark and cheerless, receiving neither light nor air direct from the outside'.[48] Furthermore, the cubicles were filthy, overcrowded with inhabitants, and cluttered with goods (Plates 6 and 7). Table 3.1, which summarizes Professor Simpson's detailed survey of 20-3 Sago Street and 18 Sago Lane, shows that the average space allotted to each occupant in the tenement block ranged from only 25 to 68 square feet. Such estimates did not take into account space taken up within the cubicles themselves by goods and other paraphernalia essential to the occupants' trades. Thus, Professor Simpson discovered a general goods hawker in one of the cubicles sleeping on a bench 18 inches below a shelf containing boots, shoes, pipes, whistles, soaps, towels, lamps, and tooth powder, whilst in another cubicle, a fruit and food hawker shared his tiny space with apples, eggs, preserved pig, mangoes (many overripe and rotten), and other kinds of fruit.[49] For the Chinese coolies who rented out these cubicles, spaces for work, sleep, and storage were undifferentiated and in many cases, one and the same. The sick breathed the same air and shared the same space as the healthy, and all activities occurred amidst

FIGURE 3.8
20-3 Sago Street and 18 Sago Lane: A Four-Storey Tenement Enlarged by Horizontal and Vertical Accretion

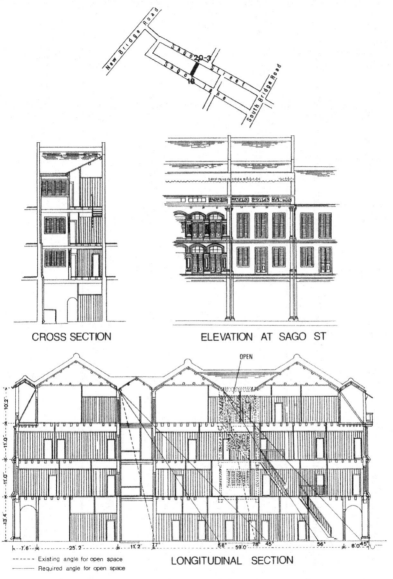

Source: Simpson, *Report on the Sanitary Condition,* pp. 14–15.

TABLE 3.1

Occupant Density and Average Cubicle Space per Occupant in a Typical Chinese Tenement House
(20-3 Sago Street and 18 Sago Lane)

Street/Floor	Total Number of Occupants	Total Number of Occupied Cubicles	Total Area of Cubicles	Average Number of Occupants per Cubicle	Average Space per Occupant (sq. ft.)
Sago Street/Ground	19	10	592.5	1.9	31
Sago Street/First	14	6	697.0	2.3	50
Sago Street/Second	15	7	776.0	2.1	52
Sago Street/Third	16	6	591.0	2.7	37
Sago Lane/Ground	11	4	271.0	2.8	25
Sago Lane/First	11	6	457.5	1.8	41
Sago Lane/Second	7	3	477.5	2.3	68
Sago Lane/Third	6	2	176.0	3.0	29
Total	99	44	4,038.5		
Average				2.3	41

Source: Simpson, Report on the Sanitary Condition, pp. 17–24.

'immense quantities of rubbish and filth'.[50] It is clear from the municipal health officer's earlier description of 'native' housing that such 'an
appalling state of affairs in the life below the surface'[51] was not peculiar
to the houses surveyed by Professor Simpson but typical of the older
parts of Chinatown:

The houses in the native quarter, and particularly those in the older parts of
town, exhibit many sanitary defects. Blocks of houses are built back to back.
Large rooms are frequently divided up by the erection of wooden partitions
into cells [of] only a few square feet, and above these lofts are frequently constructed, access to which is obtained by movable ladders.... Such houses are
frequently overcrowded. Little, if any, light ever enters; lamps are kept burning
in many cases day and night; ventilation there is none, and the air space is
reduced by the personal belongings of the inmates and by collections of odds
and ends of rubbish.[52]

Third, the situation of the houses within the building block and in
relation to one another further exacerbated the already deleterious conditions. In the traditional Chinese quarter, both narrow and wide
blocks were filled up and packed with buildings, which, in the narrow
blocks formed back-to-back houses, and, in the wider blocks, created
congeries of buildings abutting one upon another as in a piece of
mosaic work (Figure 3.9). Such an arrangement intensified the
obstruction to the admission of light and the free circulation of air
already hindered by the design of the individual houses themselves.[53]
In some of these blocks, the amount of open space to the covered-over
area was only 6–7 per cent (Plate 8), instead of 33 per cent, the
required health minimum.[54]

By the 1920s, phthisis was the 'new champion' in chalking up the
largest number of deaths within the municipal area annually (Plate 9).
At the crux of municipal understanding of phthisis, a disease which
continued to hold the city to ransom even after the Second World
War,[55] lay the belief that a panacea could be found for the disease by
ensuring that each individual is assured of an adequate supply of pure
air and light. Such a view is clearly epitomized by the principal civil
medical officer, Dr D. K. McDowell's propositions that 'wherever the
amount of air to the individual is increased when deficient, the death
rate from tubercle of lung has been greatly reduced', and that '[i]f all
inhabited rooms were properly lighted the health of the community
would improve materially'.[56] Municipal sanitary reforms were hence
motivated by the quest for better health through the banishment of
overcrowding, darkness, stale air, filth, and clutter from Asian
tenements.

From the vantage point of the municipal authorities and medical
men in Singapore, the entire catalogue of Asian and especially Chinese
domestic practices and habits—'overcrowding, the careless and filthy
habits ... with regard to the collection and removal of refuse, the urin-
ating and defaecating in house drains and in the neighbourhood of
wells, the almost universal habit of spitting, ... disregard of the dangers
of infection'[57] and 'their scant love for water and soap'[58]—signified all

FIGURE 3.9

A Typical Mosaic of Back-to-back and Abutting Houses in Chinatown

Source: Simpson, *Report on the Sanitary Condition,* pp. 26–7.

those features which seemed to set the Chinese irrevocably apart as a most insanitary race characterized by 'incurably filthy and disorderly habits'.[59] While the habits and customs of other Asian communities attracted less attention, they were in no way considered more acceptable or sanitary. Dr D. J. Galloway, one of the chief medical practitioners in Singapore, for example, alleged that there was 'nothing in the [Mohammedan] religion which tends towards sanitation as we [European medical men] interpret the word' and instead felt that much in that religion militated against hygiene, including 'the strict injunctions regarding the disposal of faeces and urine, the frequent ablutions (both tending to contaminate the water supply), the absence of scavenging efforts of the domestic pig and dog, and ... most conspicuously, the ritual of washing of the bodies of the dead, whether they have died of contagious disease or not'.[60] In the colonial explanatory scheme, these traits were attributed to the basic nature of Asians and their intrinsic racial peculiarities rather than to the inequalities and contradictions inherent in colonial society itself. One of the local dailies, for example, alleged that responsibility for the insanitary conditions of

Singapore did not lie at the door of the municipal and government officials but instead, 'blame [lay] primarily with the Chinese residents, who [were] filthy in their habits beyond all European conception of filthiness'.[61] 'There [was] not a law known to the student of hygiene which they [did] not break', the paper continued, 'and outraged nature [took] its toll periodically in epidemic death-rate'.[62] Disease was hence the *natural* consequence of racial characteristics and could be divorced from the social and political context of colonialism. By concentrating attention on the Asian environment as the cause of disease and the focus of public health intervention, the Municipal Health Department eschewed arguments which related mortality and disease to poverty and economic deprivation, or which questioned the colonial socio-economic structure. The characterization of Asians as 'incurably filthy' or the 'Chinaman' as 'the most insanitary of mortals'[63] also obviated the colonial government's responsibility to provide sufficient financial outlay for costly sanitary facilities in the city, for it was argued that even if the government were willing to make the 'financial sacrifice', 'oriental stubbornness would in all probability continue to cling on to old habits and customs, and hang on to dirt'.[64] Yet it is clear that by labelling and stereotyping the Asian as 'befogged', 'stubborn', or 'prejudiced', and in calling insanitation a 'hereditary Chinese instinct', this was in itself a clue that Europe had not completely dominated the colonial relationship and that there were limits to the extent to which the colonized could be made to fit imperial visions.

Municipal Strategies of Disease and Sanitary Control: Surveillance

Imbued with an unquestioned confidence in the supremacy of Western sanitary science on the one hand, and an antipathy for Asian domestic practices on the other, the Municipal Health Department contrived various sanitary strategies to improve the health of the city. The accumulation of a corpus of Western epidemiological knowledge as well as increasing local evidence of disease-impregnated Asian environments, which seemed corroborated with every official investigation, provided grist to the sanitary mill. Knowledge, scientific, statistical, and anecdotal, is intimately related to power (as Foucault has emphasized through his *pouvoir/savoir* couplet), both being constructed on the basis of concrete 'technologies' and utilized in specific 'strategies'.[65] These strategies to sanitize Singapore were not divorced from discourse and social policy in metropolitan countries. In English Victorian cities, concern with the degradation of the urban poor had often been expressed in the language of bourgeoisie morality: physical miasmas were said to correspond to moral miasmas and sanitary legislation often provided the means to attack overcrowded slums, which were identified as the 'rookeries' or 'haunts' of the 'dangerous classes' and 'hot-beds' of social decay, 'cholera, crime and Chartism'.[66] The main thrust of strategies to root out physical and moral disease from the cities had been through spatial reorganization and environmental reform, bringing decay originally hidden and concealed within the 'lower orders' into the ambit of official surveillance and the public gaze.[67] The campaign for sanitation was thus largely concerned with order, openness, visibility, ventilation, and the spatial demarcation of different activities. These sanitary preoccupations and strategies were to have significant parallels in the colonial context.

In colonial Singapore, public health was perceived largely in terms of racial categories and the conflation of physical disease with Asian degeneracy was assumed and accepted without question. Strategies to sanitize the city were characterized by a tendency to disparage Asian customary practices in managing health, disease, and the environment and to privilege either a replacement of these practices by Western-inspired ones, or the subjection of Asian practices to the rigours of inspection, regulation, and disciplinary action. Among the strategies frequently discussed in health reports was that of replacing *in toto* Asian practices of sewage disposal, water supply, and medical care which had developed under an era of *laissez-faire* with municipally-administered systems.[68] A second strategy which was given prominence by Professor Simpson's 1907 report, advocated the re-ordering of built form in congested parts of the city inhabited by the labouring classes through area reconstruction, backlane schemes, and the introduction of open spaces into the closely woven fabric of the city. In practice, at least up to the first decade of the twentieth century, the advancement of these strategies was held in check by several constraints, the most immediate being the heavy financial commitment involved in any of these schemes.[69] As wholesale replacement of Asian systems and recon-struction of the Asian built environment were financially unviable, the municipal authorities depended on a third strategy which required rela-tively small financial outlay but instead had the force of requiring Asians to reform their habits, customs, and daily management of the environment using their own resources. This was the strategy of sur-veillance, through which the sanitary conditions of the city were to be secured.

From the municipal perspective, satisfactory regulation of the envir-onment depended on effective surveillance to break down the barriers which kept the physical and immoral disorder of the Asian classes 'secluded from superior inspection and common observation'.[70] It was a prevalent and peculiarly Victorian belief that by breaking down the barriers and rendering the labouring classes 'visible', they would be transformed by careful policing and appropriate punishment on the one hand, and by the inculcation of good and clean habits through the 'benign gaze' of their superiors on the other.[71] The success of surveil-lance as a strategy of control depended on whether it was able to work in a manner that was pervasive, meticulous, and uniform, or in Foucault's terms, on its ability to infiltrate, re-order, and colonize.

In colonial Singapore, the practice of surveillance involved the char-acterization of the population using racial categories in order to facilitate classification of information; the accumulation of detailed information pertaining to the Asian population from population censuses, special surveys, and numerical returns generated through the daily grind of the colonial bureaucracy; and the constant supervision of various aspects of Asian life to ensure the enforcement of municipal by-laws. These aspects of surveillance, as Anthony Giddens argues, 'are closely related in principle as well as frequently in practice in virtue of the fact that the collection, synthesis and analysis of information about the members of a society can either be an aid to, or constitute a direct mode of, surveillance over their activities and attitudes.[72]

According to the municipal health officer, Dr Middleton, effective surveillance to arrest the diffusion of diseases and ensure conformity to sanitary standards depended in the first instance on the systematic collection of birth and death statistics. Only thus could the occurrence of infectious diseases be detected and 'the localization of insanitary conditions insured'.[73] Provision was first made for the registration of births and deaths by Ordinance VIII of 1868. Under this ordinance, in the case of a death not medically certified, a native policeman visited the house, inspected the body, made enquiries of friends, and submitted his report to the officer in charge of his station who then forwarded all reports to the Registrar-General. Apart from the deliberate suppression of information, the system was notoriously unreliable as police constables were not medically trained to ascertain the cause of death.[74] In 1896, in an effort to improve the system, two qualified apothecaries were appointed as deputy registrars in charge of examining corpses and ascertaining the causes of death.[75] As an additional safeguard against 'dangerous infectious diseases', the Municipal Ordinance of 1896 also provided for the compulsory notification of diseases such as cholera, enteric fever, bubonic plague, smallpox, diphtheria, and erisepelas.[76] Details of reported cases of infectious diseases, in terms of the locality of occurrence, the race, sex, and occupation of the victim, whether treated at home or in hospital, and whether the case resulted in recovery or death were meticulously recorded. These records were expected to facilitate immediate, effective measures to arrest the spread of any infectious disease as well as to provide guidelines for sanitary work in the aftermath of the outbreak to prevent a recrudescence of the disease.[77] Given the extensive prevalence of insanitary dwellings in many parts of the municipality and limited resources allotted to sanitary work, the incidence of 'dangerous infectious diseases' such as cholera provided an 'index of salubrity' to distinguish the 'worst' housing for the immediate attention of the sanitary staff.

In 1895, in order to facilitate municipal supervision of Asian life, the municipality was organized into six sanitary districts which formed the basis for the compilation of disease data and returns of nuisances and sanitary offences. By 1908, these had burgeoned into eighteen districts. Sanitary inspectors were assigned to each district and armed with powers conferred by a proliferating array of municipal by-laws to carry out inspections of premises, post notices, serve court summons on trangressors of municipal sanitary standards (Table 3.2), and remove persons suspected to be suffering from infectious diseases (Plate 10). Substantive sections of the Municipal Ordinance of 1887, 1896, and subsequent amendments were devoted to the control and policing of various aspects of Asian domestic practices, including scavenging, the removal of night-soil and household filth, the organization of space and the proportion of people to space within buildings, the maintenance of premises, privies and drains in a sanitary state, the proper use of water, and the prevention of the spread of infectious diseases. Surveillance also went hand in hand with a meticulous system of fines and penalties calculated to punish the transgression of any detail of municipal sanitary law. It was hence through the sanitary inspectors and police officers that the hierarchy of supervision and inspection, which ultimately culminated in the Governor, penetrated the hidden recesses of the Asian house or shop.

While the system of surveillance was effected on a day-to-day basis through municipal officers proceeding on their rounds and the police setting a watch for sanitary transgressions in public places, it could also be galvanized into more vigorous action when confronted with the threat of epidemic disease. Surveillance was intensified in localities suspected of harbouring infectious diseases by taking sanitary inspectors off their usual rounds and detailing them for house-to-house inspection, the removal of infected patients, and the disinfection of houses.[78] Sanitary officers were also empowered to remove cubicles, compartments, lofts, galleries, outhouses, and other structures in order to facilitate the exposure of disease and the cleansing and disinfection of premises.[79] As effective supervision was predicated on accurate information, special regulations could also be enacted for 'supplying to the Municipal Health Officer the names of all persons residing in infected premises, with the addresses of those who went to live elsewhere'.[80] The municipal system of sanitary surveillance, with its concomitant powers of description, penetration, and re-ordering, was hence most keenly felt by the plebeian classes during the onslaught of infectious diseases.

Manpower assigned to the enforcement of municipal sanitary law was, however, severely limited. In 1894, the sanitary staff comprised one health officer, six or seven inspectors, and a similar number of peons and detectives.[81] The post of deputy municipal health officer was created and sanctioned in 1897; and by 1909, the number of sanitary inspectors had increased to eighteen and an additional health officer appointed.[82] In addition, the Municipal Commission contributed annually towards the maintenance of the police force and in return was entitled to assistance from the six hundred odd police officers employed within the municipality.[83] However, in the enforcement of 'sanitary order and decency',[84] the efficacy of the police officers fell far short of anticipation. The municipal president, Alex Gentle, complained that offences against municipal law were habitually committed in the sight of the police and had gone unnoticed,

TABLE 3.2.

Municipal Summons Cases: Filthy Premises, Filthy Latrines, and Dumping of Rubbish

Name/Address Where Offence Occurred	Charge	Plea/Statement of Defendant	Sentence
Vellupillay 23 Jalan Sultan (principal tenant)	Filthy premises	Claimed that the house was not dirty	$3 + costs
Kaloon 116 Prinsep Street	Filthy premises Non-removal of night-soil	Stated that the house was not his, and he only slept in five-foot-way	Cautioned and discharged
Mohd Unos 62 Queen Street (principal tenant)	Filthy premises	Claimed that the sanitary inspector did not inspect his house or latrine	$10 + costs
Ang Hong 20 Mansur Street	Deposition of rubbish on the street	Stated that he was not the man in question	$2 + costs
Koh Tiah Hu 59 Orchard Road	Deposition of rubbish on the street	Stated that rubbish was deposited in a cart	$3 + costs
Alagapa Chetty	Filthy premises	Claimed that filth was caused by others	$3 + costs
Muna Govinda	Depositing night-soil and urine into street drain	Called witnesses to testify that night-soil was caused by children	Dismissed
J. Frankel 26–7 Orchard Road	Filthy premises	Stated that he was not the occupier but only rented and sublet the premises	Dismissed
Tan Ee Yong 213–5 New Bridge Road	Filthy latrine	Claimed that night-soil bucket had been renewed	$2 + costs

(continued)

TABLE 3.2 (*continued*)

Name/Address Where Offence Occurred	Charge	Plea/Statement of Defendant	Sentence
Kader Meydin 22 Orchard Road	Filthy premises	Stated that he was only a subtenant and not the principal	Dismissed
Yunus 32 Queen Street (principal tenant)	Filthy premises Previous conviction	Stated that sanitary inspector threatened summons unless he paid bribe-money	$5 + costs
Saleh bin Mohd 72, 73 Queen Street	Filthy premises Non-compliance with limewash notice	Claimed that premises were clean	$8 + costs for first charge, second charge dismissed as notice was not properly served
Abdul Gafar 76 Queen Street	Filthy premises Non-compliance with limewash notice	Claimed that premises were cleaned and limewash already done by contractor, and that sanitary inspector asked for bribe-money	$3 + costs for first charge, second charge dismissed
Alibux 54 Queen Street	Filthy premises Non-compliance with limewash notice	Claimed that premises were white-washed and generally clean	$3 + costs for first charge, second charge dismissed
Tan Kim Tin	Filthy premises Non-compliance with limewash notice	Pleaded guilty to first charge, claimed that he had already complied with the second	$3 + costs
Asummullah 63 Queen Street	Non-removal of night-soil Non-compliance with limewash notice	Claimed that a paid carrier removed night-soil daily and that he had already complied with limewash notice	$2 + costs on first charge, second charge postponed

Name/Address	Charge	Statement	Sentence
Govindasamy	Filthy premises	Pleaded guilty	$3 + costs
Nasumullah 71 Queen Street	Filthy latrine Previous conviction	Claimed that a paid carrier removed night-soil and cleaned latrine daily	$5 + costs
Govinevehn Cattleshed in Anderson Road	Drain obstructed with dung and urine	Dung was accumulated for European plantation	Costs only
Tan Ee Yong 213–5 New Bridge Road	Filthy latrine	Pleaded guilty	$2 + costs
Ahmad Din 15, 16 Tanglin Road	Filthy drain, not properly graded	Claimed that drain was flushed daily and was not dirty	Costs only
Si Mui 50–1 Tanjong Pagar Road (principal tenant of coolie house)	Filthy latrine Three previous convictions	Claimed that house was not dirty; bribe-money had been paid to sanitary inspector each month	$8 + costs
Peng Hee Keng 156 Rochor Rd	Filthy premises	Had nothing to say	$3 + costs
Yew Chew 100–1 Tanjong Pagar Road (principal tenant)	Non-removal of night-soil from latrine	Called witness to testify that night-soil had been removed earlier that morning	Dismissed
Chor Po Chwee House next to 61 Rochor Road (principal tenant)	Non-compliance with nuisance notice Filthy premises	Asked for time for first charge Claimed that premises were cleaned daily	$5 + costs

(continued)

TABLE 3.2 (continued)

Name/Address Where Offence Occurred	Charge	Plea/Statement of Defendant	Sentence
Gonapathy Pillay 34 Cross Street	Filthy premises Non-compliance with limewash notice	Did not know whether premises were clean as he was not there Asked for time regarding limewash	$5 + costs for first charge, 7 days given to complete limewash
Lim Hua Say (stallholder at Ellenborough Market)	Throwing out rubbish on the steps of Ellenborough Market	Claimed that he was sick and had a bad leg; left basket of rubbish on the steps and inspector kicked it over	$2 + costs
Syed Ahmad bin Gayah (landlord), Go Koh Chai (tenant) 17 houses along Lorong Krian	Filthy premises, drains and latrines	Landlord and tenant each held the other responsible	Landlord held responsible and fined $5 + costs
Kawana Usof 71 Orchard Road	Filthy premises	There was no dirt	$3 + costs
Sina Samalay 93–7 Queen Street	Filthy premises (first and ground floor coated with dirt and cowdung)	House was quite clean	$3 + costs
Manager of Khai Sun & Co. Canal Road	Allowing coolie to deposit rubbish in backlane (portion of envelope with name of defendant's firm stamped on it submitted as evidence)	Rubbish was always deposited in dustbox and emptied in municipal place in Wayang Street Pleaded guilty on court's advice	$2 + costs
Swee King 84 Jalan Sultan	Depositing rubbish in public road	Paper was blown there by the wind	Costs only

Kanoomal 57 High Street	Deposition of rubbish on the street	Rubbish was thrown into municipal box, not on the street	$5 + costs
Chia Chai Ming 5 Upper Nankin Street (principal tenant)	Filthy premises	Ground floor is sublet to others, calls upon evidence of subtenant that premises were clean	$5 + costs
Lau Sih Ping	Insanitary dwelling unfit for human habitation	Pleaded guilty	Costs Closing order (7 days)
Chua Yam Siah 23 Carpenter Street (principal tenant)	Filthy premises	Pleaded guilty	$5 + costs

Sources: Municipal Summons Note Book, 5 July 1912–13 December 1912; Municipal Summons Note Book, 16 December 1912–6 May 1913; Note Book of Municipal Cases No. 1, 3 March 1921–1 April 1924.

hence perpetuating the impression already common among the labouring classes that sanitary offences were 'no impropriety at all, or practically condoned'.[85] Police assistance in guarding houses under quarantine in cases of infectious diseases was also far from effective. The usual means adopted was to station a police constable near the house to control ingress and egress. Such a method of quarantining was practically impossible in the case of places with large numbers of occupants such as coolie depôts and lodging houses, and according to Gentle, 'even in other more hopeful positions, the plan [was] useless', largely because the constables selected for the service neither '[understood] the intention of their orders' nor '[had] intelligence enough to carry them out'.[86] In a particularly notorious incident, forty-one out of forty-seven contacts placed under quarantine regulations at 121 Queen Street as a result of the occurrence of plague managed to escape the eye of five police constables on quarantine guard.[87] The Inspector-General of Police himself admitted that a large number of men would be necessary to draw a complete cordon around an infected house and that instead of checking the spread of infectious disease, house quarantine only served to introduce the disease into police stations as it was 'useless to expect that men on duty ... [would] avoid having unnecessary communication with the inmates'.[88] That the police force were notoriously inefficient in enforcing municipal law might have made the system of surveillance less intrusive, but it also made it more arbitrary.

The arbitrary nature of the system of surveillance was further compounded by the fact that the municipal authorities were entitled to impose dominant Western-inspired definitions of what was to be taken as sanitary and insanitary without necessarily producing palpable proof of any threat or injury to the health of the people. The Municipal Ordinance of 1896, for example, conferred upon sanitary inspectors the authority to declare 'an act, omission or thing' a 'nuisance' subject to prosecution if it was in their perception 'occasioning or likely to occasion injury, annoyance, offence, harm, danger or damage to the sense of sight, smell or hearing'.[89] The essence of a 'nuisance', an elastic term which encompassed many possibilities from filthy premises to undesirable animals,[90] was hence intimately related to its degree of offensiveness in the eyes of the municipal minions, mainly Western-trained European and Eurasian men, rather than its actual threat to health.[91] It was hence an ideological construct imposed by the municipal authorities and was seldom similarly perceived or invested with equal significance by the Asian plebeian classes. Notions of dirt and purity were divorced from Asian perceptions and in practice, dictated by the arbitrary whims of sanitary inspectors involved. Hence, not only did the system of sanitary surveillance attempt to penetrate the Asian urban environment with the power of the municipal gaze, it tried to do so using Western-inspired notions and categories which were assumed, on an a priori basis, to be superior to those held by the inhabitants themselves.

Up to the end of the first decade of the twentieth century, municipal strategies of sanitary control hinged crucially on this system of surveillance coupled with the threat of prosecutions and fines. During this period, the municipal authorities by and large sidestepped the necessity of large-scale sanitary schemes which depended heavily on the municipal fund, such as a municipal system of night-soil collection, a modern sewage system, the compulsory clearance and rebuilding of extensive insanitary areas and the draining of swampy land to suppress malaria. The next section deals with the response of the Asian plebeian classes to municipal sanitary surveillance.

The Asian Plebeian Response to Municipal Sanitary Control

The success of municipal sanitary regulation depended not only on the vigour with which municipal strategies were pursued but also on the strategic conduct of the Asian plebeian classes, the principal subjects bearing the brunt of sanitary reform. From the municipal perspective, the urgency of sanitary improvement on the part of the Asians needed no further justification other than the alarmingly high mortality and morbidity rates. Impediments to the achievement of municipal sanitary ideals were invariably attributed to an ingrained Asian recalcitrance or what the municipal health officer, Dr Middleton, denigratingly described as the 'ignorance, apathy, superstition and even active opposition [of the Asians] to speedy accomplishment of sanitary improvement'.[92] From the perspective of the labouring classes, however, municipal purposes, in particular the sanitary cause, were little appreciated as means of improving civic life or mitigating the scourge of diseases in the city. In terms of its objectives, legislative framework and institutional style, the Municipal Commission, patterned after the municipal authorities of British towns and cities, was a form of control alien to the majority of the Asian plebeian classes. For the Chinese, civic authority and control were more firmly invested in the myriad clan, dialect, and territorial organizations to which they belonged. In 1896, Alex Gentle clearly discerned the wide disparity between municipal and Asian views of sanitary reform when he expressed forlorn hopes that the proposed amendments to the Municipal Ordinance would 'convince the ignorant people ... that sanitary rules [were] not the outcome of fitful zeal, or worse still, attempts at extortion on the part of a few proposed subordinate officers [of the Commission], but the settled purposes of the Government of the Colony, not to be evaded or resisted, but to be consistently enforced for the public good'.[93]

In discussing the difficulties encountered in improving health in the colonies, contemporary colonial literature tended to emphasize tensions between the 'ignorance' of the non-European population and the heroic endeavours of the Western-trained medical man. Seldom was the viability of alternative systems of health care acknowledged. The following comment is typical of the colonial view:

The struggle for victory over disease [in Malaya] is enhanced by the ignorance of the population. ... The first idea of the European who falls ill is to call in the doctor. With Asiatics it is the last. In those few words is presented the whole position. They make clear that day after day, year in and year out, the Government medical or health man wages a war of Titans. The giant battalions of disease, of ignorance, of distrust set themselves in array against him, and he is never anywhere else than in the forefront of battle.[94]

In elucidating the Asian response to municipal sanitary control, it is first argued that the plebeian response was not one of 'ignorance, apathy and superstition' but a rational reaction based on different interpretations, perceptions, and priorities derived from alternative systems of medicine and health care. Standard assumptions of mass passivity, quiescence, or ignorance must be revised, and that, indeed, even passive non-compliance on the part of large numbers of Asians could become a 'power' to be reckoned with. In certain cases, this could become strong enough to force the municipal authorities to recast their goals and priorities and modify their strategies in line with more realistic expectations.

Chinese Medical Resources and Chinese Attitudes towards Health and Disease

Ethnic medical systems provide a set of contextual parameters within which ethnic communities interpret matters of health, disease, death, and its relation to the wider environment. Among the immigrant Chinese for example, the strength of their own medical traditions and practices (which propagated a vibrant set of medical concepts at variance from those of the municipal authorities) was a crucial resource drawn upon in response to the imposition of municipal sanitary control which increasingly circumscribed their daily routines. The Chinese who came to the Nanyang brought with them traditions and practices embedded in a classical medical system of great antiquity which had survived several centuries of evolution in China. As opposed to folk medicine based on popular ideas about the causes and treatment of disease and simple remedies supplied by non-professional practitioners, a distinct Chinese medical system possessing 'a predominantly rational theoretical basis, a large corpus of medical classics and a secular class of physicians' had emerged by the Spring and Autumn period (770–476 BC).[95] In terms of organization, classical Chinese medicine had acquired a distinctive structure comprising a 'court' sector ministering to the Imperial family and its retinue and a 'public' sector serving the ordinary populace. The latter sector was divided into a 'professional' category comprising those who had served some form of medical apprenticeship, and a 'pavement' category who peddled popular remedies and prescriptions for a livelihood.[96]

Among the Chinese in colonial Singapore, a 'folk' conception of the cause of disease coexisted alongside a medical system based on traditional classics and an extensive pharmacopoeia. Among the coolie class in particular, the cause of disease was conceived in specific, concrete, and personalized terms: sickness and epidemics were often attributed to the malevolence of *gui* or evil spirits which had to be appeased by sacrificial offerings or driven off and overcome using ritualistic powers. Thus, in response to an outbreak of cholera in 1907 among rickshaw coolies residing in the Rochor district depôts, hundreds of coolies joined in ceremonies and processions, accompanied by drums, instruments, and the letting off of crackers to drive away the cholera demons.[97] At the end of a ten-day crusade,

Four ghost ships were made of bamboo and coloured paper, in the shape of Chinese junks ... and filled with paper money, rice, fruit and candles.... About forty *ricksha* coolies were dressed as demons with grotesquely painted faces and horns fastened on to their heads, and carried spears and spiked clubs. These formed the escort for the fleet. The three larger vessels were placed in the road way while certain monks began their incantation.... After a couple of hours, a fourth boat was brought out. It was pounced upon by the head monk and taken into the depôt, in which it was believed that the cholera demons were still lingering. The head monk ... exhorted the demons to get on board the vessel and leave the house. Excitement was great when the little boat got as far as the door and then turned back and careered wildly round the front room. After about twenty minutes of these antics the boat was coaxed into the road, and the devils represented by black, red, blue, green and yellow effigies were put on board the largest junk. Then the procession in a wild, mad rush, made its way to a landing-stage on the Rochor River, where crackers were let off, the head monk waved his sword in the direction of the river and exhorted the devils to depart, and the paper vessels were burned.[98]

While the 'folk' view of disease persisted throughout the nineteenth and early twentieth centuries, it was clear that by the turn of the century, various elements derived from the traditional organizational structure of Chinese medicine were also well established in Singapore to serve the rapidly expanding Chinese population.[99] The structure of the Chinese medical delivery system in Singapore comprised first, at the institutionalized level, a number of charitable medical organizations set up by wealthy Chinese merchants to provide free medical advice and treatment by Chinese physicians for indigent persons. The pioneer among these organizations was the Thong Chai Yee Say (later known as the Thong Chai Medical Institution) established in 1867.[100] It was first situated along North Canal Road and later moved to 3 Wayang Street,[101] both sites within the traditional core area of the Chinese. The Institution engaged physicians from China who provided free consultation and treatment to all Chinese regardless of clan or dialect affiliations. Medical prescriptions written out by the Institution's physicians also entitled patients to free medicines and drugs from a number of Chinese pharmacies which were reimbursed on a monthly basis by the Institution.[102] According to an 1892 petition from the trustees of the Institution to the government requesting a land grant,[103] the Institution served a total of 39,196 patients between 1884 and 1891. No records of patient numbers are available for the last decade of the nineteenth century and the first two decades of the twentieth century. However, it is reasonable to postulate that the number of patients rapidly increased during these decades as Chinese immigration rose to unprecedented levels. According to Li Bai Gai (Plate 11),[104] a prominent Chinese physician who first served for five years in the Thong Chai Medical Institution from 1901, he alone treated a total of 6,125 patients within three months when he returned to the Institution in 1922.[105] A second institution, the Sian Chay Ee Siah, was established in 1901 along Victoria Street to minister to the needs of the burgeoning Chinese population in the Rochor area.[106] In 1910, the first inpatient hospital providing Chinese medical care, the Kwong Wai Shui Hospital, was founded by Cantonese belonging to the Kwang Chou, Hui Chou, and Chao Ching prefectures at the former site of the Tan Tock Seng Hospital along Serangoon Road.[107] According to the provisions of the constitution at the time of the Hospital's inauguration, inpatients were restricted to Cantonese belonging to the three prefectures but out-patient facilities were open to all races.[108]

Another important element in the structure of Chinese medicine in colonial Singapore was the traditional Chinese medical hall or phar-macy (Plate 12). In 1883, a census taken by the British authorities recorded a total of 139 Chinese chemists and druggists in the Straits Settlements.[109] Between 1870 and 1928, there were at least fifty-eight Chinese medical halls established in various parts of Singapore by the Hakkas, the principal dialect group associated with the retailing of Chinese medical supplies (Table 3.3).[110] Besides dispensing Chinese herbs, drugs, and medicaments, medical halls also provided readily available advice on a wide range of ailments. For example, the man-aging proprietor of an unnamed Chinese dispensary testified during an inquest involving a case of oxalic acid poisoning that his range of duties included treating ordinary disease such as fever, diarrhoea, and dysentery, dispensing medicines to clients who patronized his shop, and attending to patients in their homes when sent for.[111] Besides dis-pensing medicines, Chinese druggists hence played a vital role in the delivery of a more complete system of medical care including consult-ation, diagnosis, and treatment. This was frowned upon by Western practitioners schooled under a system which drew hard and fast lines between specializations. Dr T. C. Mugliston, for example, condemned 'men who [kept] drug shops and who [did] not confine themselves to putting up the prescriptions of duly qualified practitioners' and recom-mended legislative action to remove them.[112]

While medical institutions and pharmacies catered to the general public, the care of the sick and aged among the Chinese was often based on the clan organization. Most Chinese clan and mutual benefit associations[113] owned buildings which were used not only for the conduct of the associations' business but also to accommodate the sick and aged amongst their members. Several clan associations ran recu-peration centres or sick-receiving houses which provided free food, shelter, and medical care for the diseased and chronically ill. For

TABLE 3.3
Chinese (Hakka) Medical Halls Established in Singapore, 1870–1928

Medical Hall	Address	Date of Establishment
1. Lau wan shan	Philip Street	1870s
2. Yong an tang	131 Tanjong Pagar Road	1891
3. Zhang yu he yao hang	Unknown	1890s
4. Wan an he	Philip Street	1890s
5. Wan an zhan	New Market Street	1890s
6. Zhou lan ji	Upper Nankin Street	1900s
7. He zhang chun	Philip Street	1900s
8. Wan tai he	Unknown	1900s
9. Xin wan shan	New Market Street	1900s
10. Tian shen tang	Bukit Ho Swee	1907
11. Shen he tang	313 Silat Gate (Telok Blangah)	1907
12. Pei shan yao ju	77 Rochor Road	1907
13. Wan shan tang	91 Cross Street	1908
14. Xing ling chun	139 Tanjong Pagar Road	1908
15. Pei an yao ju	39 Merchant Road	1908
16. Fu tai tang	12 Joo Chiat Road	1910
17. Ren shou tang	1324 Hougang 5½ mile	1912
18. He zhang chun yao hang	12 & 13 Philip Street	1913
19. Hua qiang yao fang	320 Lavender Street	1913
20. Da chuan he	844F Hougang 6 mile	1915
21. Song bo yao hang	14 Upper Nankin Street	1918
22. Wan an zhan	158 Tanjong Pagar Road	1918
23. Bu shen yao ju	841 Pasir Panjang Road	1918
24. Zi shen tang	82 Jalan Besar	1918
25. Wan shan quan	210 Selegie Road	1918
26. Wan an long	233 Orchard Road	1918
27. Yu an he	145–15 Holland Road 5 mile	1918
28. Ren an tang	280–1 Hougang 6 mile	1918
29. Wan sheng long	21–1 Bukit Timah Road 7½ mile	1918
30. Xing hua yao fang	Yishun 10 mile	1918
31. Ji shen tang	1 Thomson Road 5 mile	1918
32. Xing shen he	1346 Hougang 5½ mile	1920
33. An shen he	290 Geylang Road	1922
34. Pei yuan tang	109 Hill Street	1923
35. Wan shan zan	1341 Hougang 5½ mile	1923
36. Wan shan xing ji	150 Tanjong Pagar Road	1924
37. Da dong yao hang	91 Cross Street	1925
38. Da shen he yao hang	157 Havelock Road	1925
39. Wan shen tang	47 Kampong Bahru	1925
40. Nan shan zhan	109 Pasir Panjang 5½ mile	1925
41. Wan shan ji	789 South Bridge Road	1925
42. Chun shen tang	185 Jalan Besar	1925
43. Yu an tang	76 East Coast Road	1925
44. He rong chun	77A Paya Lebar Road	1925
45. Yan shou tang	89 Rochor Road	1926
46. Zhen yuan chun	145–15 Holland Road 5 mile	1926

(*continued*)

TABLE 3.3 (*continued*)

Medical Hall	Address	Date of Establishment
47. Wan xin zhan	23–1 Bukit Timah Road 7 mile	1927
48. Ren ai zhan	128 Boat Quay	1928
49. Wan shan zheng zhan	685 South Bridge Road	1928
50. Yong chun tang	153 Rochor Road	1928
51. Ren de tang	244 Rochor Road	1928
52. Wan ai he	28 Weld Road	1928
53. Wan shen chun	304 Balestier Road	1928
54. Fu chun tang	16 Kim Kiat Avenue	1928
55. Wan he xiang	193 Lorong 24 Geylang	1928
56. Wan shen he	37 Joo Chiat Road	1928
57. Fu chun tang	17 Joo Chiat Road	1928
58. Wan shen zhan	786 Changi 10 mile	1928

Source: 'List of Hakka Medical Halls', compiled by Chen Chin Ah of the Thong Chai Institute of Medical Research from *Xin Jia Bo Cha Yang Hui Guan Bai Nian Ji Nian Kan* [Souvenir Magazine of the Centenary Celebration of the Singapore Char Yong Association], Singapore, 1958.

example, the Chung Shan Association (Cantonese), which was first established in 1821, raised sufficient funds to purchase several houses along Peking (Pekin) Street to open a recuperation centre for its members right in the heart of Chinatown.[114] The Kiung Chow Association, the principal Hainanese association in Singapore, established the Lok Tin Kee, a recuperation centre for invalids, the aged, and the homeless amongst its members in 1902 at 30 Sungei Road[115] whilst the Char Yong Association, the largest Hakka district association, opened the Hakka Free Hospital for the reception of the destitute sick from the Taipu district in 1906. These centres were funded and managed by clan members with minimal support or interference from the municipality until 1913 when by-laws were passed to regulate the condition of sick-receiving houses.[116]

Finally, at the most basic level, the Chinese medical delivery system which emerged in colonial Singapore comprised freelance physicians or *chung-i* who operated from clan associations, temples, the market-place, or their own homes. These included those who had either received medical training by working in hospitals and pharmacies or served apprenticeships under family members or more experienced physicians. For example, a Hokkien *chung-i* testified that he had been employed in a Chinese medical hall before he became a *chung-i*. Having no college training, he had however learnt to treat fever and 'wind' from his father.[117]

In the absence of a municipal system of medical care and poor relief, the Chinese medical delivery system comprising medical institutions, pharmacies, clan-based recuperation centres, and *chung-i* filled a much felt void in the lives of the Chinese plebeian classes.[118] As a medical system based on different conceptions of disease, it also placed a different construction on the relation between the maintenance of health and the management of the environment. Whilst Western sanitary science advocated the removal of filth, the disinfection and ventilation of houses, and the isolation of the sick as essential preventive measures in stemming the tide of disease, Chinese medical theory did not necessarily imbue these measures with similar significance. Traditional Chinese pathological theory distinguished between internal and external causes of disease.[119] Illness was considered to result from a lack of moderation, either in the external sphere (for example, excessive dryness, damp, or cold) or in the internal sphere (for example, excessive joy, anger, melancholy, or trauma). These imbalances were then interpreted in terms of disharmony between the *yin* and the *yang*, the primogenial forces which governed the universe.[120] Diagnosis was based on detailed descriptions of symptoms and prescribed medications drawn from a rich pharmacopoeia. The Chinese physician Li Bai Gai, for example, classified common ailments in Singapore into fourteen categories, each characterized by meticulously observed symptoms.[121] Unlike Western sanitary science which regarded disease as the product of germs which were hence enemies to be conquered by 'scientific' means, traditional Chinese medicine focused on correcting imbalances and strengthening the body's resistance rather than attacking a pathogenic invader.[122] For example, in the treatment of beriberi[123] or *jiao qi*, Chinese medical men stressed sleeping upstairs to avoid damp, taking more beans and potatoes instead of rice, using purgative rather than tonic drugs, and cultivating a healthy mindset as a 'mind diseased [might] be prejudicial to the patient'.[124] Diseases marked by fever, diarrhoea, and rashes were considered 'hot' diseases, while those characterized by wasting, shivering, and apathy were 'cold'.[125] Each type of disease required a remedy belonging to the opposite category: for example, measles, a disease associated with an excess of heat and poison was treated with *mao-cao* (a type of herb) and sugar cane juice, both considered 'cold' and 'clean' remedies; whilst chronic bronchitis, a 'cold' illness was treated with pigeons and ginseng, wine and sticky rice pudding, and other 'hot' remedies.[126] Similarly, according to a nineteenth-century Chinese authority, Xu Zi Mo, 'remedies for warming and stimulating the vessels' were appropriate in the case of cholera or *huo luan* which was regarded to have arisen from 'morbific cold'.[127] The ideal disease-free state for the Chinese was one of balance. From the Western medical perspective, however, the hot/cold classification of disease represented but one of many 'curious and superstitious native ideas of diseases and their treatment' and proof of the 'astonishing ignorance displayed by even the most intelligent amongst [the Chinese] in matters concerning sanitation'.[128]

In general, the colonial government tolerated the establishment of non-Western medical systems as it obviated the responsibility of welfare provision among ethnic communities such as the Chinese.[129] At the end of the nineteenth century, however, there was increasing pressure from the European public and in particular the medical profession in Singapore to limit or control various forms of Asian medical practices. In 1887, one of the local English dailies launched an attack on the 'evil' of 'the Chinese science of medicine and the art of pharmacy', urging the authorities not to be 'content with protecting John Chinaman from the grime, squalor and smells so dear to him', but also 'to set about preventing him from helping himself to a premature grave by doctoring himself with the medicines so abundant in Chinese apothecaries' shops, the managers of which make no scruples about putting medicines of which they know little into bodies of which they know still less'.[130] The dangers posed by a system of medicine entirely in the hands of 'persons of an inferior race whose moral standpoint [was] not as high as that endeavoured to be attained in Western nations'[131] was one of the first major issues on the agenda of the Straits Medical Association formed in 1890.[132] Armed with an array of evidence showing that unqualified practice was a menace to public health and the root of a vast amount of unnecessary suffering and death, the Association pressed the government to register medical practitioners and suppress unqualified practice.[133] In this, they were partially successful: a Medical Registration Bill was passed in 1905 requiring persons without recognized medical degrees to submit themselves for examination by a medical council in order to establish competence. The main targets of legislative suppression appeared to be 'quacks' who borrowed methods and pilfered medications from Western orthodox medicine. Exemptions, however, were made in favour of 'those persons who confined themselves to the practice of any of the ancient Asiatic system of therapeutics'.[134] In the same year, a bill to regulate and restrict the sale of poisons in native bazaars was also passed.[135] In general, the medical fraternity was highly critical of any form of government support for indigenous medicine, claiming that the development of three different systems (Chinese, Malay, and Indian) would result in 'chaos', unlike 'scientific investigation by Western methods'.[136] Supremely conscious of the superiority of Western medical discourse, they had little interest in entering into a dialogue with Asian medicine and instead viewed all Asian medicine as innately inferior, confused classical with folk medicine, and branded all as irrational and superstitious.

Despite the opposition of the Western medical fraternity, traditional Chinese medicine continued to flourish and remained highly popular and accessible in colonial Singapore, particularly among the immigrant Chinese.[137] Given the vital role it played in everyday life, the bulk of the Chinese plebeian class remained unsympathetic towards Western-imposed concepts of disease propagation and the importance of sanitation.[138] Their hostility towards municipal sanitary measures was hence not the result of ignorance or apathy but stemmed from a different conception of how disease could be banished and health pursued. The next section examines specific Asian strategies used in countering municipal sanitary control.

Asian Counter-strategies against
Municipal Sanitary Control

Although prolonged strikes and revolutionary uprisings against the colonial authorities did not loom very large in the period under consideration, the Asian plebeian classes were capable of knowledgeable and at times skilled strategies to prevent control over their daily practices and management of the environment from being totally wrested from them.[139] In countering the expanding scope of institutional power over various aspects of their quotidian routine, the most common strategies were the 'passive' rather than 'active' means of inflecting municipal control. These usually took the form of apparent acquiescence to municipal demands as a façade for unobtrusive non-compliance. The municipal reports and summons notebooks abound in examples of such forms of 'passive resistance'.[140]

In March 1895, vexed by the inefficient scavenging of the city, the municipal commissioners ordered the distribution of notices among people living in the district between the Singapore River and Kreta Ayer to prohibit the deposition of household refuse on the streets. All occupiers were required to provide themselves with dustboxes placed outside their houses. This was to prevent refuse from being left to fester in the streets and also to facilitate scavenging when the municipal dustcarts came on their rounds. At the end of the year, the municipal president, Alex Gentle, reported that few householders had complied with the notice and the use of dustboxes or baskets was in most streets 'an exception rather than the rule'.[141] The acting municipal engineer protested that 'no [scavenging] scheme [could] be expected to keep the Town clean when all classes of the community persist[ed] in the practice of throwing into the streets at all hours of the day, office, house and kitchen refuse, which even when swept into heaps [was] so exposed that it [was] scattered by fowls and the wind'.[142] In 1896, he reiterated that the order issued regarding the proper disposal of refuse appeared 'a dead letter' and 'rubbish [was] freely scattered on the roads daily'.[143] As a rule, offences against the order could not be brought home to the offender as he had to be caught in the act of throwing out refuse before the municipal authorities could prosecute.[144] Hence, unless 'a tell-tale envelope betray[ed] its origin',[145] the ubiquity of the offence and the anonymity of the offender amidst the Asian masses rendered municipal surveillance ineffective. Even when a sanitary offence could be traced to a particular house, it was often difficult to single out the actual culprits. For example, in a municipal summons case, conservancy overseer Michael Gabriel found night-soil 'overflowing with maggots' coming down a drain at the back of a sundry shop at Telok Blangah Road. The case against the owner of the shop had to be withdrawn as the upper floors of the shophouse were inhabited by a large number of coolies 'for whom [the shopowner] could not be held responsible'.[146] In another case, conservancy overseer Ambrose W. Lewis charged M. Govinda, a hackney-carriage syce living at 69 Orchard Road with emptying a bucket of night-soil and urine into a drain at the back of the house.[147] In his defence, Govinda denied responsibility for the condition of the drain and claimed that the night-soil was already in the drain when the overseer directed his attention to it. He was able to produce a witness—another hackney-carriage syce living at the same address—to testify that children had defecated into the drain and as a result, the case was dismissed. At the police court, municipal summons cases were considered minor offences or 'odd jobs' often delegated to be heard by the most junior magistrates[148] and dismissed after a perfunctory hearing and the occasional fine. Summons cases for sanitary offences culled from court notebooks[149] listed in Table 3.2 show that sentences for such offences ranged rather arbitrarily from dismissal to a maximum fine of $10.00, with an average of $3.00.

In most cases of filthy premises, illegal erection of cubicles, overcrowding, and the non-removal of night-soil, given the difficulty of distinguishing the guilty from the innocent, summonses were frequently served on the principal tenants in the hope that they would find ways and means of exerting control over subtenants to abate nuisances. At best, however, this resulted in a perfunctory cleansing of premises, minor repairs, and the temporary removal of partitions rather than more permanent sanitary improvements. Once municipal vigilance was averted, '[c]ubicles pulled down by order of the Sanitary Department [were] frequently put up again at the earliest opportunity, improvements effected to drains and latrines [were] not properly maintained and soon [in a state of] disrepair'.[150] Municipal efforts to improve sanitation were hence often reduced to a 'Sisyphus task'.[151] Minor fines were unlikely to deter principal tenants from erecting cubicles and overcrowding premises as their gains from increased rents were more than adequate to offset occasional fines. Regulations against overcrowding were especially difficult to enforce as overcrowding occurred chiefly at night. Except in the case of suspected infectious disease, municipal officers only had powers of entry under certain conditions between sunrise and sunset. According to Dr Middleton, night inspection was impracticable for not only would it entail the employment of a night subordinate staff which had to be carefully supervised to ensure that duties were properly carried out, but such a measure would also inevitably stir up 'complaints, more or less well founded, against the staff for invading sleeping apartments'.[152] Furthermore, Dr Middleton doubted the efficacy of night surveillance and felt that it would only lead to more 'prosecutions, convictions, fines—and little of any improvement'.[153] In 1909, with the prevalence of overcrowding unabated, Dr Middleton reiterated the futility of prosecution as a means of preventing overcrowding:

If an occupier were prosecuted often enough to instill into him an observance of the law it would only lead to those who were put out of the overcrowded houses seeking shelter elsewhere and adding to the numbers of inmates in houses that already contained their share, or sleeping on the five-foot-ways where, if they ran less risk of contracting Phthisis they would run more of contracting other diseases, e.g., Bronchitis and Pneumonia.[154]

Despite the proliferation of municipal laws to strengthen the system of surveillance, the Asian plebeian classes were capable of exploiting 'openings' in the system and thwarting the realization of sanitary objectives. Actively to challenge the imposition of sanitary laws was often beyond their power and resources, but by alternating temporary observance with rampant disregard of the laws, and by executing non-compliance and non-cooperation in anonymity, they were sometimes able to escape the grasp of sanitary law.

The efficacy of sanitary reform depended not only on the stringency of sanitary law but also on the quality of its enforcers and those directly involved in working out its implications on a day-to-day basis. Given the meagreness of municipal pay, police constables, sanitary inspectors, and peons, the so-called 'backbone of sanitation',[155] were often open to the offer of bribes from those who were willing to pay a small sum to avoid trouble with the police courts. The health officer's annual reports and municipal minutes record a handful of cases where inspectors had been dismissed for receiving illegal gratification. However, evidence gleaned from the police notebooks suggests that corruption was far more widespread than the number of dismissals indicated. There were, for example, several instances where defendants prosecuted for sanitary transgressions in turn accused the inspectors and police officers of accepting bribes amounting virtually to a regular income and threatening trouble if they were not given pecuniary satisfaction.[156] One hawker went as far as to claim that 'it [was] customary to give money to police'.[157] Sanitary ideals at higher levels of municipal administration were hence seldom translated into equally visionary zeal at the level of enforcement. In practice, sanitary issues were often subject to pecuniary negotiation between enforcer and offender.

Another range of Asian counter-strategies—those of dissimulation, evasion, and concealment—were mostly clearly demonstrated in their attempts to evade municipal disease control. Dr Middleton warned that 'in dealing with cases of infectious disease, the sanitary staff [had] to be on the alert to all the dodges practised by the natives to conceal such cases'.[158] Although the failure to report the presence of a 'dangerous infectious disease' carried a fine of up to $25.00, such a penalty was far outweighed by the practical trouble incurred by occupants if a 'dangerous infectious disease' were traced to a particular dwelling. The resulting inconveniences included domiciliary visits by sanitary inspectors, isolation, the disinfection and destruction of contacts' belongings and occasionally the dwellings concerned,[159] the imposition of quarantine regulations,[160] all of which measures were perceived by the Asian population to be irksome interferences with their livelihood routines.[161] The removal of victims from the home to an orderly, sanitized hospital regime was also highly unpopular.[162] There were frequent complaints that hospitals were socially isolating, that there was no strict segregation of sex, race, and religion, that conditions were overcrowded and inferior to what 'the better class of Asiatics were accustomed to in their own houses', and that Asians were not allowed treatment 'according to their own customs' by their own 'medicine men'.[163] The Asians were hence keenly conscious of the consequences of detection and avoided it at all costs. The municipal health officer confessed himself nonplussed by the persistent reluctance of the Asians to report cases of infectious disease and the 'cheerfulness' with which they paid their fines when caught.[164] Concealment of infectious cases was rife, as evidenced by the large number of cases discovered only after death.[165] Victims of infectious disease were frequently smuggled out to the outskirts of the town[166] and attempts to trace the disease to its source were often frustrated by the reluctance of patients and friends alike to reveal addresses.[167] The 'persistent efforts of Asiatics to conceal the existence of [infectious diseases], and the deliberate mis-statements made to put Sanitary Officers off the track' were not confined to 'the more ignorant classes of the population, but were 'freely indulged in by the more intelligent and educated classes of the Asiatic community' including those who worked as clerks and storekeepers in European offices.[168] Strategies resorted to by the Asians to avoid detection were highly imaginative, as illustrated by the following catalogue:

The existence of cases of infectious disease [were] carefully concealed, the patients surreptitiously removed in *jinrikishas* or *gharries* to the hospital or to sick receiving houses. Or they [might] be taken in a moribund condition or after death and deposited on the street or any convenient piece of vacant ground from which they [were] removed to the hospital or cemetery by the Police. Every device is resorted to, to prevent the authorities from tracing the houses from which such cases were removed, such as changing *jinrikishas* two or three times between the house and the hospital, giving false addresses, or declaring the patient had newly arrived in the town and had been picked up on the street or 5-foot-ways.[169]

In 1909, the municipal commissioners attempted to counter Asian dislike of quarantine by paying each contact sent to the Quarantine Station at St John's Island 15 cents for each day of quarantine.[170] The experiment was, however, scrapped after a year on the basis of the health officer's report that it had failed to encourage the Asians to come forward more readily with information pertaining to the existence or source of infectious disease.[171]

Deaths suspected to have resulted from 'dangerous infectious diseases' also entailed post-mortem examination which was highly disliked and seen as 'an intolerable interference with the religions and sentiments of different classes of the [Asian] [c]ommunity'.[172] As such, they were frequently misreported as resulting from other innocuous causes. An investigation into the accuracy of death returns in 1896 revealed that no less than eighty-seven cases of infectious diseases (mainly cholera and enteric fever) were within six weeks returned under other names.[173]

As the cause of death in a large number of cases (an average of 65 per cent of deaths each year in the first decade of the twentieth century)[174] was not medically certified but dependent on a perfunctory inspection of the corpse and inquiries from friends, it was inevitable that a significant proportion of misinformation evaded detection. This, according to Dr Middleton, was the vital flaw in municipal disease detection and control for 'where the cause of death [was] obscure and there were no means for satisfactorily determining them [sic], there must be many missing links ... in [the] chain connecting many of the cases with each other'.[175] As such, the municipal health officer felt that the inaccuracy of death returns militated against any attempt to introduce 'special measures' aimed at particular diseases and that 'it [would] only be possible to proceed on general lines and introduce such measures as experience in other places ha[d] shown to be instrumental in benefiting and preserving the public health'.[176] It was hence clear that municipal surveillance could only be effective if it were predicated on an efficient turnover of information. By the purposeful withholding, corrupting, and falsifying of information and by setting a high premium on it, the Asians were not only instrumental in thwarting attempts to trace the origin and course of infectious disease, they also played a part in rendering it difficult for the Health Department to devise more stringent and precise control measures.

The Dialectics of Power

In their daily conflicts with the municipal authorities over the sanitary condition of the environment, the Asian plebeian classes were neither powerless nor inert. Whilst they lack formal power, the strength of their own systems of health and medicine rendered them less accessible to hegemonic practice. They also possessed a 'power' contained in the anonymity of each individual, the opaqueness of Asian society to municipal understanding, and specific strategies of evading, countering, and inflecting municipal control. How effective were these forms of 'passive resistance' in protecting or advancing the interests of the Asian masses in the long run? In James Scott's discussion on the 'weapons of the weak' in the context of peasant resistance in rural Kedah, he cautioned against over-romanticizing 'people's power'.[177] Asian counter-strategies were not capable of revolutionizing the colonial pecking order nor introducing large-scale structural changes into society. However, these strategies might effect immediate, de facto gains for the plebeian classes, and occasionally, passive non-compliance on a large scale and over a substantive period might force the colonial authorities into revising their goals and strategies. About a quarter of a century after the Municipal Ordinance of 1887 came into effect, a commission set up to enquire into the efficacy of various aspects of municipal administration including sanitary regulation concluded that the system of sanitary surveillance had failed to improve health conditions in the city: 'In a place like this Colony ... where there is so much hostility to sanitation, and so little belief in its utility on the part of the bulk of the population, it is not surprising to find that the effect of these sanitary measures is not reflected in the mortality returns, and that they have not given many visible beneficial results, but that on the other hand they have only created dissatisfaction and discontent.'[178]

The commission acknowledged the crucial role the plebeian classes played in thwarting municipal sanitary measures: 'All the good effects which may be expected from the measures taken by the Health Officer are, as a rule, counterbalanced by the action or inaction of the persons whom they are intended to benefit.... It is of very little practical use to depend on the assistance of the portion of the public accustomed to insanitary surroundings, or to expect any help from them in the execution of sanitary reforms.'[179] Undoubtedly, it was easier to impose the form of a municipal government, to legislate for municipal powers, or even provide municipal facilities, than to instil civic consciousness or create a civic order through the strategy of surveillance among a transient urban population riven by potentially divisive forces of race, language, religion, dialect, clan, and neighbourhood.[180]

Given the difficulty experienced in forcing compliance on an unwilling Asian public, the commission concluded that 'larger scale' measures which were 'more automatic in their operation' were necessary if any improvement in sanitary conditions were to be realized.[181] It urged the Health Department to 'initiate ... schemes outside its routine duties' even if these involved 'questions of finance and engineering' including large-scale anti-malarial work, the establishment of public dispensaries, and the putting into effect of some of Professor Simpson's proposals for improving the health of the city.[182]

The report of the enquiry commission marked a significant turning-point in municipal goals and strategies. After 1910, whilst the system of surveillance was not abandoned, the municipal authorities increasingly turned to other strategies such as the replacement of Asian systems and the reconstruction of built form to secure the sanitation of the city. In the second decade of the twentieth century, several schemes which had hitherto been left languishing on the shelf for years as a result of municipal abstention were finally inaugurated. In 1911, a malarial committee with $20,000 at its disposal was formed[183] in order to supervise the infilling of mosquito-infested swamps.[184] In the same year, a modern water-carriage sewerage system to replace the Asian pail system was prepared and adopted.[185] In 1912, a hospital for the isolation of infectious diseases opened[186] and from 1913 under the provisions of a new Municipal Ordinance, a 2 per cent Improvement Rate was levied to provide a fund devoted to sanitary works. The 'power' of the Asian masses, and their counter-strategies of evasion, concealment and non-compliance, were at least in indirect ways, partly responsible for forcing the hand of the municipal commissioners. In Foucault's terms, just as disciplinary institutions employ a panoply of meticulous, minute techniques embodying the 'micro-physics' of power, the subjects of discipline were themselves capable of 'small acts of cunning endowed with a great power of diffusion', which were 'all the more real and effective because they [were] formed right at the point where relations of power [were] exercised'.[187]

Power over the sanitary conditions of the urban landscape was hence neither the prerogative of the municipal authority nor the people, but had to be continually negotiated in the quotidian flow of activities. Power was not the intrinsic property of any one group, but instead operated through an ensemble of strategies, of tactics and techniques. The encounter between the rulers and the ruled was a dialectical one, and the relationship required dynamic adjustments and readjustments. The former's control of the central ground was seldom directly contested but concessions could be wrested at the margins as the latter continually stretched the bounds of the possible and chiselled away at the corners of existing structures in what Scott describes as 'long-run campaigns of attrition'.[188] The change from surveillance to direct environmental control on the part of the municipality was, at least in part, a testimony to the success of this campaign by the 'powerless' masses.

1. *Straits Times*, 17 February 1887.

2. J. R. R. Tolkien, *The Lord of the Rings*, London: Unwin Hyman, 1978 (first published in 1954), p. 287.

3. Kenneth Macleod, 'The Scope and Aim of the [Tropical Diseases] Section's Work', *Journal of Tropical Medicine*, August (1900): 18.

4. William McGregor, 'An Address on Some Problems of Tropical Medicine', *Journal of Tropical Medicine*, October (1900): 63.

5. ARSM, 1895, 'Appendix Q: Registration of Deaths', p. 140.

6. *Straits Times*, 21 February 1896.

7. F. B. Smith, *The People's Health, 1830–1910*, London: Croom Helm, 1979, pp. 195–6.

8. Felix Driver, 'Moral Geographies: Social Science and the Urban Environment in Mid-nineteenth Century England', *Transactions, Institute of British Geographers, New Series*, 13 (1988): 278.

9. Michel Foucault, 'The Eye of Power', in Colin Gordon (ed.), *Michel Foucault: Power/Knowledge, Selected Interviews and Other Writings, 1972–1977*, Brighton: Harvester Press, 1980, pp. 150–1.

10. Michel Foucault, 'The Politics of Health in the Eighteenth Century', in Gordon, *Michel Foucault*, p. 175.

11. Malcolm Watson, quoted in Gordon Harrison, *Mosquitoes, Malaria and Man*, London: John Murray, 1978, p. 138.

12. J. A. Cockburn, 'Sanitary Progress during the Fifty Years, 1876–1926—Overseas Dominions and Colonial Aspect', *Journal of the Royal Sanitary Institute*, 47 (1926): 89.

13. Ibid., pp. 89–90.

14. James Kirk, 'Analysis of One Hundred and Fifty Cases of Local Fever', *Journal of the Malayan Branch of the British Medical Association*, January (1904): 51.

15. In 1891, Singapore recorded a death rate of 31.88 per thousand compared to 26.0 in Calcutta, 40.0 in Madras, 25.69 in Bombay, 28.0 in Rangoon, and 24.5 in Colombo. The death rate of infants under one year of age in Singapore was 338.44 compared to 273.3 in Calcutta and 264.45 in Bombay ([C. E.] Dumbleton, 'The Need of Reform in the Present System of Registration of Deaths in the Straits Settlements', *Journal of the Straits Medical Association*, 4 (1892/1893): 70). In 1905, Singapore

recorded a death rate of 43.7 per thousand compared to 48.6 in Bombay and 58.7 in Madras. However, as the local press observed, 'whereas Madras and Bombay [paid] [a] yearly toll in thousands of deaths from plague, Singapore's abnormal rate [was] maintained from year to year under what we must regard as the normal conditions under which the common people [were] housed' (*Straits Times*, 19 June 1907). It was also reported that Singapore's mortality rate was more than twice those of principal towns with the highest death rates in Ireland, England, and Scotland.

16. Dumbleton, 'The Need of Reform', p. 70.

17. Ibid.

18. The term 'zymotic disease' apparently originated with the English statistician William Farr, the first 'Compiler of Abstracts' in the Registrar-General's Office in England. Farr considered that zymotic diseases acted as 'an index of salubrity' and saw much utility in keeping them as a distinct class of disease as they had two unifying features: first, their enormous capacity to decimate and to threaten public health; and second, their responsiveness to stringent sanitary control (Margaret Pelling, *Cholera, Fever and English Medicine, 1825–1865*, Oxford: Oxford University Press, 1978, p. 82; F. M. M. Lewes, 'Dr Marc D'Espine's Statistical Nosology', *Medical History*, 32 (1988): 310).

19. Between 1896 and 1910, phthisis averaged 1,313.8 deaths per annum, beriberi 1,137.3, and remittent fever 934.2. For every death from a 'dangerous infectious disease', there were 14 deaths from fevers, phthisis, and beriberi (W. R. C. Middleton, 'The Working of the Births and Deaths Registration Ordinance', *Malaya Medical Journal*, 9, 3 (1911): 47, 50).

20. The view that health was chiefly related to climatic factors, topography, and vegetation was embodied in a number of 'medical topographies' of Singapore written mainly in the first half of the nineteenth century. These were the first systematic investigations into the causes of ill-health and disease in Singapore (T. M. Ward and J. P. Grant, *Official Papers on the Medical Statistics and Topography of Malacca and the Prince of Wales' Island, and on the Prevailing Diseases of the Tenasserim Coast*, Pinang: Government Press, 1830; 'Report on the Diseases of Singapore', *Madras Quarterly Medical Journal*, 1 (1839): 59–77; Robert Little, 'On the Medical Topography of Singapore Particularly in Its Marshes and Malaria', *Journal of the Indian Archipelago and Eastern Asia*, 2 (1848): 449–94; A. Graham, *Medical Topography of Singapore and Sarawak*, Edinburgh: Murray and Gibb, 1852).

21. J. A. Bethune Cook, *Sunny Singapore*, London: Elliot Stock, 1907, p. 31.

22. Ernest Hart, *The Nurseries of Cholera: Its Diffusion and Its Extinction*, London: Smith, Elder & Co., 1894, p. iv; ARSM, 1896, 'Supplement: Report on a Special Investigation into the System of Death Registration in Force in Singapore, with Some Observations on Local Conditions Adversely Affecting the Health of the Population', pp. 12–13.

23. ARSM, 1895, Health Officer's Department, p. 44.

24. Experimental proof of the association of beriberi with the continuous consumption of polished, decorticated white rice as a staple article of diet was only established at the beginning of the second decade of the twentieth century by H. Fraser and A. T. Stanton, both researchers at the Institute of Medical Research in Kuala Lumpur (Henry Fraser and A. T. Stanton, 'The Cause of Beri-Beri', *Lancet*, 1 (12 March 1910): 733–4; 'The Etiology of Beri-Beri', *Lancet*, 2 (17 December 1910): 1755–7; and *The Etiology of Beriberi*, Studies from the Institute for Medical Research No. 12, Singapore: Kelly & Walsh, 1911).

25. Max F. Simon, 'Some Remarks on the Nature and Causes of So-called "Beriberi" or Peripheral Neuritis, in the Tropics, and on Its Place in the "Nomenclature of Disease"', *Journal of the Straits Medical Association*, 3 (1891/2): 59.

26. Ibid., p. 57.

27. Ibid., p. 59.

28. 'Report on a Special Investigation into the System of Death Registration', p. 12.

29. ARSM, 1907, Health Officer's Report, p. 2; MPMCOM, 27 March 1908, 22 May 1908, and 26 September 1919; MGCM, 10 October 1919.

30. 'Report on a Special Investigation into the System of Death Registration', p. 13.

31. Ibid., pp. 9–10.

32. 'Registration of Deaths', p. 135.

33. For the years with severe cholera outbreaks (1895, 1896, 1900, 1902, 1906, 1907, 1911, and 1914), the Chinese accounted for 81–97 per cent of notified cases. These figures were comparatively high in view of the fact that the Chinese accounted for about 75 per cent of the total municipal population during this period.

34. 'Report on the Special Investigation into the System of Death Registration', p. 13.

35. T. C. Allbutt (ed.), *A System of Medicine*, Vol. 2, London: Macmillan and Co., 1897, p. 28.

36. David J. Galloway, 'Observations on the Death Rate', *Journal of the Malayan Branch of the British Medical Association*, January (1907): 3.

37. ARSM, 1909, Health Officer's Report, p. 38.

38. Nair, 'Etiology and Early Diagnosis of Pulmonary Tuberculosis', *Malaya Medical Journal and Estate Sanitation*, 1, 2 (1926): 16.

39. These figures were extremely high considering that between 1908 and 1913, death rates for respiratory tuberculosis in European cities only ranged between 0.89 (Hull) and 3.50 (Paris) per thousand living (Gerard Kearns, 'Zivilis or Hygaeia: Urban Public Health and the Epidemiologic Transition', in Richard Lawton (ed.), *The Rise and Fall of Great Cities*, London: Belhaven Press, 1989, p. 102).

40. Simpson was so-named because of his appearance in various parts of the world 'whenever there was "dirty weather" in the sanitary sense' to advise on the course of remedial action (Andrew Balfour and Henry H. Scott, *Health Problems of the Empire: Past, Present and Future*, London: W. Collins, Sons & Co., 1924, p. 149). He was Professor of Hygiene at King's College, London, and had previously served for many years as the municipal health officer of Calcutta. He had also been called upon to investigate the outbreak of plague in Cape Town and Hong Kong prior to spending three months from mid-May to mid-August 1906 in Singapore.

41. W. J. Simpson, *Report on the Sanitary Condition of Singapore*, London, Waterlow & Sons, 1907, p. 11.

42. The houses which Professor Simpson assumed to be principal foci of tuberculosis were in fact 'houses for the dying' where tubercular victims in a moribund condition were taken and deposited. These houses hence had an artificially inflated number of deaths occurring in them (D. J. Galloway, 'Notes on Tuberculosis', in *Proceedings and Report of the Commission Appointed to Inquire into the Cause of the Present Housing Difficulties in Singapore and the Steps Which Should Be Taken to Remedy Such Difficulties*, 1918, Vol. 2, p. C9).

43. Simpson, *Report on the Sanitary Condition*, p. 11.

44. Ibid., p. 12.

45. Ibid.

46. Ibid.

47. Ibid., p. 28.

48. Ibid., p. 14.

49. Ibid., pp. 15–16.

50. Ibid., p. 16.

51. F. J. Hallifax, 'Municipal Government', in Walter Makepeace, Gilbert E. Brooke, and Ronald St. John Braddell, *One Hundred Years of Singapore*, Vol. 1, London: John Murray, 1921, p. 322.

52. W. R. C. Middleton, 'The Sanitation of Singapore', *Journal of State Medicine*, 8 (1900): 697–8.

53. Simpson, *Report on the Sanitary Condition*, pp. 25–6.

54. Ibid., p. 27.

55. *Report of Singapore Housing Committee, 1947*, Singapore: Government Printing Office, 1948, p. 1.

56. ARSM, 1905, 'Appendix L: Correspondence with Government on the Subject of High Death Rate in Singapore', p. 107.

57. Ibid., p. 108.

58. K. Hintze, 'Sanitäre Verhältnisse und Enrichtungen in den Straits Settlements and Federated Malay States (hinterindien)', *Archiv für Schiffs- und Tropen-Hygiene*, 10, 17 (1906): 526.

59. ARSM, 1896, p. 17.

60. W. J. R. Simpson (and discussants), 'Discussion on Sanitation of Villages and Small Towns, with Special Reference to Efficiency and Cheapness', *British Medical Journal*, 2 (1911): 1276.

61. *Straits Times*, 4 July 1907.

62. Ibid.

63. Galloway, 'Observations on the Death Rate', p. 4.

64. Hintze, 'Sanitäre Verhältnisse und Einrichtungen', p. 525.

65. Michel Foucault, 'Power and Strategies', in Gordon, *Michel Foucault*, p. 142. The importance of knowledge as a basis for structuring social action and conflict is also discussed in Nigel Thrift, 'Flies and Germs: A Geography of Knowledge', in Derek Gregory and John Urry (eds.), *Social Relations and Spatial Structure*, Basingstoke: Macmillan, 1985. Thrift argues that knowledge is differentially distributed among social groups over space and in time and is not perfectly communicable among groups. The availability of particular types of knowledge sensitizes the actors and inspires them towards specific forms of social action. In the context of colonial Singapore, it is argued here that municipal strategies of sanitary control were both inspired and constrained by prevailing knowledge of disease propagation and local evidence, both statistical and impressionistic, of the relationship between diseases and Asian practices. Such knowledge is not objective but ideologically drawn upon.

66. Gareth Stedman Jones, *Outcast London: A Study of the Relationship between Classes in Victorian Society*, London: Peregrine Books, 1984, p. 167.

67. Driver, 'Moral Geographies', pp. 279–81.

68. See for example, ARSM, 1896, Health Officer's Department, p. 86.

69. The question of reordering the built form of the city is examined in greater detail in Chapter 4 while Chapter 5 looks at the attempts to introduce municipal water and sewerage schemes.

70. Edwin Chadwick, quoted in Peter Stallybrass and Allon White, *The Politics and Poetics of Transgression*, London: Methuen, 1986, p. 126.

71. Stallybrass and White, *The Politics and Poetics of Transgression*, p. 135.

72. Anthony Giddens, *A Contemporary Critique of Historical Materialism. Volume One: Power, Property and the State*, Basingstoke: Macmillan, 1981, p. 169.

73. Middleton, 'The Working of the Births and Deaths Registration Ordinance', p. 33.

74. An 1896 investigation into the system of death registration revealed that out of 1,265 deaths registered within six weeks, there were 423 (33.4 per cent) erroneous returns including the incorrect reporting of beriberi, phthisis, cholera, enteric fever, bronchitis, and dysentery as well as a number of other diseases ('Report on a Special Investigation into the System of Death Registration', pp. 19–20).

75. ARSM, 1896, Appendix O, pp. 151–4.

76. Contagious diseases such as sexually transmitted diseases were separately administered under the powers of the Contagious Diseases Ordinance of 1870 which introduced a system of registration and inspection of brothels and prostitutes initially under the auspices of the Registrar-General's department but transferred to the Chinese Protectorate in 1881. This ordinance was repealed in 1887 largely because of the campaign waged by moral reformers, feminists, missionaries, and civil libertarians in England against state regulation of prostitution and what was seen as licensing a social evil. Registration of licensed brothels was continued after 1887 but this was also abolished in 1894. See James Francis Warren, 'Prostitution and the Politics of Venereal Disease: Singapore, 1870–98', *Journal of Southeast Asian Studies*, 21 (1990): 360–83; and *Ah Ku and Karayuki-san: Prostitution in Singapore, 1870–1940*, Singapore: Oxford University Press, 1993, for a discussion of the uneven history of state regulation of prostitution, the mixture of motives and attitudes of policy-makers, and the impact on and response of prostitutes and brothel keepers to these policies.

77. 'Ordinance [IX] of 1887: An Ordinance for Consolidating and Amending the Law Relating to Municipal Government', in *Straits Settlements Ordinance, 1887*, Singapore: Government Printing Office, 1888, pp. 35–115; 'Ordinance [XV] of 1896: An Ordinance to Amend and Consolidate the Law with Regard to Municipalities', in C. G. Garrad, *The Acts and Ordinances of the Legislative Council of the Straits Settlements from the 1st April 1867 to 7th March 1898*, Vol. 2, London: Eyre & Spottiswoode, 1898, pp. 1474–571.

78. ARSM, 1901, Health Officer's Department, p. 95; ARSM, 1902, Health Officer's Department, p. 122.

79. ARSM, 1894, 'Appendix F: Sanitary Regulations', pp. 58–60.

80. ARSM, 1901, Health Officer's Department, p. 95.

81. ARSM, 1894, 'Appendix J: Contribution to the Cost of Police Force: Gentle to Colonial Secretary, 21 September 1894', p. 68.

82. PLCSS, 1910, 'Report of the Municipal Enquiry Commission', p. C199.

83. Garrad, *The Acts and Ordinances*, Vol. 2, p. 1556 (Section 268) and p. 1559 (Section 318).

84. ARSM, 1894, 'Contribution to Cost of Police Force: Gentle to Colonial Secretary, 21 September 1894', p. 68.

85. Ibid.

86. ARSM, 1893, 'Appendix Q: Dangerous Infectious Diseases: Gentle to Colonial Secretary, 30 October 1893', p. 47; MPMCOM, 2 June 1904.

87. MMFGPC, 8 July 1910.

88. ARSM, 1899, 'Appendix I: Hospital Accommodation and Police as Quarantine Officers', p. 101.

89. Garrad, *The Acts and Ordinances*, p. 1476.

90. See Garrad, *The Acts and Ordinances*, pp. 1534–5 for a comprehensive list of 'nuisances'.

91. To take an example, in campaigning against the use of night-soil in vegetable gardens within the city, municipal officers had found their hands tied because it was impossible to provide medical proof that such use of manure was 'injurious to health', a condition required by the 1887 Municipal Ordinance for the purposes of prosecution (*Straits Settlements Ordinances, 1887*, Singapore: Government Printing Office, p. 80, section 183). In the 1896 amending ordinance, the word 'nuisance' was substituted for 'injurious to health' (Garrad, *The Acts and Ordinances*, p. 1539, section 211), thus allowing the municipal commissioners greater flexibility in taking action against the use of manure on the basis of 'a disagreeable smell' rather than having to prove 'injury to health' (*Straits Times*, 24 July 1896).

92. ARSM, 1905, 'Appendix L: Correspondence with Government on the Subject of High Death Rate', p. 120.

93. ARSM, 1896, p. 11.

94. Cuthbert Woodville Harrison, *Some Notes on the Government Services in British Malaya*, London: Malayan Information Agency, 1929, pp. 108–9.

95. J. A. Jewell, 'Theoretical Basis of Chinese Traditional Medicine' in S. M. Hillier and J. A. Jewell, *Health Care and Traditional Medicine in China 1800–1982*, London: Routledge & Kegan Paul, 1983, p. 232.

96. S. M. Hillier and J. A. Jewell, 'Chinese Traditional Medicine and Modern Western Medicine: Integration and Separation in China', in Hillier and Jewell, *Health Care and Traditional Medicine*, pp. 306–7.

97. *Straits Times*, 20 August 1907.

98. Song Ong Siang, *One Hundred Years' History of the Chinese*, Singapore: Oxford University Press, 1984 (first published in 1902), p. 420. Similar processions lasting a few days and completely obstructing the principal thoroughfares leading to Commercial Square were held in 1862 during an outbreak of cholera (Charles B. Buckley, *An Anecdotal History of Old Times in Singapore*, Singapore: Oxford University Press, 1984 (first published in 1902), p. 688; *Straits Times*, 5 April 1862.

99. According to Chinese perceptions of illness, the view that disease was caused by evil spirits (thus requiring ritualistic exorcism) and the view that disease was a result of imbalances within the body (hence necessitating medical remedies) need not be

contradictory but were in fact complementary. According to Ahern, the Chinese distinguished between disease caused by 'being hit' by evil spirits and illnesses from 'within the body'. In the former, the gods and their emissaries had to be applied to for assistance but in the latter, doctors were often consulted. These categories were, however, rather fluid and in some cases both solutions were resorted to (Emily M. Ahern, *Chinese Ritual and Politics*, Cambridge: Cambridge University Press, 1981, pp. 9–10).

100. Su Xiao Xian, 'Tong Ji Yi Yuan Yan Ge Shi Lue' [A Brief History of the Thong Chai Medical Institution], in *Tong Ji Yi Yuan Da Sha Luo Cheng Ji Nian Te Kan* [Souvenir Magazine of the Opening Ceremony of the Newly Completed Thong Chai Medical Institution Building], Singapore, 1979, p. 117.

101. A piece of land comprising Crown leases Nos. 1334 and 749 situated at Wayang Street was granted to the trustees of the Thong Chai Yee Say on 14 November 1892 according to 'Straits Settlements Statutory Land Grant No. 3650' in the possession of the Thong Chai Medical Institution.

102. Li Song, 'Xin Jia Bo Zhong Yi Yao De Fa Zhan 1349–1983' [The Development of Chinese Medicine in Singapore 1349–1983], *Singapore Journal of Traditional Chinese Medicine*, 10 (1983): 48.

103. 'Petition from Gan Eng Seng, Seong Kheng Tong, Lim Kheng Hoo and Jew Chu Wan to the Government for a Land Grant, 1892', in the possession of the Thong Chai Medical Institution.

104. Li Bai Gai (or Li Peh Khai), a scholar under the Imperial Examination system during the Ching Dynasty sailed for Singapore in 1900. He devoted himself to the study of Chinese medicine and served as principal consultant in the Thong Chai Medical Institution for five years before setting up his own practice, the Thong Lin Medical Hall. He was a staunch defender of Chinese medicine, founding member of the United Chinese Medical and Druggists Association, and editor of the Chinese Medical Monthly (Chin Chan Wei, 'Short Biography of the Author', in Li Bai Gai, *Yi Hai Wen Lan* [Chinese Medical Discourses], ed. Hsu Yun-tsiao, Singapore: Lai Kai Joo, 1976, pp. 1–2).

105. Li Bai Gai, 'Yuan zhen ji su' [A Brief Description of Diagnosis at the Thong Chai Medical Institution], in *Tong Ji Yi Yuan Da Sha Luo Cheng Ji Nian Te Kan*, p. 155.

106. Li Song, 'Xin Jia Bo Zhong Yi Yao De Fa Zhan', p. 49.

107. Peng Song Toh, *Directory of Associations in Singapore 1982–1983*, Singapore: Historical Culture Publishers, 1983, p. P36; 'Brief Account of the Kwong Wai Shiu Hospital', compiled by the Secretary of the Hospital, 1988.

108. 'Brief Account of the Kwong Wai Shiu Hospital'.

109. Ooi Giok Ling, 'Conservation-Dissolution: A Case-Study of Chinese Medicine in Peninsular Malaysia', Ph.D thesis, Australian National University, 1982, p. 142.

110. List compiled by Mr Chen Chin Ah of the Thong Chai Institute of Medical Research from *Xin Jia Bo Cha Yang Hui Guan Bai Nian Ji Nian Kan* [Souvenir Magazine of the Centenary Celebration of the Singapore Char Yong Association], Singapore, 1958.

111. T. C. Mugliston, 'Unqualified Practice in Singapore', *Journal of the Straits Medical Association*, 5 (1893/4): 78.

112. Ibid., pp. 71–3.

113. Estimates of the number of traditional Chinese voluntary associations with welfare functions existing at the end of the nineteenth century vary from about sixty-eight to eighty-six (James Daniel Vaughan, *The Manners and Customs of the Chinese of the Straits Settlements*, Singapore: Oxford University Press, 1985 (first published in 1879); Maurice F. Freedman, 'Immigrants and Associations: Chinese in Nineteenth-century Singapore', in G. W. Skinner (ed.), *The Study of Chinese Society: Essays of M. Freedman*, Stanford: Stanford University Press, 1979, p. 76; Thomas Tsu Wee Tan, 'Singapore Modernisation: A Study of Traditional Chinese Voluntary Associations in Social Change', Ph.D. thesis, University of Virginia, 1983, p. 85.

114. *Xin Jia Bo Guang Hui Zhao Bi Shan Ting Qing Zhu 118 Zhou Nian Ji Nian Te Kan* [118th Anniversary Souvenir Magazine of the Singapore Guang Hui Zhao Bi Shan Ting], Singapore, 1988, unpaginated.

115. *Xin Jia Bo Qiong Zhou Hui Guan Da Sha Luo Cheng Ji Nian Te Kan* [Souvenir Magazine of the Opening of the Kheng Chiu Building], Singapore, 1965, p. 63.

116. ARSM, 1913, Health Officer's Report, p. 39. In 1913, of the twenty-three sick-receiving houses inspected by the municipal health officer, only two were considered fit for the accommodation of the sick. There were, however, difficulties in dealing with unfit houses which did not meet municipal standards as those under the supervision of a medical practitioner were exempt from the operation of the by-laws, even if supervision were purely nominal.

117. Mugliston, 'Unqualified Practice in Singapore', p. 79.

118. The sustained popularity of Chinese medicine in Singapore today can be seen from a 1978 survey of voluntary institutions providing traditional Chinese medical consultations and medicine. It is estimated that nineteen such institutions accounted for 25 per cent of the total number of patient visits to government outpatient clinics, the official form of health care based on Western practices. (Suzanne Chan Ho, Kwok Chan Lun, and W. K. Cheng Hin Ng, 'The Role of Chinese Traditional Medical Practice as a Form of Health Care in Singapore—II. Some Characteristics of Providers', *American Journal of Chinese Medicine*, 11, 1 (1983): 20). A related study in 1980 returned twenty-six traditional Chinese medical clinics distributed throughout the island and an estimated 1,000 traditional practitioners (Suzanne Chan Ho, Kwok Chan Lun, and W. K. Cheng Hin Ng, 'The Role of Chinese Traditional Medical Practice as a Form of Health Care in Singapore—III. Conditions, Illness Behaviour and Medical Preferences of Patients of Institutional Clinics', *Social Science and Medicine*, 18 (1984): 745).

119. Pathogenic factors recognized in traditional Chinese medicine were not the pathogens of Western medicine. While the Chinese concept traditionally included the idea of an 'epidemic pathogenic agent' (a concept most closely resembling the microbe or bacteria), it remains a very small part of the general class of pathogenic factors. Instead, in Chinese medicine, pathogenic factors or 'evil *qi*' are 'harmful substances that disturb the relative internal balance of the human body or the balance between the body and the external environment'. These are divided into three categories: 'exogenous' factors such as climatic excesses in wind, cold, heat, dampness, dryness, and fire; 'endogenous' factors such as the excessive emotions of joy, anger, sadness, pensiveness, grief, fear, and fright; and a miscellaneous category including pestilence, intemperate sexuality, and animal bites (Liu Yangchi, *The Essential Book of Traditional Chinese Medicine*, Vol. 1, New York: Columbia University Press, 1988, pp. 141–66).

120. Jewell, 'Theoretical Basis of Chinese Traditional Medicine', pp. 232–4.

121. Li Bai Gai, 'Yuan Zhen Ji Su', pp. 155–6.

122. Jewell, 'Theoretical Basis of Chinese Traditional Medicine', p. 234.

123. At the turn of the century, medical men in Singapore still supported the thesis that beriberi was a 'filth' disease caused by germs and inseparable from insanitary conditions. See previous discussion.

124. R. M. Gibson, 'Beri-beri in Hong Kong', *The Journal of Tropical Medicine*, 4 (1901): 98.

125. K. Gould-Martin, 'Hot Cold Clean Poison and Dirt: Chinese Folk Medical Categories', *Social Science and Medicine*, 12, 1B (1978): 39.

126. Ibid., p. 43.

127. K. Chimin Wong and Wu Lien-Teh, *History of Chinese Medicine being a Chronicle of Medical Happenings in China from Ancient Times to the Present Period*, Tientsin: Tientsin Press, 1932, pp. 107 and 240.

128. T. C. Avetoom, 'Curious and Superstitious Native Ideas of Causation of Diseases and Their Treatment', *Journal of the Malayan Branch of the British Medical Association*, December (1905): 6.

129. Ooi, 'Conservation-Dissolution', p. 110.

130. *Straits Times*, 19 July 1887.

131. Mugliston, 'Unqualified Practice in Singapore', p. 68.

132. David J. Galloway, 'Introductory Address', *Journal of the Straits Medical Association*, 1 (1890): 23–4.

133. Mugliston, 'Unqualified Practice in Singapore', pp. 68–85; 'Association Intelligence: Malaya Branch', *The British Medical Journal*, 2 (1895): 1520; David J.

Galloway, W. R. C. Middleton, and John Ritchie, 'Report of Committee of Malaya Branch of the British Medical Association Appointed to Suggest Improvements to the Medical Registration Bill', *Journal of the Malayan Branch of the British Medical Association,* January (1907): 118–23.

134. 'Medical Progress in the Straits Settlements', *Journal of the Malayan Branch of the British Medical Association,* December (1905): unpaginated. It was argued that 'Doctahs China' should be tolerated as long as they confined themselves to their own 'active medicines' such as 'dragon's blood [a colouring substance derived from the ripe fruit of a species of rattan], bat's dung, boar's tusks, boa constrictor's lungs and virgin's urine' which were 'most likely innocuous if extremely nasty'. However, since 'they [had] not learnt the more accurate remedies of Europe', they should not encroach on 'any remedy in the Pharmacopoeia Britannica' (Mugliston, 'Unqualified Practice in Singapore', p. 72).

135. 'Medical Progress in the Straits Settlements', unpaginated.

136. 'Ayurvedic Medicine—An Ancient System', *Malaya Medical Journal,* 2, 3 (1927): 114–15.

137. Cf. Lenore Manderson, 'Race, Colonial Mentality and Public Health in Early Twentieth Century Malaya', in Peter J. Rimmer and Lisa M. Allen (eds.), *The Underside of Malaysian History,* Singapore: Singapore University Press, 1990, p. 193.

138. In analysing the Asiatic response to the imposition of sanitary surveillance, it is not the intention here to impute a static, unitary consciousness embracing all Asians. Indeed, a fundamental difference in attitude could be distinguished among the Asiatic educated élite which included several Western-trained medical men such as Dr Lim Boon Keng, who collaborated with municipal officers in investigating the causes of Asian insanitation. Others like Tan Kah Kee, a wealthy rubber magnate and philanthropist who urged the China government to emulate Western sanitary and housing reforms in Singapore, were convinced of the benefits of 'sanitary science' without completely breaking faith with Chinese concepts of health and medicine (see Chen Jia Geng (Tan Kah Kee), 'Zhu Wu Yu Wei Sheng' [Housing and Hygiene] in Chen Jia Geng, *Nan Qiao Hui Yi Lu* [Recollections of an Overseas Chinese in Southeast Asia], Hong Kong: Cao Yuan Chu Ban She, 1979, pp. 384–92). The attitude of the editors of *The Straits Chinese Magazine* (the organ of progressive Straits-born Chinese opinion) was equally equivocal: whilst accepting the basic premise that sanitation was vital to public health, they continued to champion certain Chinese practices such as the use of night-soil as manure (see Chapter 5). It was largely through the Asian educated élite that the colonial state hoped to persuade the plebeian masses to come to terms with Western medical care and sanitary reform. In time, given the Chinese bent towards syncretism, Western ideas of health and medical remedies were easily incorporated within an existing pluralistic repertory. However, whilst Western medicine was gradually accepted as an alternative to, rather than a complete replacement for, Chinese medicine, the intrusive penetration of the 'medical police' continued to be disliked and resisted.

139. Anthony Giddens criticizes Foucault's analysis of surveillance for not adequately acknowledging 'that those subject to the power of dominant groups themselves are knowledgeable agents, who resist, blunt or actively alter the conditions of life that others seek to thrust upon them'. Surveillance as 'a subtle, calculated technology of subjection' applied to society is not total, and "the docile bodies" which Foucault says discipline produces 'turn out very often to be not so docile after all' (Giddens, *A Contemporary Critique of Historical Materialism,* pp. 170–2).

140. This is a phrase used by the municipal president, Alex Gentle (ARSM, 1895, p. 15).

141. ARSM, 1895, p. 15.

142. ARSM, 1895, Municipal Engineering Department, p. 23.

143. ARSM, 1896, Municipal Engineering Department, p. 28.

144. ARSM, 1895, p. 16.

145. Ibid.

146. Municipal Summons Note Book, 5 July 1912–13 December 1912, Case 5821 (6 August 1912).

147. Ibid., Case 6506 (10 September 1912).

148. The experiences of young cadets appointed as supernumerary magistrates to deal with municipal summons cases in the early twentieth century are recounted in Edward Wilmot F. Gilman, 'Personal Recollections, Speeches and Newscuttings, c.1920–1931 while a Civil Servant in Malaya', Mss.Ind.Ocn.s.127; and Frank Kershaw Wilson, 'Letters Home, January 1915–December 1916 while Administrative Cadet', Singapore, Mss.Ind.Ocn.s.162.

149. The survival of court notebooks which focus entirely on summons cases under the Municipal Ordinance and municipal by-laws is extremely fragmentary. After careful search, only three of these notebooks have been uncovered in the pre-1930 era: Municipal Summons Note Book, 5 July 1912–13 December 1912; Municipal Summons Note Book, 16 December 1912–6 May 1913; and Note Book of Municipal Cases No. 1, 3 March 1921–1 April 1924. The cases listed in Table 3.2 represent the bulk of sanitary offence cases extracted from the three notebooks. It is evident that the list is far from complete and that a large number of cases are not extant.

150. 'Report of the Municipal Enquiry Commission', p. C200.

151. Simpson, *Report on the Sanitary Conditions*, p. 33.

152. 'Correspondence with Government on the Subject of High Death Rate', p. 118.

153. Ibid.

154. ARSM, 1909, Health Officer's Report, p. 38.

155. Andrew Balfour, *War against Tropical Disease*, London: Bailliere, Tindall & Cox, 1920, p. 17.

156. See, for example, Municipal Summons Note Book 5 July 1912–13 December 1912, Case 6578 (2 October 1912) and Case 9166 (13 December 1912). The widespread incidence of bribery, extortion, and corruption within the police force is also attested to by the evidence of senior police officers as well as a number of anonymous petitions from the public accusing specific officers of corrupt practices (CO273/259: Swettenham to Chamberlain, 13 December 1900).

157. Note Book of District Court Cases No. 2 of 1923, Cases 115, 116, and 117 (29 May 1923).

158. Middleton, 'The Sanitation of Singapore', p. 702.

159. MPMCOM, 18 December 1908 and 26 February 1909.

160. During an outbreak of smallpox in the first quarter of 1899, the average number of houses in quarantine at any one time was 7.29 and the average duration of quarantine, seven to ten days (ARSM, 1899, 'Appendix I: Hospital Accommodation and Police as Quarantine Officers', p. 101). Contacts were also sometimes sent to St John's Island Quarantine Station on the orders of the municipal health officer in the late nineteenth century and the first decade of the twentieth century. In 1912, Dr Peter Fowlie, a municipal commissioner, put forward the case that quarantine and the isolation of contacts only served to encourage concealment of cases and somewhat extravagantly claimed that 'if isolation [were] abolished, every case of cholera [would] be reported'. He was supported by Chia Keng Chin, one of the Chinese commissioners, but the other members of the Board favoured some form of isolation as a necessary safeguard against the spread of infectious disease (MMFGPSYC, 24 September 1912; MPMCOM, 27 September 1912). In 1915, a new Infectious Disease Ordinance allowed contacts to be isolated in their own houses or for the health officer to require them to report at the Municipal Office instead of quarantine on St John's Island. By the early twenties, isolation was rarely resorted to and only beggars and those who could not be relied upon to report at the Municipal Office were sent to St John's Island. The usual procedure was for contacts to report at the Municipal Office periodically (MPMCOM, 27 May 1921).

161. Quarantine measures were highly repulsive to the Asian communities for many reasons. Complaints were lodged that owners were not always compensated for property destroyed and that quarantined persons placed under police supervision were treated like criminals by the native police (*Straits Times*, 10 August 1907).

162. In the late nineteenth century, Europeans and Eurasians were sent to the dangerous infectious disease wards of the General Hospital while patients of all other nationalities were accommodated at the Dangerous Infectious Hospital at Balestier Road.

163. 'Hospital Accommodation and Police as Quarantine Officers', pp. 104–6; MMSYC, 2 September 1910; MMFGPSC, 24 September 1912; MPMCOM, 28 October 1921; 24 February 1921.

164. ARSM, 1899, Health Officer's Department, p. 35.

165. For example, in 1902, a particularly virulent year, 19 per cent, 39 per cent, and 58 per cent of reported smallpox, enteric fever, and cholera cases respectively were discovered after death (ARSM, 1902, Health Officer's Department, pp. 118–21). In 1911, 114 or almost half the total number of reported cholera cases were 'concealed cases' discovered after death owing to the necessity of getting a burial permit, and another thirty-eight were found dumped in various parts of town (ARSM, 1911, p. 11). In the early 1920s, corpses of those who died from infectious diseases were still 'found with fair frequency in various parts of the town' and the municipal commissioners resorted to offering a reward of $50.00 for information leading to the discovery of the residence in which death occurred (MPMCOM, 29 April 1921). Concealment of cases however, continued unabated.

166. MPMCOM, 2 June 1904; ARSM, 1909, Health Officer's Report, p. 5.

167. ARSM, 1902, Health Officer's Department, p. 118; MPMCOM, 25 February 1927.

168. ARSM, 1910, Health Officer's Report, p. 6.

169. 'Special Investigation into the System of Death Registration', p. 10.

170. MPMCOM, 30 April 1909.

171. MMSYC, 1 April 1910.

172. 'The Working of the Births and Deaths Registration Ordinance', p. 50.

173. 'Report on a Special Investigation into the System of Death Registration', p. 8.

174. In comparison, only about 2.5 per cent of deaths in the United Kingdom remained uncertified in the early 1900s (ARSM, 1904, Health Officer's Department, p. 24).

175. ARSM, 1905, Health Officer's Department, p. 6.

176. ARSM, 1904, Health Officer's Department, p. 24.

177. James C. Scott, *Weapons of the Weak: Everyday Forms of Peasant Resistance*, New Haven: Yale University Press, 1985, pp. 29–30.

178. 'Report of the Municipal Enquiry Commission', p. C199.

179. Ibid., p. C200.

180. In late nineteenth-century Japan, surveillance, quarantine measures, and compulsory disinfection were also heavily employed as relatively low cost public health techniques to control the spread of infectious diseases. Johansson and Mosk argue that these strategies were relatively successful in suppressing communicable diseases largely because of Japan's high educational levels and also because of its distinctive institutional and cultural legacy which facilitated acceptance of the national government's right and duty to interfere with individual freedom in order to improve collective welfare. The success of surveillance in securing sanitary conditions was dependent on a high level of receptiveness and co-operation among the local population (S. Ryan Johansson and Carl Mosk, 'Exposure, Resistance and Life Expectancy: Disease and Death during the Economic Development of Japan, 1900–1960', *Population Studies*, 41 (1987): 220–1). In contrast, the relative failure of surveillance as a sanitary strategy in turn-of-the-century Singapore could well be accounted for by the unavailability of education and the lack of strong allegiance to the colonial government. Whilst Japan had already achieved 90.0 per cent literacy rates by the turn of the century, Singapore had literacy rates of only 36.6 per cent for all languages and 8.3 per cent for English in 1921 (J. E. Nathan, *The Census of British Malaya, 1921*, London: Waterlow & Sons, 1922, pp. 322, 332).

181. 'Report of the Municipal Enquiry Commission', p. C200.

182. Ibid., pp. C200–2.

183. ARSM, 1911, Health Officer's Report, p. 38.

184. Such a strategy concentrated on eradicating the vector of disease in order to break the aetiological chain. As Michael Worboys has observed for British territories as a whole, given the view that the native population was incapable of playing an active role in disease prevention, the 'ideal' strategy was one which 'did not require their co-

operation or even their involvement' (Worboys, 'British Colonial Medicine and Tropical Imperialism', p. 165).

185. ARSM, 1911, Health Officer's Report, p. 42.

186. ARSM, 1912, Health Officer's Report, p. 6.

187. Foucault, *Discipline and Punish*, p. 139; Foucault, 'Power and Strategies', p. 142.

188. Scott, *Weapons of the Weak*, p. 298.

4
Shaping the Built Form of the City: From the Regulation of House Form to Urban Planning

Singapore the City of Mean Streets and high land values needs rebuilding on a vast scale.[1]

Much of the work of housing reform may be interpreted as an attempt to change social behaviour via physical change which was often frustrated or deflected by a conflict between the donor's and recipient's expectation of how the environment should be made effective.[2]

An Alternative History of Housing in Singapore?

THE (largely unwritten) history of working-class housing in nineteenth- and early twentieth-century Singapore has mainly been approached as a descriptive exercise tracing the efforts of the authorities to solve what were perceived as 'housing problems'.[3] The view taken is often from the vantage point of the town planner rather than through the eyes of the slum dweller, and as such there has been much less concern in the literature with the contemporary social meanings of house and urban forms for the inhabitants of the city,[4] let alone whether the 'improvements' to housing imposed by the authorities were accepted or resisted. This chapter examines the interaction between municipal efforts to shape the built form of housing space within the city and Asian attempts to adapt and modify the existing built environment to render it 'effective' for their own purposes. It analyses some of the tensions and readjustments as the aims of the 'managers of the potential environment'[5] encountered the expectations of its users in the process of shaping the built form of the city. From the municipal perspective, the Asian housing question was inextricably conflated with problems of overcrowding, insanitation, disease, and the 'inexorable logic of Asiatic ignorance'. Municipal discourse focused on regulating the spatial built form of the city in order to combat disease and insanitary housing conditions, and the difficulties involved in translating such ideals into concrete strategies in the realm of practices and effects. Such an approach to the housing question was contrasted by the perspective of the Asian communities themselves, who possessed their own adaptive strategies to utilize limited housing space in an overcrowded city.

The Municipal Perspective: The Evidence for Overcrowding and Insanitary Housing Conditions

In Singapore as in other burgeoning nineteenth- and early twentieth-century cities, the question of housing the masses which supported its growth was one of the most persistent and perplexing of urban problems. The unremitting pressure of a rapidly increasing population on a dwindling supply of space in which people could live within reach of their workplace grew more and more acute with each upsurge in immigration. Overcrowding did not simply describe a state of affairs, but became the mechanism by which the urban economy sustained a market for menial and more or less casual labour. Subdivided tenements, makeshift cubicles, and back-to-back houses were a crucial part of the urban infrastructure by which the coolie population could be physically absorbed.

Whilst there was a deepening 'crisis of habitability'[6] in the city, neither the colonial government, the municipality, nor the private sector were prepared to shoulder the expense of providing housing for the Asian labouring classes.[7] The trend in both private and public ambition, and of capital, did not favour the construction of working-class housing which yielded far lower returns compared to investments in rubber, tin, and commerce.[8] As Jim Warren has argued, 'It was easier and more profitable in Singapore to construct quarters for government servants and the employees of big firms, and of monumental buildings like the Post Office in 1928, the massive Municipal Building in 1929 and Clifford Pier, than to clean [up the city]. [It was] easier to embellish its face than sound its depths.'[9] Instead, investigations into the question of Asian working-class housing tended to focus on issues such as overcrowding, insanitation, and 'house diseases' which could be 'solved' through regulating the spatial built form of the city.

By the second decade of the twentieth century, the relationship between defective, overcrowded housing conditions and the incidence of disease in urban Singapore was well elaborated by a number of census and health reports. Professor Simpson's 1907 report discussed in Chapter 3 provided graphic details and benchmark statistics of the dismal housing conditions in Singapore which were reiterated and authenticated by subsequent reports. Medical men in Singapore continued to identify 'the crux of the tuberculosis question' with the housing conditions of Chinese town dwellers, in particular 'craftsmen, carpenters, builders, painters, engineers, artificers, goldsmiths, peddlers, etc.' who formed the 'cubicle population' of urban Singapore.[10] While overcrowding *per se* was not seen as necessarily injurious to health, it was an unmitigated 'evil' where 'it occur[red] in insanitary dwellings'.[11] In Singapore, alleged the municipal health officer, Dr W. R. C. Middleton, the bulk of houses on the south side of the Singapore River

('Chinatown') as well as houses in several areas on the north side were *ipso facto* insanitary, largely because they were built back-to-back, an unhealthy condition aggravated by the subdivision of floor areas on each storey into cubicles to absorb excessive numbers of occupants.[12] The deleterious effects of back-to-back housing, one-room apartments, and overcrowded conditions and their association with high rates of diseases like phthisis and pulmonary diseases, gastro-enteritis, infantile diarrhoea, and infectious diseases were already well substantiated by the investigations of Dr Darra Mair[13] and Dr A. K. Chalmers[14] in English and Scottish cities respectively, and whilst the comparison of such cities with Singapore was not 'scientifically accurate', it was 'for practical purposes ... sufficiently permissible and convincing'.[15]

Although no comprehensive, city-wide survey of housing conditions was made, the twin phenomena of overcrowding and insanitation were considered self-evident and widespread enough in urban Singapore to allow details to be extrapolated from a number of 'typical' sample surveys. Table 4.1 summarizes the findings of several areal housing surveys conducted between 1906 and 1917. Block densities varied between 635 and 1,304 persons per acre while house densities ranged from 18.7 to 44.5 persons per house. While the fragmentary nature of the data does not allow accurate analysis of variations in housing conditions through time, a comparison of the statistics for the Hokien Street Block between 1908 and 1917 indicates that house and areal densities had increased by roughly 15 and 30 per cent respectively within a decade. This is in agreement with the general finding that the overall municipal population had increased more rapidly than the number of houses (Table 4.2; Figure 4.1), resulting in an overall increase in person-to-house density for the municipal area as a whole (Table 4.3).

From the municipal perspective, overcrowding was not merely a question of high population density but also a matter of spatial arrangement and built form. This has two aspects: first, in terms of the overcrowding of persons in houses, the spatial distribution of occupants within the house is as crucial as the actual person-to-house density. According to the Municipal Health Department, housing conditions in Singapore seldom fell short of the legal yardstick for overcrowding, which was a minimum of 350 cubic feet of clear internal space per adult (two children under ten counting as one adult).[16] However, whilst 'legal overcrowding' in terms of the cubic contents of each house rarely occurred, the health officer was convinced that actual overcrowding was rampant within cubicles particularly at night because 'the distribution of people within each house [was] wrong'.[17] Not only was a large part of the cubic space within the typical shophouse used for storage of 'miscellaneous collections of rubbish and other articles' or for business purposes, thereby severely reducing the actual 'living space', the 350 cubic feet rule was fundamentally flawed as a method of measuring legal space essentially because it took no account of the spatial distribution of people between cubicles and rooms.[18] As the

TABLE 4.1

Sample Surveys of Housing Conditions in Inner Municipal Areas,
1906–1917

Housing Conditions	Sago Street House 1906[a]	Upper Chin Chew Street Block, 1906[b]
Area (sq. ft.)	4,038.5[†]	239,580
Population	99	4,819
Number of houses	2	194
Density per acre	1,068	876
Density per house[*]	44.5	24.8

	Hokien Street Block 1908[c]	Hokien Street Block 1917[d]
Area (sq. ft.)	73,912	73,912
Population	1,078	1,403
Number of houses	61	61
Density per acre	635	826
Density per house[*]	20.0	23.0

	Pagoda Street Block 1917[e]	Johore Road Block 1917[f]
Area (sq. ft.)	29,280	57,614
Population	877	1,053
Number of houses	25	56
Density per acre	1,304	796
Density per house[*]	36.5	18.7
Number of cubicles	305	103
Number of rooms	16	58

Sources: W. J. Simpson, Report on the Sanitary Condition of Singapore, London: Waterlow and Sons, 1907, pp. 17–24; ARSM, 1909, Health Officer's Report, p. 38; 'Appendix III: Census Report', in Housing Difficulties Report, 1918, Vol. 1, pp. A71–A77; 'Memorandum N: The Representation Made by Dr W. R. C. Middleton, Municipal Health Officer, Singapore, on 30 May 1908 under Section 270(J) of the Municipal Ordinance, 1896 (as Amended by Ordinance [XXXVII] of 1907) Regarding the Insanitary Area between Macao Street and Hokien Street', in Proceedings and Report of the Commission Appointed to Inquire into the Cause of the Present Housing Difficulties in Singapore, and the Steps Which Should Be Taken to Remedy Such Difficulties, Vols. 1 and 2, Singapore: Government Printing Office, 1918 [hereafter Housing Difficulties Report].
[*]Excludes vacant houses.
[†]Total area occupied by cubicles only.
[a]The Sago Street 'double tenement house' includes two fused houses, 20-3 Sago Street and 18 Sago Lane.
[b]The Upper Chin Chew Street Block, located in 'Chinatown', comprised the whole of Upper Chin Chew Street, and one side each of Upper Nankin Street, and Upper Cross Street.
[c]The Hokien Street Block, located in 'Chinatown', is bounded by Macao Street on the north, Hokien Street on the south, part of China Street on the east, and part of South Bridge Road on the west.
[d]See note c.
[e]The Pagoda Street Block, located in 'Chinatown', is bounded by Pagoda Street, Trengganu Street, Temple Street, and the Hindu temple.
[f]The Johore Road Block, located in the 'Kampong Glam' district, is bounded by Ophir Road, Johore Road, Arab Street, and Queen Street.

TABLE 4.2

Size of Population and Number of Houses within
Municipal Limits, 1891–1917

Year	Population	Vacant Houses	Occupied Houses	Total Number of Houses
1891	155,683	1,719	17,732	19,451
1901	193,089	1,245	21,132	22,377
1911	259,610	2,406	26,196	28,602
1915	289,375	1,073	25,806	28,598
1916	296,951	860	25,997	26,857
1917	304,815	581	26,811	27,392

Source: 'Memorandum R by Dr W. R. C. Middleton, Municipal Health Officer, Singapore', in *Housing Difficulties Report*, Vol. 2, p. C76.

deputy health officer, Dr J. A. R. Glennie, explained, 'If you take [people] out of this cubicle and put them into that cubicle, and so on, you can get all the people in the house without overcrowding.... It is a difference of the arrangement of the houses that is required.'[19] Similarly, the Housing Commission set up in 1917 to investigate the causes of 'housing difficulties' in Singapore observed that 'in respect of Density of Population, ... it [was] not so much the number of persons to the acre that matters as the manner in which these persons [were] housed. What matters most of all [was] the number of rooms occupied by each family.'[20]

Second, overcrowding also took the form of an excessive crowding of houses within a limited area without sufficient provision of open space.[21] According to Dr Middleton, in congested districts such as Chinatown, the number of houses per acre ranged from twenty-one to thirty-four, well over the maximum of twelve houses to the acre consistent with the requirements of health.[22] There were no provisions in the by-laws for limiting the number of houses that might be built on an area and as a result, houses tended to be built in rows adjoining one

TABLE 4.3

Number of Persons per House within Municipal Limits, 1891–1917

Year	Persons per House
1891	8.7
1901	9.1
1911	9.9
1915	12.5
1916	11.4
1917	11.3

Source: 'Memorandum R by Dr W. R. C. Middleton, Municipal Health Officer, Singapore', in *Housing Difficulties Report*, Vol. 2, p. C76.

FIGURE 4.1
Comparison of Average Annual Percentage Changes in the Municipal
Population and the Number of House, 1891–1917

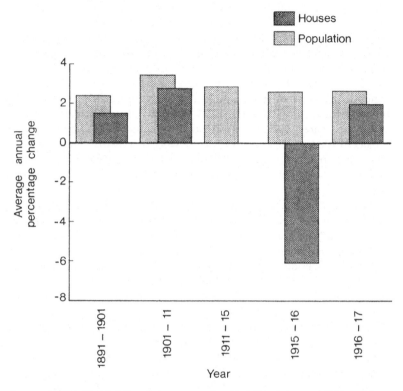

Source: ARSM, Health Officer's Department/Health Officer's Report, 1894–1916.

another and back-to-back in order to maximize the use of valuable
urban space. What resulted was an urban landscape where 'numerous
immense blocks of houses stretch[ed] from street to street, without a
single lane, alley or court of any description, by means of which a free
circulation of air [could] be ensured. Houses and godowns [were] built
side by side and back-to-back, until they constituted huge agglomerated
masses of bricks and mortar.'[23]

Overcrowding and insanitation were largely perceived as endemic
'Asiatic' problems. In the opinion of the Housing Commission
appointed in 1917, overcrowding was a problem which could be
defined and circumscribed within two well-delimited areas: the 'southern congested area' more generally known as 'Chinatown', and the
'northern congested area', often referred to as the 'Malay Quarter'.[24]
These two areas were occupied solely by the Asian communities,
whether rich, middle class, or poor.[25] F. J. Hallifax, the municipal
president, argued that although the exigencies of war had resulted in an

overall scarcity of houses for all classes, overcrowding and insanitary conditions were only associated with the 'Asiatic classes' comprising shopkeepers, artisans, and hawkers. Among Europeans, and 'better-classed Asiatics' and Eurasians, the shortage of housing accommodation had pushed rents up, making it necessary to rent houses in less convenient locations or to resort to hotels and boarding houses. However, sanitary standards among these classes were adequate and there were no signs of overcrowding.[26] Among the 'Asiatic' labouring classes, however, the tendency to overcrowd had a far more inexorable logic: 'The Asiatic of [the poorer] class has an invincible determination to live right on his work and right amongst his competitors and customers. He cares nothing for sanitation or ventilation. All he requires is a place to sleep in as near to his daily work as possible, and a more or less secure place to keep his belongings. He can get all this simply by railing off a portion of a passage or room in a house and making a cubicle.'[27]

In general, the consensus among those who surveyed or studied housing conditions in the city held that 'Asiatic ignorance and apathy' were among the chief factors for the prevalence of 'filth[y] sodden dens with neither light nor clean air to destroy the constantly accumulating germs of deadly disease'.[28] The point of departure for most discussion was the degree to which the colonial and municipal governments were responsible for countervailing 'Asiatic' insanitary standards and whether such efforts would be of any avail. On the one hand, the Housing Commission favoured a vigorous government or municipal campaign 'to educate the Asiatic in the importance of sanitation'; the commissioners being convinced that 'the Asiatic community [was] not only capable of being educated in municipal and domestic hygiene, but [was] eager to be educated. As soon as the community [understood] the position, public opinion [would] urge the introduction of measures which now, because it [did] not understand, it either oppose[d] or resent[ed].'[29]

On the other hand, there were those like Dr Glennie who argue that education was an uphill task because amongst Asian communities, 'old customs [were] difficult to eradicate'.[30] He contended that

[the] factor in the cause of overcrowding [which] has the largest influence ... is the habit of a large portion of the population of congregating together in houses. This is not entirely due to financial reasons....[31] The Asiatic does not like air in his dwelling-house. It does not matter how many windows or ventilation openings there are, he always endeavours to close them up, so that custom does away with any good that well-designed houses are intended to produce....[32] [Neither] does it matter how much room you give [Asiatics], they will huddle together, and it does not matter what you have in the way of special legislation [against overcrowding], it will be a most difficult thing to carry out that legislation.[33]

Most, however, were of an opinion similar to that expressed by the editor of *The Straits Times* who felt that though 'the blame [for in-

sanitary housing conditions] [lay] primarily with the Chinese residents, who [were] filthy in their habits beyond all European conception
of filthiness', it was up to the colonial and municipal governments to
'push ahead the improvement of the town until the reproach that it
[was] one of the most insanitary in the world [could] no longer be
made against it'.[34]

The People's Perspective: The Organization of House Space in an Overcrowded City

In colonial Singapore, the most common house type was the traditional
two- or three-storey Chinese shophouse[35] which accounted for over
half the number of dwellings in the municipal area.[36] The shophouse
was normally built in brick with common party walls shared with
adjoining houses and a tiled jack roof to relieve the heat in the house. It
comprised a narrow frontage which rarely exceeded 16 to 18 feet, and
enormous depth, normally over three or four times its width and occasionally running to almost 200 feet.[37] The reasons for the relatively
insignificant proportions of the frontage to the depth were, first, the
greater value of 'frontage area' over that more remote from the street,
and second, the mode of house construction: 'When the span or
bearing between supporting pillars is reached, another span of the same
or less amount succeeds to support the upper portion of the building.... Front walls of the building [are erected] directly upon the longitudinal timber beams which in their turn are carried upon the pillars
or columns.... In consequence, the limit of the actually unimpeded
frontage of a house is soon reached.'[38]

In the typical shophouse, it was the practice to build two or more
gabled roofs running transverse to the longitudinal axis of the shophouse, with airwells positioned at the space in between sections. Rooms
usually led into each other although where there was sufficient width,
rooms and cubicles might be reached off both sides of a central corri-
dor.[39] Part of the front room was not only used as a common room
but was the 'ritual heart of a house' containing a table facing the door
upon which were placed ancestral tablets, images of gods, and ceremo-
nial paraphernalia (Plates 5 and 10).[40] The front portion of the house
was ventilated by window openings facing the street while back rooms
received air and light from airwells, or from clerestory openings under
a jack roof.[41] In function, the shophouse combined the dual purpose of
workplace and residence within the same premises and accommodated
high population densities and intensive economic activity. Commerce,
retail, and cottage industries occupied the front portion of the ground
floor whilst the upper floor was used as a residence, a convenient set-
up considering that traditional Chinese living and business were very
much a family affair. Hence, economic enterprise and living quarters
for the Asian labouring classes were almost always physically integrated
within the same premises or closely integrated within a highly dense
environment. This was in contrast to the 'separation of spheres'[42]—the

physical separation of public space associated with productive work from the private home associated with consumption, social, and family life[43]—characteristic of the lifestyle of the typical European colonialist who 'return[ed] in the evening after his day's work [in the city]', passing along 'smooth and well-watered roads' to his suburban residence 'sequestered in its own grounds amongst beautiful and shady trees and well-kept lawns, forming a pleasing and home-like prospect'.[44] The physical integration of productive work and consumption not only minimized transport costs by abolishing the journey to work, but also allowed a large proportion (at least 100,000 or over 30 per cent according to 1921 figures) of the Chinese population who were either dependent on casual work or engaged in service sector jobs (Table 4.4) to remain in central locations (or near the harbour) within easy reach of potential employers and customers.

The essence of the shophouse–tenement tradition was the internal division of the house using partitions, which were not part of the permanent structure of the building, to create distinct compartments which could be let to individuals, families, or business users who would have the use of the common stair, kitchen, and bathroom. The system of subdividing living space into this cubicle form functioned as a mechanism by which subtenants could live within the central area of the city on urban space which commanded high rents. The cubicle system allowed the labouring classes to occupy minimal accommodation at rents which although high, could be reckoned within their means, as houses or even complete rooms could not. The temporary nature of the partitions allowed them to be readily put up or taken down and for living space to be quickly rearranged according to changing circumstances. It was a system well adapted to the needs of immigrants in search of cheap lodgings, and to the labouring classes who spent little time at home but were dependent on hawkers for cheap and quick meals and on the social life in the streets (see Chapter 7). Among the lowest class of housing types such as *rickisha* pullers' lodging houses, rooms or *pangkeng* were not partitioned into individual cubicles, but instead communally shared by a group of men usually from the same dialect group. Bed 'spaces' or 'platforms' were, however, clearly allotted and could be rented out for eight- or twelve-hour periods.[45] The involution of space is clear from the following description given by Low Ngiong Ing, a Hokchiu immigrant who came to Singapore in the 1910s, of his childhood home in lower Victoria Street, a slum area inhabited principally by *rikisha* pullers: 'they [the *rikisha* pullers] dossed in tiers, tier above tier, the higher two feet above the lower so that there must have been at least three tiers on each wall. The middle space was left as a passageway in the daytime and littered with canvas camp-beds at night.'[46]

According to Frank Leeming who has studied similar housing types in Hong Kong, the intensive use of space inherent in the cubicle and *pangkeng* systems 'represented versatility, as well as high densities and low standards'.[47] From the contemporary European perspective,

TABLE 4.4

Number of Chinese Dependent on Service Sector Jobs
or Casual Work, 1921[a]

Occupation	Males	Females	Total
Watch, clock, and chronometer makers and repairers	215	0	215
Tailors and tailors' machinists	2,323	0	2,323
Boot and shoe makers	1,338	19	1,357
Clog makers	165	0	165
Seamstresses	0	1,161	1,161
Basket makers	698	96	794
Carpenters	7,303	26	7,329
Sawyers	1,816	0	1,816
Boat and shipbuilders	543	0	543
Builders	343	0	343
Builders' labourers	499	187	686
Masons	1,639	0	1,639
House painters	1,201	0	1,201
Bullock cart drivers	2,119	0	2,119
Rikisha and handcart pullers	18,261	0	18,261
Petty officers and deck hands	1,251	0	1,251
Boatmen and lightermen	7,850	0	7,850
Dock labourers	5,571	0	5,571
Proprietors and managers of business	12,860	411	13,271
Brokers and agents	515	0	515
Salesmen and shop assistants	7,774	187	7,961
Hawkers	12,996	488	13,484
Actors	630	120	750
Musicians	97	246	343
Restaurant keepers	676	67	743
Lodging house keepers and runners	193	2	195
Barmen and waiters	1,294	0	1,294
Dhobies	920	214	1,134
Hairdressers	1,036	13	1,049
Others in personal service	0	2,192	2,192
Opium shopkeepers and attendants	582	15	597
Beggars	437	10	447
Total	93,145	5,454	98,599
Total Chinese (Singapore Settlement)	215,918	101,573	317,491

Source: J. E. Nathan, *The Census of British Malaya 1921*, London: Waterlow and Sons, 1922, pp. 289–93.

[a]Only major categories with more than 100 persons enumerated have been listed. Also excluded from the list are 53,018 women classified under 'household duties at home' and 66,529 persons with 'no occupation'. A significant proportion of persons in these two categories were likely to have been engaged in some form of unspecified casual work or belonged to households dependent on service sector jobs or casual work. The total figure listed is hence an underestimate of the actual number dependent on such occupations.

however, while it was conceded that as a 'style of Chinese architecture, if it [could] be dignified by such a title, ... it [was] strictly in consonance with [Chinese] views of ornament and decoration', and while 'the internal arrangements ... seem[ed] well adapted to the habits and customs of the people', these were, as forms of accommodation, at best 'picturesque' if not 'peculiar' and 'grotesque'.[48]

Municipal Strategies against Overcrowding:
The Regulation of Spatial Form

The failure of surveillance to combat overcrowding and insanitary housing conditions became increasingly obvious in the first two decades of the twentieth century. Not only was the legal definition of overcrowding unworkable in practice, as already seen in Chapter 3, the strategy of surveillance inscribed in municipal legislation was ineffectual against a phenomenon which was both rampant and at the same time easily concealed from the municipal gaze. A different municipal strategy, more 'automatic' in effect and less dependent on direct Asian compliance, was needed. The essence of such a strategy was summed up by Hallifax:

Anything that is to be really effective will have to be automatic. You can make it impossible to overcrowd by making your houses of such a construction that they cannot be overcrowded. Anything in the way of ... continual inspection [would be less effective]. ... [49] The personal factor should be removed as much as possible and this can only be done by aiming at making insanitary conditions automatically impossible, at any rate as far as the structure or design of a house is concerned.[50]

Thus, in order to combat overcrowding and its attendant 'house diseases', it was necessary to control the spatial arrangement and built form of housing in the city. In one sense, the control of spatial form as a technique of power was simply a reworking of the strategy of surveillance. The municipal 'inspecting gaze'[51] had shifted from overseeing the daily practices of the Asian population carried out in specific spaces (such as the house, street, market, or a public place) to controlling the dimensions, arrangements, and legibility of particular spaces (such as the house, the building block, and ultimately, the city as a whole) in order to influence the practices of those who inhabited or used such spaces.

'Sisyphean Jugglery': The Campaign against Cubicle Housing

The modification of house form had very modest beginnings when the Municipal Health Department in August 1894 embarked on a scheme to demolish or alter the internal partitioning of insanitary dwellings. Sanitary inspectors were empowered by a fresh set of regulations[52] to enter and inspect all houses between sunrise and sunset and to deal with insanitary dwellings where 'the interiors ... had been divided by wooden partitions into cells, each of a few square feet with a narrow passage running between the rows, ... [an] arrangement lead[ing] in many cases to almost complete obstruction of light and ventilation and to accumulation of filth, seldom if ever removed'.[53] The scheme aimed at sanitizing the Asian dwelling by removing or at least reducing the number of cells, cutting openings for the entrance of light and air, and cleansing and limewashing houses. It was initially pursued with much vigour, affecting several hundreds of houses annually during the closing years of the nineteenth century (Figure 4.2). However, it soon petered out after the turn of the century when it became obvious that the measure had failed to check worsening housing conditions and the rising tide of death and sickness within the city. In the face of failure, in a manner typical of municipal rationalizing, the physical darkness of house interiors was conflated with and explained by the darkened understanding in the minds of the Asians themselves. The municipal president, E. G. Broadrick, lamented that '[w]herever possible, steps are being taken by opening up new streets and removing cubicles, to bring light and air into the dark places of the town, but the work is much hindered owing to the ignorance and apathy of a great part of the Asiatic population'.[54]

It was gradually admitted that to do away with the cubicle system altogether was impossible as cubicles were 'a necessary evil'[55] given the nature of the Asian population in Singapore.

FIGURE 4.2
Number of Insanitary Houses Dealt with by the Municipal Health
Department, 1894–1916

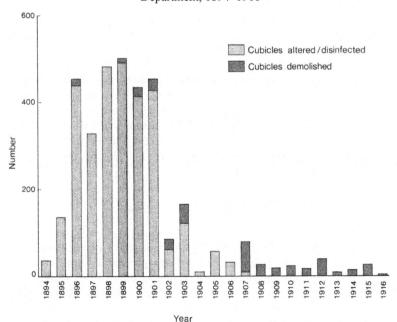

Source: ARSM, Health Officer's Department/Health Officer's Report, 1894–1916.

The occupants of [cubicles and single room dwellings] are too poor to pay for more than one or two rooms, but however poor they are, they must have privacy, even if it only amounts to a small cabin for sleeping in. This applies not only to married people but to those who are single. The latter require to keep their money and belongings in some place where they shall not be touched by others, and although a number of *ricksha* coolies and others live a common life in one large room, it is doubtful if many would live in this way if there were other accommodation.[56]

The municipal campaign against cubicle dwelling amounted to no more than 'a kind of Sisyphean jugglery'[57] for it was 'useless to pull down [cubicles] because necessity, knowing no law, puts them up again forthwith'.[58] Finally convinced of this, the Municipal Health Department substituted their campaign of 'ruthless destruction' of all compartments by one of regulation and controlled development.[59] In 1908, building by-laws were formulated permitting the erection of partitions on floors other than the ground floor.[60] However, these by-laws were so generally contravened that in 1913 it was considered that it would make 'better policy to regulate rather than prohibit [cubicle] erection'.[61] In a move considered 'retrograde' by some commissioners such as Dr Peter Fowlie, by-laws were altered to allow cubicles to be put up on all floors on the conditions that 'the Commissioners were satisfied that the health of inmates would not suffer' and approved plans of the cubicles were prominently displayed on premises.[62] In spite of this, the gradual 'liberalization' in the policy towards the cubicle system did not succeed in either regulating or checking its proliferation. In 1918, the former Chief Sanitary Inspector, T. O. Mayhew, observed that '[t]he overcrowding [of houses] is more acute now than ever it was. And the tendency to run up cubicles is on the increase'.[63] By-laws against cubicle housing were hence 'a dead letter' and two decades after the campaign began, the municipal commissioners had admittedly 'given up trying to deal with cubicles'.[64] Municipal action in demolishing cubicles was incapable of challenging the rapidity with which improvised structures of wood, cloth, canvas, matting, sack, and even tea boxes could be put up by cubicle dwellers to demarcate and protect the privacy of their living spaces. The control of internal divisions of private space within the house had proved far less 'automatic' than initially envisaged.

'Paper Layouts' and 'Paper Schemes': Backlane and
Area Reconstruction Schemes

A more ambitious effort to modify the built form of dwelling places originated with a government enquiry into prevailing high death rates in the city in 1905. Instead of attempting to control the spatial partitioning of house interiors, as envisaged by the previous scheme against cubicle housing, the new strategy aimed at a larger scale rearrangement of building blocks to combat the deleterious effects of overcrowding on the building site. Initially the Governor Sir John Anderson, who had newly arrived in the Colony in April 1904, proposed that in order to give the inhabitants 'the advantages of more light and air', land reverting to the government on the expiry of short leases might be utilized as open spaces in the more congested districts.[65] However, a committee comprising the municipal engineer, the municipal health officer, and the principal civil medical officer set up to consider the Governor's proposal recommended that instead of providing special open spaces 'to which the inhabitants of the town might go for light and air', the sanitation of the city would be better accomplished by 'bringing light and air to the dwellings in which the greater part of the lives of a large portion of the inhabitants [was] spent'.[66] To secure this, the committee proposed 'the opening up of back-lanes not less than fifteen feet wide between each block of back-to-back houses and the laying out of cross streets in the more congested districts'.[67] At the end of the year, the Legislative Council decided to invite an 'outside and independent expert' to investigate sanitary conditions in the city and the causes of the prevailing high mortality.[68] Professor Simpson, the expert who arrived in 1906, gave the stamp of approval to the Peirce–Middleton–McDowell scheme to 'open up' closely built up Asian residential areas which 'constitut[ed] regular rabbit-warrens of living humanity'[69] (Figure 4.3) and devoted a substantial part of his report enumerating the advantages of backlane and other house improvement schemes. Not only could backlanes serve to open up housing blocks to 'the blessings of light and air',[70] the scheme was 'a necessary precursor of any permanent scheme for dealing with the removal of nightsoil' (see Chapter 5).[71] By effecting a rearrangement of the building block so as to allow the 'bringing of all latrines into alignment, contiguous to an open space',[72] backlanes could also function as scavenging lanes, allowing the contents of latrines to be removed from the back of houses, and for house drains to discharge sullage water into back drains rather than passing through the interior of houses to the front.[73] Furthermore, backlanes provided convenient places for dustbins and public latrines and helped reduce the risk of the spread of fire and infectious diseases.[74] In effect, according to the Attorney-General, the backlane (Plate 13) was the latest spatial innovation in the war against disease, a spatial technique of combating 'the enemy' by using what Foucault calls 'power through transparency' or 'subjection through illumination':[75]

If our health officers are to succeed in the fight [against] sickness and disease in this city, it cannot be expected that they will do so without provision [for backlane schemes] for the reason that the enemy with which they are fighting finds shelter in the impenetrability of these dark and unopened places. He can only be reached by the power of light and air, and that can only be brought to dislodge him by this measure of driving back-lanes through these haunts.[76]

In July 1907, a bill was introduced into Legislative Council to give effect to Professor Simpson's recommendations for the opening up of insanitary areas. The chief purpose of the bill was to give the municipal

FIGURE 4.3

Section of Back-to-back Chinese Dwelling Houses in Pekin Street,
Telok Ayer Street, Church Street, and China Street

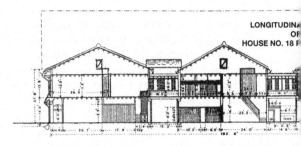

Source: Simpson, *Report on the Sanitary Condition of Singapore,* pp. 27–8.

commissioners the necessary powers to order the demolition of insanitary or obstructive buildings and the reconstruction of unhealthy areas through improvement schemes.[77] In carrying out the proposed improvements, the municipal commissioners were to either come to an amicable settlement with property owners affected by the schemes or, failing that, have the right of compulsory purchase. In terms of the financing of these improvements, it was proposed that where the property affected was held on short leaseholds (ninety-nine years or less), the government would undertake the schemes on approving the representations of the municipal health officer. It was felt that the government was in a better position to carry out such improvements as it would be able to offer the conversion of short leases into statutory grants as an effective lever in negotiations with property owners. In improvement schemes affecting other types of property, the municipal commissioners were to have enlarged borrowing powers and the right to levy an improvement rate in order to finance the schemes.

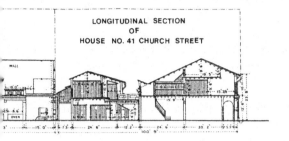

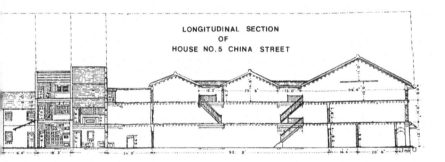

The bill was expected to initiate 'a very sweeping measure for the improvement of [the] sanitary condition [in Singapore]' when it became law.[78] Tan Jiak Kim, the Chinese member of the Legislative Council, estimated that if the new law were to be rigidly enforced, at least 60–70 per cent of the dwelling houses of the Chinese and native communities would be condemned as 'unfit for occupation'.[79] Both Tan Jiak Kim and Dr D. J. Galloway, another unofficial member of the Council, warned that unless alternative accommodation was provided in advance, improvement schemes would not only lead to an increase in the cost of living among the labouring classes and a concomitant rise in the cost of labour, but would also result either in a great number of the poor being 'thrown out into the streets without any shelter' or in the reproduction of 'the evils [of] their previous habitations' in other locations.[80] These ominous prognostications were, however, summarily brushed aside and the government chose to assume that new accommodation for the labouring classes would automatically be provided by market forces when property owners perceived 'the advantages of building elsewhere and improving their houses when they see that the town must spread by the improvements to be made by [the new law]'.[81] Conceived in the magnanimous spirit that 'money ... must come second to the question of sanitation and health',[82] the bill was passed at the end of 1907 'without a single dissentient voice' in an equally optimistic mood envisaging that it would effect 'a very large measure of improvement in the health conditions of Singapore, with a minimum of friction'.[83]

Such rhetoric quickly proved to be merely 'brave words' and a decade later, it had to be admitted that these 'high aims' had failed.[84] The alteration of the built form of housing blocks proved far more complex and costly than had been envisaged. In the first place, the scheme for 'the reconstruction of unhealthy areas' comprising Crown land held on short leases was essentially stillborn. In May 1908, Dr Middleton submitted to the government a representation made under the section of the Municipal Ordinance dealing with 'the reconstruction of unhealthy areas' regarding an insanitary district in the heart of 'Chinatown' bounded by Hokien Street, Macao Street, part of China Street, and part of South Bridge Road.[85] The area comprised sixty back-to-back shophouses of two and three storeys mainly built on Crown land held on 99-year leaseholds. The block exemplified most of the grosser features of insanitary areas: narrow, deep house plots on an overbuilt site with an almost complete absence of open spaces; disproportionate window to floor space (1 to 18.7 instead of the recommended 1 to 10) resulting in poor ventilation; defective drainage with insufficient fall causing 'a back flow of reeking filth into living and sleeping rooms'; floors and woodwork in a dismal state of repair 'providing a refuge for filth and vermin'; overcrowded cubicles; and the lack of proper bathroom, latrine, and kitchen accommodation.[86] Dr Middleton represented that these sanitary defects could not be effectively remedied by any means other than through 'an Improvement Scheme for the rearrangement and reconstruction of the houses' to pave the way for demolishing obstructive structures, driving backlanes in between the backs of houses, and rebuilding the rear of dwellings to include sufficient open space in conformity to building by-laws.[87] The municipal health officer also recommended the acquisition and reservation of adequate space for a hawkers' shelter in order to 'place within reach of residents in the locality a supply of properly cooked food'.[88]

A government order authorizing this 'pioneer' improvement scheme was duly published in the *Government Gazette* on 21 May 1909 but revoked four months later.[89] Despite a flurry of anxious letters from the municipality to the government pressing for the implementation of the scheme, the latter remained intransigently evasive on the issue, refusing both to execute the scheme or to disclose its reasons for revoking the original order. It was later surmised that the main deterrent was the cost of the scheme. In 1918, the Commissioner of Lands, J. Lornie,

1 Sanitizing the city: conservancy was a major municipal function from the early days of the Municipal Board. (National Archives, Singapore)

2 Alex Gentle, President of the Municipal Commission in the 1890s, depicted here setting out on a promising if controversial course of municipal reform. (*Straits Produce,* July 1894:6)

3   Municipal commissioners, 1890. (Hallifax, 'Municipal Government', facing p. 322)

4   Municipal officers, 1915. (Hallifax, 'Municipal Government', facing p. 324)

5 Photograph taken by Professor Simpson to show the 'dark central passage' on the lower floor of a tenement house with the entrances of cubicles 'indicated by the position of figures in the passage'. (Simpson, *Report on the Sanitary Condition,* pp. 24–5)

6 A cubicle dweller living in 'a dark and verminous room in Chinatown'. (Fraser, *The Work of Singapore Improvement Trust,* facing p. 9)

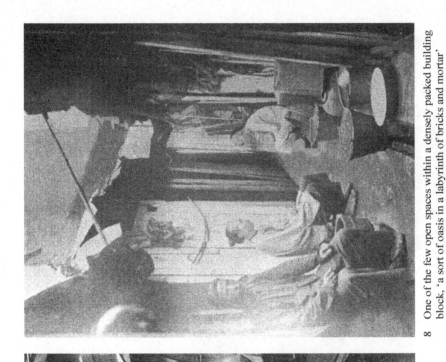

7 Photograph taken by Professor Simpson to show 'the interior of a cubicle and its furnishings'. (Simpson, *Report on the sanitary Condition*, pp. 25–6)

8 One of the few open spaces within a densely packed building block, 'a sort of oasis in a labyrinth of bricks and mortar' (Simpson, *Report on the Sanitary Condition*, pp. 26–7)

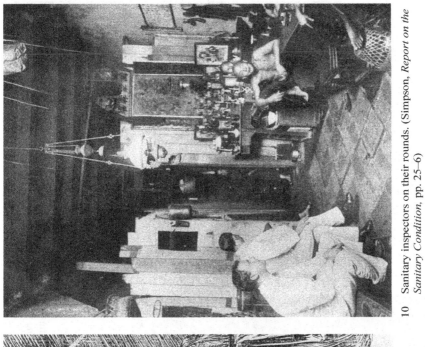

9 A cartoon entitled 'Sorrowfully dedicated to the thousands whose lives have been lost owing to bad housing conditions'. (*Dream Awhile: Cartoons from Straits Produce*, [1933])

10 Sanitary inspectors on their rounds. (Simpson, *Report on the Sanitary Condition*, pp. 25–6)

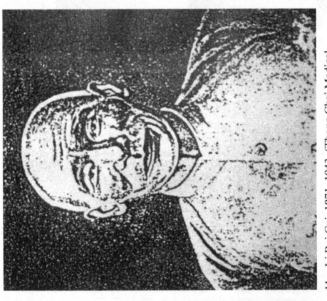

11 Li Bai Gai, 1871–1943. (Thong Chai Medical
Institution, Singapore)

12 Eu Yang Seng, a company specializing in the import,
export, and retail of Chinese medicine according to
prescription.

13 Backlanes, 'the latest innovation in the war against disease': (a) demolition of the rear of back-to-back houses in Cross Street and (b) the newly constructed backlane. (National Archives, Singapore)

14 Orchard Road, a well-shaded avenue leading to the European suburbs. (National Archives, Singapore)

15 South Bridge Road, one of the main thoroughfares through Chinatown known to the Chinese as the 'big horse carriage road'. (National Archives, Singapore)

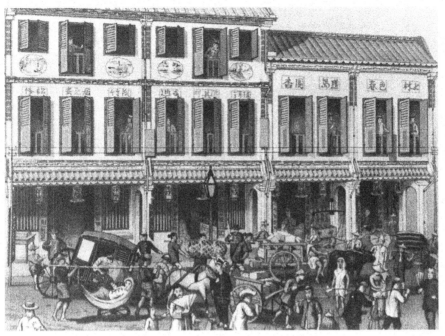

16 A somewhat artificial model of North Boat Quay Presented in 1886 at the Colonial and Indian Exhibition in London. (Schlegel, 'A Singapore Streetscene', pp. 122–4 and Plate ix)

17 A more realistic street scene at Cross Street and South Bridge Road, c.1910. (National Archives, Singapore)

18   Verandahs as the 'great crockery and hardware repositories'. (*Souvenir of Singapore*, 1905, p. 34)

19   In the morning, vendors selling fresh vegetables often lined the sides of streets in working-class areas. (National Archives, Singapore)

20 Verandahs and sides of streets were often 'colonized' by food hawkers and their customers. (National Archives, Singapore)

21 An itinerant Chinese pork hawker laying down his wares on the street to serve a customer. (National Archives, Singapore)

22 An itinerant Chinese
    barber serving a customer
    on the verandah.
    (National Archives,
    Singapore)

23 An itinerant Bengali bread
    seller traversing the length
    of the verandah to sell his
    wares. (National Archives,
    Singapore)

24 Trengganu Street, *c.* 1919.
    Although relatively deserted
    as roadside stalls were
    prohibited along this street,
    itinerant hawkers were still
    in evidence. (National
    Archives, Singapore)

25 Trengganu Street in the
    early 1990s. Verandahs and
    the sides of streets are
    liberally used to store and
    display goods.

26 A funeral party accompanied by elaborate ritual paraphernalia gathered on the verandah
before the start of the funeral procession, *c.* 1900. (National Archives, Singapore)

27 Collyer Quay, *c.* 1900. (National Archives, Singapore)

28　Raffles Place, formerly called 'Commercial Square', *c.* 1900. (National Archives, Singapore)

29　The Hokkien Seh Ong burial grounds at Kheam Hock Road.

30 The Bukit Brown Municipal Chinese Cemetery at Kheam Hock Road/Sime Road.

31 One of the better-kept tombs on Bukit Brown contrasted by its neighbour which is almost completely overgrown.

offered the following explanation as to why the government had retracted from its original position: 'The reason why the Hokien Street scheme fell through was because of a section of the Acquisition of Land for Public Purpose Ordinance [under which the land for the scheme was to be bought] which says that if any part of a house is taken, the owner may require [the government] to take the whole, and the Government feared that the result would be that the whole block would have to be purchased.'[90] Magnanimity and the enlightened concern for the health of the labouring classes which had inspired the 1907 bill had rapidly evaporated on encountering the prospect of an actual large disbursement of funds. It was clear that when it came to bearing the cost of improving the people's health, the colonial government was unwilling to grasp the nettle of finance.

Having suffered bureaucratic vacillation and what was construed as a deliberate rebuff at the hands of the government, the municipal commissioners resolved in 1913 that no further representations regarding unhealthy areas would be made under the same section of the Ordinance.[91] Hence, not only did 'matters remain exactly where they were' in the specific case of the Hokien Street–Macao Street block, 'not a dollar of Government money [was expended] in improving a single house on Crown land held under 99-years' leases'.[92]

On their part, the municipal commissioners attempted to uphold their end of the bargain by proceeding with backlane schemes on land held other than on 99-year leases. Yet, from 1907 to 1913, little was accomplished to drive backlanes through congested areas. Although several schemes for backlanes set out on paper by the municipal engineer were approved by the Municipal Board,[93] they represented prospective rather than immediate plans, to be executed in the event that the property was to be rebuilt. In other words, they were no more than 'paper lay-outs'.[94] Seldom did the municipality have the resources to deal with an entire block of houses as a coherent whole; instead, the system depended on property owners sending in plans for rebuilding or on piecemeal purchases of 'a bit here and a bit there'.[95] Again, cost proved to be the major prohibitive factor: not only was the value of most properties to be acquired for backlane schemes extremely high, the commissioners also blamed the 'extravagant demands' of property owners for compensation in inflating the cost of these schemes.[96] In total, only seventy-six houses were altered to provide backlanes between 1910 and 1913, of which few were situated in the congested districts of 'Chinatown'.[97]

When the amending Municipal Ordinance XIII of 1913 was passed, the basis of compensation was altered under section 137 in order to facilitate cheaper acquisition of land for backlane schemes.[98] Even then, as the Housing Commission observed in 1918, the results were extremely meagre:

From the first to the last, in the whole of Singapore City—as the result of the general feeling aroused in 1907 by Dr Simpson's report—there have only been

FIGURE 4.4

Spatial Distribution of Backlane Schemes, Including Those Completed,
in Hand, and Where Negotiation Was Being Carried Out, 1917

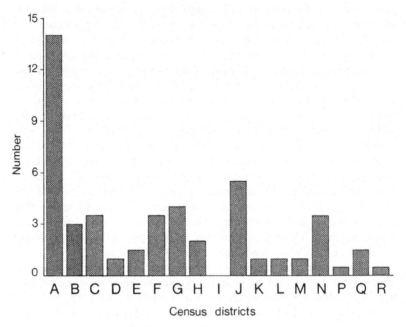

Source: Housing Difficulties Report, Vol. 2, pp. B216 and B249.

22 completed back-lane schemes under section 137 of the Ordinance. Only
426 houses in all have been affected. Of this number, only 238 houses have
obtained communication with the back-lane, and 188 are still unconnected.
Only 156 houses are affected by the 10 schemes now in hand; whilst 512
houses will be affected by 14 schemes in which acquisition proceedings are
under negotiation. And that is all that 'Schemes' have done up to date.[99]

By 1917, although backlane schemes were concentrated in districts
where Asians were heavily represented such as census districts A, C,
F, G, and J (Figure 4.4), Dr Middleton himself confessed that backlanes 'ha[d] not affected the worst parts of the town' and that he
doubted whether the task of opening up all congested areas would be
accomplished in a century if the rate of progress was not radically
improved.[100]

The failure of backlane schemes to procure the 'blessings of air and
light' for congested areas was, however, not simply a question of inadequate financing but also because they failed to secure the co-operation
of both Asian houseowners and occupiers. In a typical case, in order to
open up the block of back-to-back houses shown in Figure 4.5A, the
municipal commissioners proceeded by acquiring and demolishing the

FIGURE 4.5
Opening up a Block of Back-to-back Houses through a Backlane Scheme

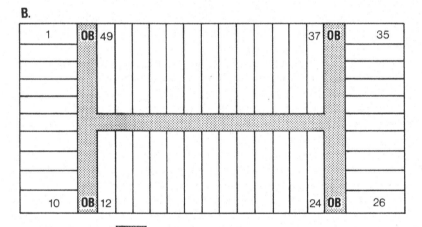

Proposed backlanes

OB   Obstructive buildings

Source: Housing Difficulties Report, Vol. 1, pp. A33–A34.

four obstructive buildings as shown in Figure 4.5B, thereby providing access to backlanes for houses 1–10 and 26–35. Beyond this point, the prospect of constructing the connecting backlane for houses 12–24 and 37–49 was far more daunting as it required the forcible entry of a demolition party into these houses and the destruction of their rear portions, a procedure bound to court opposition, legal tussles, and ruinous compensation bills. In practice, whether a backlane could be formed was dependent on whether houseowners acquiesced in rebuilding, and even if this happened, through communication along the entire backlane system could not be effected as long as any one owner refused to rebuild. Furthermore, even where a backlane was already provided, as in the case of houses 1–10 and 26–35, proprietors often took no action to connect their premises with it. At this point,

the Municipal Health Officer sets to work to cajole or bully them into connecting their premises with the back-lane. It is thoroughly in keeping with the whole ridiculous procedure that the bullying is administered by the Municipal Health Officer by means of serving the owners with 'Nuisance Notices'. If the landlord, yielding to cajolery or overcome by Nuisance Notices, decides to connect his back premises (kitchen, bathroom and latrine) with the back-lane and submits a plan accordingly, he learns still more of the subtility [sic] of the Municipal Ordinance. His plan is rejected ... unless he *also* reconstructs his building in such a manner as to give up a third of it for an open space.[101]

The illusion of improvement created by marking out clean, straight lines of interconnecting backlanes cutting through irregular conglomerations of slum properties on plans of the city lost much of its credibility when these schemes were actually attempted in practice. The spatial form of housing blocks, created by casual and unsystematic accretions and additions by individual property owners over the decades, proved far more resistant to systematic modification than had been anticipated.

It was also ironic that municipal building by-laws designed to safeguard building standards were perceived by Asian property owners as deterrents to rebuilding and improving house form. When an owner desired to alter a particular part of his house, a plan of the whole property had to be submitted to the municipality for approval, on which the authorities would 'mark a red line' across for a proposed backlane. Hence, in order to make even minor alterations such as putting in a doorway or altering a front staircase or a back kitchen, the owner would also have to give up land for a backlane and set his house back in order to conform to the rule that one-third of his property had to be left as open space. These by-laws only applied to those who proposed to build or rebuild and did not affect 'the man who [was] content to let his property remain as it was when he bought it'.[102] As such, not only were the by-laws 'bitterly resented' and a source of 'very considerable friction between the public and the Municipality', they were to a large extent self-defeating as they inhibited rather than encouraged the improvement of house form.[103]

From the perspective of Asian occupiers, municipal attempts to alter the house form through backlane and improvement schemes, intended to introduce the 'power of light and air' into the dark, diseased interiors of Asian tenements, only served to compound the problems of house scarcity and to disrupt the extended family system. As Seah Liang Seah, who had served both as municipal commissioner and unofficial member of the Legislative Council, explained to the Housing Commission in 1918,

Professor Simpson's back-lane system caused difficulty regarding houses because one family who formerly occupied one house now had to separate into two houses. Formerly, in a house they had two rooms, a front room and a back room, with an airwell and a passage between, but since the back-lane system has been introduced, the Municipality will only allow one room. . . . Now except for a kitchen at the back it is all front part and open space. [There is] no air well. . . . That is hard on the people who occupy many of the houses. Formerly, a family of husband and wife and two children could occupy the front part, and say a mother-in-law and sister-in-law the back, and downstairs the coolie could be accommodated. Now there is only one room for the family, and the mother-in-law and sister-in-law must find another house.[104]

The airwell was a vital element of the Chinese shophouse for several reasons. As a structural element, it served not only to admit light and a constant draught to keep the house cool in a hot and humid climate but to divide each storey into distinct semi-private compartments occupied by different generational tiers or separate families sharing the same house.[105]

The municipal authorities, however, did not acknowledge the significance of airwells in Chinese house design. Instead, Dr Middleton considered them 'rather objectionable', and in cases of reconstruction, insisted on closing them and providing an open air space equivalent to one-third of the total area at the rear of the house, the airwell not counting as contributing to the amount of open space demanded.[106] Figures 4.6 and 4.7 illustrate the contrast between the traditional shophouse plan based on the airwell concept (1886 plan) and the new municipally sanctioned plan, where the airwell had been replaced by an 'open area' at the rear of the house and a 'proposed back lane' (1921 plan). From the municipal perspective, an opening encapsulated as a private space within the house was far harder to regulate than open space located outside the house structure itself, partly as a backlane accessible to municipal servants. Indeed, Dr Middleton alleged that the airwell within the Chinese house was often roofed over and cluttered up, serving little purpose as a ventilation opening.[107] In effect, by re-arranging the spatial layout of open and enclosed spaces within the house plot, backlane schemes converted essentially private spaces into public places accessible to the municipal gaze. They imposed a prescribed 'legible order' on the building block by opening up and exposing what was previously hidden, private, and inaccessible. These schemes aroused strong resentment amongst occupiers as they represented an erosion of private control of usable space. In time, however, the space occupied by backlanes were occasionally 're-colonized' by occupiers and others for private purposes such as the dumping of rubbish, storage of goods, setting up of hawker stalls, and the erection of lean-tos and even cowsheds.[108] Hence, even where the municipal authorities had successfully 'opened up' a building block through a backlane scheme, there was no guarantee that the 'air space' would not be 'diminished' by 'illegal encroachments'.[109] The backlane provided another arena where conflicts over the structure and use of space had to be negotiated.[110]

Besides reducing the amount of space enclosed as private territory within the house plot, backlane schemes also entailed the reconstruction of the rear portion of houses in a manner objectionable to certain

158

FIGURE 4.6

Approved Plan (No. 161 of 1886) of Seven Shophouses to Be Built in
Japan Street, the Property of Moona E.L.C. Chitty

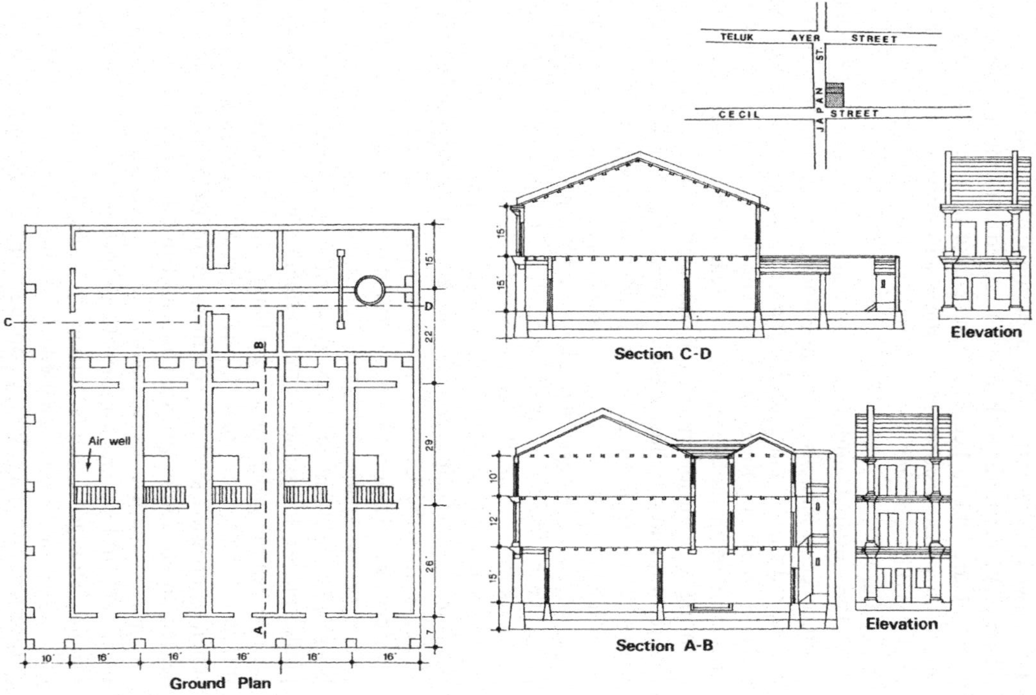

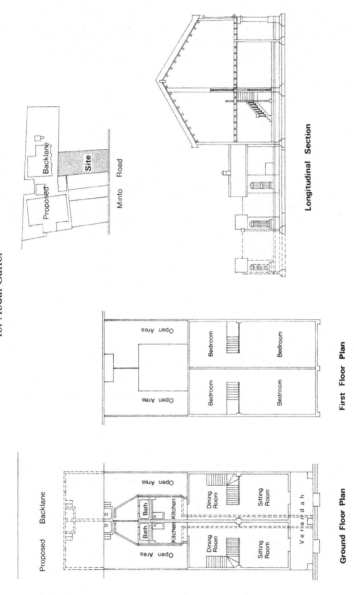

FIGURE 4.7
Approved Plan (No. 88 of 1921) of Two Shophouses at Minto Road
for Abdul Gaffor

Asian communities. Municipal building by-laws required that recon-
structed terraces had bathrooms and kitchens built back-to-back on
either side of the wall between two houses. According to Dr H. S.
Moonshi, a Muslim Indian member of the Municipal Board, this prescribed spatial order imposed by building by-laws was objectionable to
Asian tenants as it did not 'give much privacy'.[111] Instead, it would be
preferable from the Asian perspective if kitchen and bathroom terraces
could be built on one side of the wall only. Despite what was municipally perceived as overcrowded conditions, the claims of privacy were
important elements in the arrangement of domestic space among the
labouring classes. Not only was this evident in the persistent erection of
cubicles to provide enclosed private spaces, but for certain groups such
as the Muslim and Hindu communities, considerations of ritual pollution and the gender roles of women were particularly crucial to the
organization of domestic space. Cooking was perceived as the most
sacred domestic activity other than purely religious ones, while defecat-
ing, which required bathing afterwards, was the most polluting.[112] For
orthodox Hindu and Muslim families, kitchens formed an important
element of women's territory and had to be organized in such a way as
to protect purdah or the seclusion of women from public view. As
such, kitchens and bathrooms had to be closely isolated as private
spaces impenetrable to the gaze of outsiders. Dr Moonshi's suggestion
that building by-laws be modified to suit Asian cultural practices was,
however, rejected by the municipal president on the grounds that 'the
setting of terraces back-to-back [was] a very important health factor' as
it greatly increase[d] light and ventilation' and was an arrangement
'universally practised in up-to-date Municipalities'.[113] The attainment
of privacy was in no way to interfere with what were municipally prescribed as 'health factors'.

'Salus Populi Suprema Lex': The Move towards Town Planning

By the 1920s, the vision of a healthy and efficient city had expanded in
scale and complexity. It was no longer considered sufficient to deal
with the problems of urban malaise through piecemeal projects focusing on selected housing blocks. Instead, the improvement of the city
had to be envisioned and planned as a co-ordinated project encom-
passing the entire city and its multifarious functions as an organic
whole. In his presidential address to the Fifth Biennial Congress of the
Far Eastern Association of Tropical Medicine held in Singapore in
1923, Dr A. L. Hoops summed up what was becoming the main thrust
in the campaign against urban disease in tropical cities:

Most large cities have in the past grown up without law and order; year by
year more people have pressed into them. The remedy is clear: town planning,
zoning, the building of enough houses in residential areas to relieve overcrowding; the provision of cheap transport to enable the poor to live in these areas,
and to travel to and from their work; modern sanitary arrangements. These

measures must precede the destruction or reconstruction of old insanitary tenements and the provision of playgrounds.[114]

These concerns were part of the emerging movement for urban planning which had gathered momentum first in Britain and then throughout the Empire. In early twentieth-century Britain, the vision of a modern 'planned' environment had grown out of Victorian housing, social, and land reform movements which had sought to ameliorate the environmental malaise of the nineteenth-century industrial city. The focus of British urban planning was the improvement of environmental quality through a rational ordering of space. Urban planning, codified in legislation and given statutory force in the Housing and Town Planning Act of 1909, was mobilized into a coherent movement in tandem with the Garden City project, and in so doing, changed 'a century-long process of sanitarianism and bye-law control into one of land-use control and comprehensive urban management'.[115] The vision of urban planning, inextricable from its ideological assumptions that socio-economic problems could be met by physicalist solutions, was 'exported' to British dependencies on the colonial fringe.[116] Colonial cities like Singapore accepted the tenets of urban planning as 'a form of technical expertise' by which local environments could be modelled and controlled in accordance with an assumed 'public good'.[117] Indeed, though Singapore had evolved from an 'island swamp' to 'a great tropical city', the persistent image of miasma-generating tropical environments and its associated unhealthiness further vindicated the importance of urban planning:

The whole island of Singapore is destined to become urban, and because its geographical location is tropical, the planning of its streets, the provision of its open spaces ... are matters more vitally important than they would be in a temperate zone. Great main arteries, broad and handsome, should be formed, ... lined so that the cleansing and refreshing breezes that rise on the sea may sweep along them unrestrained, sweetening and purifying as they go.[118]

The 1918 Housing Commission Report had concluded that 'the improvement and development of Singapore must take place upon broader and more general lines than ha[d] hitherto been the case'.[119] It recommended that if town improvement were to be conceived in broader terms, it could no longer be entrusted to the municipal authorities which, apart from being already overburdened with multifarious day-to-day functions, did not possess the technical expertise to plan and lay out the city according to modern planning principles. Neither could the task of improving the city and providing sanitary dwellings for the urban poor depend on private enterprise or philanthropic agencies for 'the Tan Tock Seng[120] of the present generation of Singapore Chinese ha[d] not yet appeared'.[121] It was thus necessary to create new machinery in the form of an Improvement Trust, similar to those already in existence in the larger towns of India, to supervise the improvement and future expansion of Singapore.[122] In mid-1919, it

was decided to engage a technical expert with experience in town planning to study the conditions in the city and draft the legislation essential for an effective 'Town Planning, Housing and Improvement
Trust'.[123] In May 1920, Captain E. P. Richards, an experienced civil
engineer, arrived to take up the post of deputy chairman of the proposed Improvement Trust (hereafter DCIT).[124] It was immediately
realized that effective town planning had to be preceded by accurate
surveying and mapping, and the depiction of the form and spatial
layout of the entire city on the drawing board. Given 'the absence of
contour survey maps and indifferent topographical surveys [of
Singapore]', the first stage in the life of the newly created Improvement
Trust was given over to 'the preparation and bringing up to date of
such maps as existed'.[125] Accurate landuse and topographical surveys,
averred the DCIT, were the *'sine qua non* of practical investigation'
without which 'there can be no rapid and extensive detail planning and
replanning, quickly organised improvement and remedy of present
order or congestion, ... or precision planning for the future'.[126] A
'Unified Plan of Singapore', on which were demarcated areas for
various specific uses—shophouse development, different classes of
housing, trade, offensive and non-offensive industries, public buildings,
open spaces and playing grounds—was prepared so as to fit in various
improvement, housing and other schemes.[127]

The mood of the early 1920s was one of strong faith in town planning and legislative solutions as the panacea to cure the evils of congested cities and set them 'on the path of ordered growth'.[128] The
testimony of successful town planning in European cities inspired the
remedy for the urban ills of Singapore:

By their superior [planning] laws, [Continental cities] have got far ahead in all
the greater aspects of orderly civic growth and planning. Because of this, their
post-war housing is a simple problem. Britain, however, gave the world the
single family dwelling—the type that all civilization aims for today, and we
gave the world perfect water supply, sanitation, sewerage, drainage and the
macadam road. Of the dwelling itself, and of dealing with slums, the English
have learnt most, but the Continent rapidly absorbed our good points.... It is
now our turn to make full use of Continental and other knowledge for the
benefit of Singapore.[129]

Undaunted by what had in the past been intractable problems of urban
growth in Singapore, the DCIT alleged that once the essential powers
for town planning had been acquired, it '[would] be accomplished
within our own lifetimes, a great increase of trade facility and elbow-
room for business, with life-saving on a big scale by creating accessible,
healthy and attractive residential areas for every class, by better
housing, and by the steady but equitable elimination of slums—at reasonable cost'.[130] Given the magnitude of the problem, these were
enthusiastic words indeed. Their sweeping optimism disclosed a sense
of expectation, of standing at the cutting edge of a campaign to
conquer slums and squalor. Town planning was the ultimate solution

to urban malaise, for not only was it expected to achieve results, it was able to do so on a large scale, within a relatively short time, at an affordable cost, and in a manner which 'hangs the scales very fairly between the true right of the Individual, and the true rights of the Community'.[131]

One of the principal tenets of effective town planning was embodied in the concept of 'zoning' or 'districting'. In the words of the DCIT, '"zoning" is a careful and experienced division of the whole of a city into many areas, each of which is specially suitable—by reasons of both present and likely future conditions—for a particular use and purpose'.[132] Zoning was considered a practical spatial tool to facilitate a nested hierarchy of planning controls: the city was first divided broadly into areas 'zoned' for specific landuses and then further subdivided into smaller areas governed by specific building by-laws relating to building height and open space. Each part could be controlled separately and at the same time subsumed under the dictates of a general plan of the entire city. By facilitating the 'scientific' and co-ordinated manipulation of urban space, zoning, according to the DCIT, could 'prevent further disordered growth, remedy past mistakes and present congestion, and cause ... business-like, ordered, commonsense expansion and amenity'.[133] An effective use of zoning regulations could also increase the predictability of landuse and stabilize land values in the city. Zoning was, according to Walter H. Collyer, the Acting DCIT who succeeded Captain Richards in 1924, advantageous to both the city authority and the individual property developer: the former was better able to tailor the provision of roads and other public facilities to the needs of each zone more economically, whilst the latter was protected from the depreciation of property values resulting from the invasion of objectionable forms of landuse.[134] Zoning hence allowed 'orderly' development, building regulations, and planned landuses to be introduced into various parts of the city inhabited by different classes according to the standards deemed appropriate to each area. This implied that planning was by no means an egalitarian process, but one which conformed to the basic division of colonial society between ruler and ruled, Asian and European, rich and poor. By regulating the type of housing permitted in each zone, such as the proposed prohibition of shophouse development in 'bungalow' or 'terrace house' areas, zoning facilitated the inscription of an implicit apartheid on the colonial urban landscape.

In the draft town improvement bill submitted to the government in 1923, the DCIT devoted a large section to zoning regulations empowering the Improvement Board to set apart areas for specific landuses and to specify building lines and areal layouts.[135] However, a committee appointed to report on the draft bill in 1924 considered the proposed zoning regulations far too inflexible and impracticable to cope with the rapidly changing urban environment in Singapore. A town, declared the committee, '[could] not be forced to grow as an Espalier Fruit tree [was] trained on a wall'.[136] Instead, it recommended that the municipal commissioners should continue to regulate landuses and trades by the prohibition of specific uses in specified areas.[137] It was felt that zoning powers were too large and unwieldy, and constituted an unwarranted interference with the private rights of owners which would adversely affect the property market. A means of landuse control which could be selectively imposed through the operation of by-laws was preferable to one which was rigidly prescriptive in a general way.

Trimmed of its zoning powers, the final Improvement Bill passed in 1927[138] was deprived of its cutting edge in effecting changes in urban landuse. Whilst the Improvement Trust could prepare schemes laying out roads, open spaces, and building lines, it had no power to dictate the type of buildings or landuse suitable for a particular area. As such, the General Improvement Plan drawn up was in fact, 'in no sense a complete development plan, but merely a record, built up in piecemeal fashion, of existing development, together with layouts for proposed future development which had received statutory approval'.[139]

The newly independent Improvement Trust also inherited from the municipality the responsibilities for condemning 'houses unfit for human habitation' and the eradication of slum property in the city.[140] The improvement scheme was regarded as 'the most satisfactory method of dealing with slums'[141] and although several schemes were either initiated or taken over from the municipality in the closing years of the 1920s,[142] the policy of the Trust after 1930 shifted towards 'an intensification of the back-lane programme ... as the best means of opening up insanitary blocks of back-to-back houses to light, air and municipal services'.[143] Again, improvement schemes requiring the buying up of large blocks of slums for demolition and reconstruction proved far too expensive as compensation had to be paid not simply for the houses affected but also for the site value, a condition seen to be not only prohibitive but distinctly unfair. The Trust's legal adviser, Roland Braddell, for example, considered it 'a crime against the community that a landlord received compensation based on rentals of slum property derived from overcrowded insanitary houses'.[144] In the long term, while a backlane programme proved a more feasible way of opening up insanitary blocks, its actual value in solving the slum problem was rather doubtful. First, renewing the rear portions of old and dilapidated houses gave a new lease of life to property which was obsolete and overdue for demolition and rebuilding, thereby perpetuating rather than ameliorating the slum problem. Second, by reconstructing the rear portion of a house for backlane purposes, living accommodation was cut down by almost half in many cases, thus creating rehousing problems and aggravating overcrowding.[145] During its early years, the Trust had no power to build except where expressly laid down in an improvement scheme and was under no obligation to provide housing except for persons dishoused as a result of the execution of an improvement scheme.[146] As such, the lack of public provision of working-class housing to ameliorate the unremitting pressure on the existing housing stock caused by rapid demographic increase as well as the reduction of living space resulting from backlane improvements meant that the Improvement Trust, as in the case of its predecessor (the municipality), fought a rapidly losing battle against the problems of overcrowding, congestion, and the proliferation of slum properties.

Another major concern of the town planning agency in Singapore was the need to provide open spaces as 'breathing lungs' scattered throughout congested areas. The 1918 Housing Commission Report had found Singapore 'lamentably deficient in open spaces and parks', for only two places—the 80-acre Botanical Gardens located in Tanglin, the élite European suburb, and the People's Park, an open space of 7 acres adjoining the southern congested area—could possibly be included in any definition of 'park'.[147] It was the responsibility of the town planning agency to create open spaces of all shapes and scales, whether small playgrounds for children, playing fields for organized games, parks and ornamental gardens, or private spaces in the interstices between buildings throughout the city.[148] It was also desirable to provide open spaces in the centres of deep blocks of buildings so as to reduce the depth of plots and improve the conditions of insanitary dwellings (Figure 4.8).[149] Such open spaces in the centres of solid blocks of masonry were particularly valuable not only for purposes of sanitation but also as recreational grounds for both children and adults living in the surrounding tenements, who in the absence of recreational spaces were either confined to their 'dark airless houses' or compelled to transfer their recreational activities to the streets.[150] Both these alternatives were undesirable; the former was inimical to health whilst the latter added to the congestion of the streets and hindered urban circulation (see Chapter 7). Furthermore, public parks were considered to have immense aesthetic and educational value in enhancing civic life. The editor of *The Straits Times*, for example, urged that public parks, 'kept strictly clean, made graceful and attractive' should be liberally provided in various parts of the city, for they 'would not only serve as lungs but would help to awaken a sense of appreciation of beauty and neatness among the dirt-scarred masses who live in viler holes than the average pig'.[151]

Planning for open spaces was, however, only the beginning of a long-drawn process of actually creating 'lungs' for the city. These spaces could only be created when rebuilding occurred or in conjunction with an improvement scheme. Although a number of plans for public open spaces to be reserved in congested areas was sanctioned by both the municipal commissioners and the government in the early 1920s,[152] the actual work of creating these spaces was only initiated in 1925 in connection with the improvement of two city blocks. The first opportunity to create an open space arose as a result of the rebuilding occasioned by a burnt-out foundry which occupied the central portion of a 7½-acre block in a northern congested area bounded by Beach Road, Middle Road, North Bridge Road, and Tan Quee Lan Street, described by the DCIT as 'a rabbit warren of dangerously insanitary

FIGURE 4.8
Plan Showing the Demolition of Buildings and the Creation of an Open Space
in the Centre of a Deep, Irregular Building Block

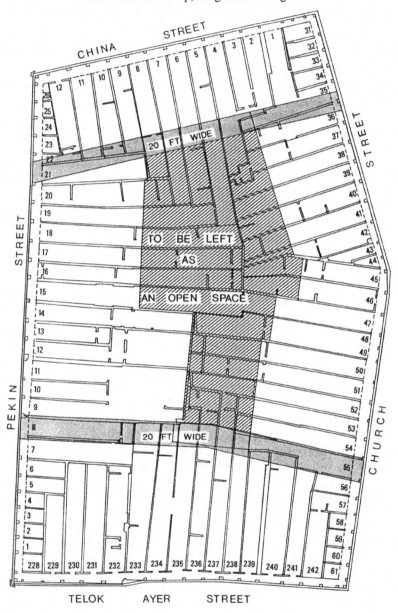

Source: Simpson, Report on the Sanitary Condition, pp. 36–7.

structures containing over 500 people' and 'one of the worst places in the city'.[153] Two acres in the centre of the block was purchased with the aid of government funding and reserved as an open recreation space. A second project initiated in the same year involved putting in backlanes and opening up an acre of open space in the heart of a 3½-acre 'island of solid masonry' in the southern congested area bounded by Chuliah Street, Philip Street, Church Street, and Synagogue Street.[154] These two schemes were completed in the late 1920s and early 1930s, but progress on creating other open spaces was exceedingly slow. In the words of the DCIT, 'doctoring a sick city [was] a long and painful process'.[155]

Transforming the Built Form of the City: Intentions versus Effects

The overriding, even obsessive, concern with health was the driving force behind much of the colonial preoccupation with modifying and improving the built form of urban Singapore. The creation of physically healthy environments, defined according to the medical and cultural criteria of metropolitan power, became a major objective pursued with little consideration of, and at times much disdain for, the strategies of spatial management and arrangement of house form used by the colonized in adapting to the constraints of the colonial urban environment. In his evaluation of colonial urban planning policy, A. D. King argues that

indigenous definitions of health states, the means for achieving them and the environments in which they existed were replaced by those of the incoming power in a total ecological transformation. Thus, vital statistics from metropolitan society are used as the reference point to 'measure' health states in the colonial population, historically and socially derived concepts of 'overcrowding' developed in metropolitan society are applied to the indigenous environment, irrespective of cultural context or the larger economic and political situation. In the interest of 'health', new environments are created—[houses] ... surrounded by 'light and air', 'open space', gardens, and recreational areas in total disregard of the religious, social, symbolic or political meaning of built environments as expressed in indigenous villages and towns.[156]

In transforming Singapore from a city of 'mean streets', 'airless cubicles', and 'dark house interiors' to one which conformed with the municipal vision of a healthy, salubrious, open-textured city, however, there was constant slippage between intentions and effects. First, the attempts by the municipal authorities and the Improvement Trust to modify the built form of the city according to metropolitan norms were fraught with difficulties because intentions were not sufficiently matched by financial commitment. Various 'improvement schemes' designed by health and town planning experts aimed at introducing 'the blessings of light and air' into working-class housing in particular and the built environment of the city as a whole were seldom implemented on a scale and at a rate sufficient to significantly alter the

spatial built form of the city. Second, schemes which appeared financially more feasible such as the demolition and regulation of cubicles or
the construction of backlanes failed to secure the intended effects as
they were often resisted by Asian property owners and occupiers who
viewed these schemes as unwarranted intervention with the organization of Asian house space. The municipal aim of creating a healthy city
hence faltered on two main counts: first, in abrogating its responsibilities to provide alternative accommodation to absorb the rapid influx of
immigrants into the city as well as those displaced by its 'sanitizing'
efforts; and second, in its failure, while attempting to modify built
form, to take into account Asian traditions of building and organizing
space which had taken root in the city. The project of 'sanitizing' and
'disciplining' the form of housing space in the city was hence one of
conflict and negotiation fought out among those in authority who
aimed at imposing particular 'norms and forms' on the built environment, the fraction of capital who built and owned house property, and
those who actually lived and used housing space within the city.

1. W. C. Oman, 'A Plea for the Registration of Architects in Singapore', *Journal of the Singapore Society of Architects*, 1, 4 (1924): 32.

2. Martin J. Daunton, 'Public Place and Private Space: The Victorian City and the Working-class Household', in Derek Fraser and Anthony Sutcliffe (eds.), *The Pursuit of Urban History*, London: Edward Arnold, 1983, p. 213.

3. For example, see *Report of the Housing Committee of Singapore, 1947*, Singapore: Government Printing Office, 1948, pp. 1–4; Singapore Housing and Development Board, *Homes for the People: A Review of Public Housing*, Singapore: Straits Times Press, (1965/6), pp. 8–20; Stephen H. K. Yeh (ed.), *Public Housing in Singapore*, Singapore: Singapore University Press, 1975; Riaz Hassan, *Families in Flats: A Study in Low Income Families in Public Housing*, Singapore: Singapore University Press, 1977, pp. 3–4; Teo Siew Eng and Victor R. Savage, 'Singapore Landscape: A Historical Overview of Housing Change', *Journal of Tropical Geography*, 6 (1985): 48–63; Aline K. Wong and Stephen H. K. Yeh (eds.), *Housing a Nation*, Singapore: Maruzen and Housing and Development Board, 1985. The small number of studies which explicitly examines the ideological basis of public housing in the post-war era include Manuel Castells, Lee Goh, and R. Yin-Wang Kwok, *The Shek Kip Mei Syndrome: Economic Development and Public Housing in Hong Kong and Singapore*, London: Pion, 1990; Chua Beng-Huat, 'Not Depoliticized but Ideologically Successful: The Public Housing Programme in Singapore', *International Journal of Urban and Regional Research*, 15 (1991): 24–41.

4. An exception, albeit one which focuses on post-war rather than pre-war housing conditions, is Kaye's sociological study of Chinese households living in a densely popu-lated area (Barrington Kaye, *Upper Nankin Street, Singapore: A Sociological Study of Chinese Households Living in a Densely Populated Area*, Singapore: University of Malaya Press, 1960).

5. Daunton, 'Public Place and Private Space', pp. 213–14.

6. James Francis Warren, *Rickshaw Coolie: A People's History of Singapore (1880–1940)*, Singapore: Oxford University Press, 1986, p. 194.

7. In British cities such as London and Glasgow, with the failure of philanthropic capitalism to provide adequate low cost dwellings for the working classes, municipal authorities intervened to provide subsidized housing. Before the outbreak of the First

World War, over 55,000 people were living in London County Council accommodation while about 10,000 were housed by Glasgow's Improvement Trust. Model dwelling houses organized along the lines of 'morality, sobriety, cleanliness, order and discipline' were provided in an attempt to reform the habits of the working classes through improvements in the built form of housing. Municipal housing, however, seldom benefited more than a small proportion of the working class and remained beyond the reach of the casually employed or unskilled labourer before the First World War (Anthony S. Wohl, 'The Housing of the Working Classes in London 1815–1914', in Stanley D. Chapman (ed.), *The History of Working-class Housing*, Newton Abbot: David and Charles, 1971, pp. 40–3; John Butt, 'Working-class Housing in Glasgow, 1851–1914', in Chapman, *The History of Working-class Housing*, pp. 63–4). In Singapore, although model dwellings were advocated by sanitary experts such as Professor Simpson, municipal socialism in the provision of working-class housing failed to take root largely because neither the colonial government nor the municipality were willing to bear the financial burden.

8. *Proceedings and Report of the Commission Appointed to Inquire into the Cause of the Present Housing Difficulties in Singapore, and the Steps Which Should Be Taken to Remedy Such Difficulties* (hereafter *Housing Difficulties Report*), 1918, Vol. 1, p. A12.

9. Warren, *Rickshaw Coolie*, p. 213.

10. 'Memorandum D: Notes on Tuberculosis by Dr D. J. Galloway', in *Housing Difficulties Report*, 1918, Vol. 2, pp. C8–C9; 'Memorandum G by Dr J. A. R. Glennie, Acting Municipal Health Officer, Singapore', in *Housing Difficulties Report*, 1918, Vol. 2, p. C18.

11. 'Memorandum R by Major W. R. C. Middleton, Municipal Health Officer, Singapore', in *Housing Difficulties Report*, 1918, Vol. 2, p. C78.

12. Ibid.

13. L. W. Darra Mair, 'Report on Back-to-Back Houses', *Parliamentary Papers*, 38 (1910).

14. A. K. Chalmers (Medical Officer of Health for Glasgow), *The House as a Contributory Factor in the Death-rate*, Glasgow: Royal Philosophical Society, 1913.

15. 'Memorandum R', pp. C79–C81.

16. See section 173(2) of Municipal Ordinance XV of 1896 (C. G. Garrad, *The Acts and Ordinances of the Legislative Council of the Straits Settlements from the 1st April 1867 to 7th March 1898*, Vol. 2, London: Eyre & Spottiswoode, 1898, p. 1526). This rule differed from the official definition of overcrowding used in British cities where overcrowding was fixed at more than two persons per room. If the British yardstick had been applied to Singapore, overcrowding would be 'general' throughout the city for the average number of persons per cubicle in a typical area was 2.5 ('Memorandum R', p. C78).

17. *Housing Difficulties Report*, 1918, Vol. 2, p. B2.

18. ARSM, 1909, Health Officer's Report, pp. 38–9.

19. *Housing Difficulties Report*, 1918, Vol. 2, p. B114.

20. Ibid, Vol. 1, p. A4.

21. W. J. Simpson, *Report on the Sanitary Conditions of Singapore*, London: Waterlow & Sons, 1907, p. 29; 'Memorandum R', p. C77.

22. 'Memorandum R', p. C77.

23. *Straits Times*, 12 June 1888.

24. *Housing Difficulties Report*, 1918, Vol. 1, p. A1.

25. Ibid.

26. 'Memorandum B: Notes by Mr F. J. Hallifax on the Shortage of Housing in Singapore and Its Relation to Taxation', in *Housing Difficulties Report*, 1918, Vol. 2, p. C3.

27. Ibid.

28. *Straits Times*, 6 February 1925.

29. *Housing Difficulties Report*, 1918, Vol. 1, p. A8.

30. 'Memorandum G', p. C17.

31. 'Memorandum M: A Minute on Overcrowding Addressed on 14 August 1917 by Dr J. A. R. Glennie, Acting Municipal Health Officer, Singapore to the President of the Municipal Commissioners', in *Housing Difficulties Report*, 1918, Vol. 2, p. C47.

32. 'Memorandum G', p. C17.

33. *Housing Difficulties Report*, 1918, Vol. 2, p. B114.

34. *Straits Times*, 4 July 1907.

35. In basic form, the origins of the traditional Chinese shophouse could be found in the towns and villages of southern China. However, transplantation of this basic house form into a Malayan setting had resulted in the absorption of Palladian and Renaissance influences which were particularly evident in the evolution of shophouse façades in Malaya (David G. Kohl, *Chinese Architecture in the Straits Settlements and Western Malaya: Temples, Kongsis and Houses*, Kuala Lumpur: Heinemann Educational Books, 1984, pp. 172–85).

36. In 1917, 'shophouses, shops and tenements' accounted for 55.3 per cent of a total of 27,392 dwellings within the municipality ('Memorandum R', p. C83).

37. *Straits Times*, 12 June 1888.

38. Ibid. The narrow frontage of houses was also considered auspicious according to *feng shui* practices. *Feng shui*, or the art of adapting residences to harmonize with 'cosmic breath' inherent in the landscape, is discussed in detail in relation to the siting of Chinese burial grounds in Chapter 8). This particular belief also had a pragmatic basis for in ancient China house tax was gauged according to the width or frontage of the house (Evelyn Lip, *Chinese Geomancy*, Singapore: Times Books International, 1979, p. 60).

39. Jacques Dumarçay, *The House in South-East Asia*, trans. Michael Smithies, Singapore: Oxford University Press, 1987, p. 62.

40. Ronald G. Knapp, *China's Vernacular Architecture: House Form and Culture*, Honolulu: University of Hawaii Press, 1989, pp. 43–4.

41. Kohl, *Chinese Architecture*, p. 176.

42. Leonore Davidoff and Catherine Hall, 'The Architecture of Public and Private Life: English Middle-class Society in a Provincial Town, 1780 to 1850', in Fraser and Sutcliffe, *The Pursuit of Urban History*, p. 327.

43. In nineteenth-century Britain, the industrial revolution led to a separation of the sphere of production from that of reproduction, the public domain from the private home, giving rise to what has been termed 'the divided city' (Linda McDowell, 'Towards an Understanding of the Gender Division of Urban Space', *Environment and Planning D: Society and Space*, 1 (1983): 59–72; Suzanne Mackenzie, 'Women in the City', in Richard Peet and Nigel Thrift (eds.), *New Models in Geography*, Vol. 2, London: Unwin Hyman, 1989, pp. 109–26).

44. Ambrose B. Rathborne, *Camping and Tramping in Malaya*, London: Swan, Sonnenschein & Co., 1898, pp. 5–6.

45. See Warren, *Rickshaw Coolie*, pp. 202–6, for a detailed description of the intensive use of space in the lodging houses of *rikisha* pullers.

46. Low Ngiong Ing, *Recollections: Chinese Jetsam on a Tropical Shore [and] When Singapore was Syonan-to*, Singapore: Eastern Universities Press, 1983, p. 71.

47. Frank Leeming, *Street Studies in Hong Kong: Localities in a Chinese City*, Hong Kong: Oxford University Press, 1977, p. 26.

48. *Straits Times*, 12 June 1888.

49. *Housing Difficulties Report*, 1918, Vol. 2, p. B3.

50. 'Memorandum B', p. C5.

51. Michel Foucault, 'The Eye of Power', in Colin Gordon, *Michel Foucault: Power/Knowledge, Selected Interviews and Other Writings, 1972–1977*, Brighton: Harvester Press, 1980, p. 155.

52. These were passed under the authority of section 202A of the Municipal Ordinance of 1887–9. See MPMCSM, 4 July 1894 and 18 July 1894; MPMCOM, 1 August 1894; ARSM, 1894, 'Appendix F: Sanitary Regulations', pp. 59–60.

53. ARSM, 1894, Health Officer's Department, p. 27.

54. *Housing Difficulties Report*, 1918, Vol. 1, Appendix 1, p. A61.

55. 'Memorandum D', p. C9.

56. Simpson, *Report on the Sanitary Conditions*, p. 33.

57. *Straits Times*, 28 June 1907.

58. *Housing Difficulties Report*, 1918, Vol. 1, p. A35.

59. 'Memorandum D', p. C9.

60. Ibid.

61. ARSM, 1913, Health Officer's Report, p. 41.

62. MPMCSM, 27 June 1913; ARSM, 1913, Health Officer's Report, p. 41.

63. *Housing Difficulties Report*, 1918, Vol. 2, p. B70.

64. Ibid., Vol. 1, p. A33.

65. 'Memorandum V: The Peirce–Middleton–McDowell Report of 3 March 1905', in *Housing Difficulties Report*, 1918, Vol. 2, p. C109.

66. Ibid.

67. Ibid.; ARSM, 1905, 'Appendix L: Correspondence with Government on the Subject of High Death Rate in Singapore', p. 118.

68. PLCSS, 22 December 1905, p. B228.

69. *Straits Times*, 12 June 1907.

70. PLCSS, 12 July 1907, p. B99.

71. PLCSS, 2 August 1907, p. B111.

72. Ibid.

73. Simpson, *Report on the Sanitary Conditions*, p. 35.

74. *Straits Times*, 10 June 1907.

75. Foucault, 'The Eye of Power', p. 154.

76. PLCSS, 4 August 1911, p. B102.

77. PLCSS, 12 July 1907, pp. B99–B103.

78. Ibid., p. B110.

79. Ibid.

80. Ibid., pp. B110–B111.

81. Ibid., p. B114.

82. PLCSS, 22 December 1905, p. B228.

83. PLCSS, 2 August 1907, pp. B114, B117; 20 December 1907, p. B262.

84. *Housing Difficulties Report*, 1918, Vol. 1, p. A9.

85. 'Memorandum N: The Representation Made by Dr W. R. C. Middleton, Municipal Health Officer, Singapore, on 30 May 1908, under Section 270(J) of the Municipal Ordinance, 1896 (as Amended by Ordinance XXXVII of 1907) Regarding the Insanitary Area between Macao Street and Hokien Street', in *Housing Difficulties Report*, 1918, Vol. 2, pp. C50–C64.

86. Ibid., pp. C51–C54.

87. Ibid., pp. C55–C56.

88. Ibid., p. C55; MPMCOM, 25 September 1908.

89. 'Memorandum T by Mr Roland St. J. Braddell on the Municipal Papers Relating to the Hokien-Macao Streets Improvement', in *Housing Difficulties Report*, 1918, Vol. 2, p. C100.

90. *Housing Difficulties Report*, 1918, Vol. 2, p. B55.

91. MPMCOM, 28 March 1913.

92. *Housing Difficulties Report*, 1918, Vol. 1, p. A9.

93. Approved backlane schemes were recorded in municipal minutes (see for example, MPMCOM, 23 June 1909, 20 August 1909, 21 January 1910, 18 February 1910, and 15 July 1910; MMFGPSYC, 22 April 1910; MMSYC, 1 July 1910; MPMCOM, 15 July 1910; MMSYC, 22 July 1910, 12 August 1910, and 2 September 1910; MMFGPSYC, 9 September 1910; MMSYC, 5 December 1910, 11 January 1911, 10 February 1911, 3 March 1911, 28 March 1911, 12 May 1911, 16 June 1911, 4 August 1911, and 8 September 1911; MMFGPSYC, 6 October 1911, and 24 November 1911; MMSYC, 5 September 1913).

94. *Housing Difficulties Report*, 1918, Vol. 2, p. B216.

95. Ibid., p. B215.

96. Ibid., Vol. 1, p. A10.

97. 'Memorandum Y: Particulars of Back Lanes', in *Housing Difficulties Report*, 1918, Vol. 2, p. C117.

98. Under the altered law, compensation was only payable where the land to be set aside or acquired for a backlane was covered with buildings. Where the land was not built upon, the owner must give up the space required for a backlane without compensation (PLCSS, 15 November 1912, p. B160).

99. *Housing Difficulties Report*, 1918, Vol. 1, p. A31; 'Memorandum Y', pp. C114–C117.

100. *Housing Difficulties Report*, 1918, Vol. 2, pp. B216 and B249.

101. Ibid., Vol. 1, pp. A33–A34.

102. 'Memorandum I by Mr Roland St. J. Braddell', in *Housing Difficulties Report*, 1918, Vol. 2, p. C29.

103. Ibid.

104. *Housing Difficulties Report*, 1918, Vol. 2, p. B135.

105. Ibid., Vol. 2, pp. B147–B148.

106. Ibid., Vol. 2, p. B214. Pressure to modify building by-laws so as to include air-wells in the calculation of open space had hitherto failed. In 1916, a motion introduced at a municipal meeting to include the airwell as open space was defeated. Instead, despite the protest of the Chinese members See Tiong Wah and Tan Kheam Hock, the commissioners resolved that 'the Board supported the Municipal Health Officer in condemning airwells as a means of ventilation in buildings' (MMFGPSYC, 27 June 1916 and 4 July 1916).

107. *Housing Difficulties Report*, 1918, Vol. 2, p. B214.

108. MGCM, 13 September 1918; MMSC3, 13 May 1921; Note Book of Municipal Cases No. 1, 3 March 1921–1 April 1924, Cases [unnumbered] (10 Mar 1922), 3238 (16 June 1922) and 3257 (16 June 1922); SIT 217/27, 9 March 1927: 'Back-lane off Lorong 27A at the Back of No. 539 Geylang Road'. See also Ho Kong Chong and Valerie Lim Nyuk Eun, 'Backlanes as Contested Regions: Construction and Control of Physical Space', in Chua Beng-Huat and Norman Edwards (eds.), *Public Space: Design, Use and Management*, Singapore: Centre for Advanced Studies/Singapore University Press, pp. 40–54.

109. MGCM, 13 September 1918.

110. The theme of conflict over the structure, definition, and use of urban space is discussed in greater detail in Chapter 7.

111. MPMCOM, 24 June 1921.

112. Jon Lang, 'Cultural Implications of Housing Design Policy in India', in Erve Chambers and Setha M. Low (eds.), *Housing, Culture and Design: A Comparative Perspective*, Philadelphia: University of Pennsylvannia Press, 1989, pp. 382–3.

113. MPMCOM, 24 June 1921.

114. A. L. Hoops, 'The Prevention of Disease in the Tropics', in A. L. Hoops and J. W. Scharff (eds.), *Transactions of the Fifth Biennial Congress Held at Singapore, 1923, Far Eastern Association of Tropical Medicine*, London: John Bale, Sons & Danielsson, 1924, p. 9.

115. Gordon E. Cherry, 'The Town Planning Movement and the Late Victorian City', *Transactions, Institute of British Geographers*, New Series, 4 (1979): 306–7.

116. Anthony D. King, 'Exporting Planning: The Colonial and Neo-colonial Experience', in Gordon E. Cherry (ed.), *Shaping an Urban World: Planning in the Twentieth Century*, London: Mansell, 1980, p. 209.

117. Ibid., p. 210.

118. *The Singapore Free Press Centenary Number*, 8 October 1935.

119. *Housing Difficulties Report*, 1918, Vol. 1, p. A55.

120. Tan Tock Seng was a wealthy merchant and philanthropist, best known for founding the Chinese Pauper Hospital in 1844.

121. *Housing Difficulties Report*, 1918, Vol. 1, p. A24.

122. Ibid., pp. A40–A44.

123. MGCM, 27 June 1919.

124. ARSM, 1920, 'Appendix H: Improvement Trust for Singapore [1920]', p. 125.

125. ARSM, 1920, 'Appendix G', p. 121.

126. ARSM, 1921, 'Appendix E: Improvement Trust of Singapore, Report for 1921 by Deputy Chairman', p. 93.

127. 'Improvement Trust for Singapore [1920]', p. 126.

128. ARSM, 1921, 'Appendix A: Preface to Memorandum on Laws of Other Countries', p. 101.

129. Ibid., pp. 100–1.

130. Ibid.

131. Ibid.

132. 'Improvement Trust for Singapore, Report for 1921', p. 98.

133. Ibid. p. 97.

134. SIT, 793/28, 10 August 1928, 'Zoning or Districting for Singapore: Memorandum by W. H. Collyer, Manager, Singapore Improvement Trust: Zoning or Districting for Singapore', p. 2.

135. SIT, 793/28, 10 August 1928: 'Section 63 of "A Bill for the Town Improvement and Development Ordinance, 1923"', pp. 1–2.

136. SIT, 793/28, 10 August 1928: 'Extract of Report of Committee on Draft Improvement and Development Bill'.

137. For example, in 1924, the area bounded by Tanglin Road, Orchard Road, Paterson Road, and Jervois Road (a high-class residential district where a large proportion of Europeans lived) was declared a prohibition area for offensive trades in order to preserve its amenities (ARSM, 1924, 'Appendix B: Improvement Trust for Singapore, Report for 1924 by Deputy Chairman', p. 9).

138. This became No. 10 of 1927: 'An Ordinance to Provide for the Improvement of the Town and Island of Singapore' (*Straits Settlements Ordinances, 1927*, Singapore: Government Printing Office, 1927, pp. 55–98). The ordinance signalled a change in the status of the Improvement Trust from that of a department of the municipality to an independent body with Collyer as its first manager.

139. Singapore Housing and Development Board, *Homes for the People*, p. 19.

140. The principal source of revenue was an improvement rate of 2 per cent levied on the value of all houses and lands within the municipal area and a contribution from the government out of the general revenue of a sum equal to the proceeds from the improvement rate. This could be supplemented when necessary by the flotation of loans. In addition, in 1926, under the governorship of Sir Laurence Guillemard, a sum of ten million dollars was earmarked for the clearance of slum areas (J. M. Fraser (compiler), *The Work of the Singapore Improvement Trust, 1927–1947*, Singapore: Singapore Improvement Trust, 1948, pp. 4–5).

141. SIT, 816/32, 8 January 1932: 'Chairman, Singapore Improvement Trust to Colonial Secretary, 8 January 1932'.

142. The principal schemes were those located at Albert Street, Lorong Krian, Sago Street–Smith Street, Dickenson Hill, Serangoon Road–Lavender Street, Balestier Road, Waterloo Street–Beach Road, Tanjong Katong Road–Joo Chiat Road, and the Race Course (Fraser, *The Work of the Singapore Improvement Trust*, pp. 45–6).

143. Fraser, *The Work of the Singapore Improvement Trust*, p. 10.

144. SIT, 969/27, 21 November 1927, 'Singapore Improvement Ordinance 1927': 'Legal Position under Parts VI and VII'.

145. Fraser, *The Work of the Singapore Improvement Trust*, p. 10.

146. Ibid., pp. 6–7.

147. *Housing Difficulties Report*, 1918, Vol. 1, p. A8. In the early 1920s, a second open space was created close to the southern congested area as a result of the Malaya–Borneo Exhibition. It was called the Stadium, 'a broad green space of turf swept by the sea breeze, and within easy reach on foot from miles of densely crowded, noisome, unhealthy streets and buildings' ('Improvement Trust for Singapore, Report for 1921', p. 99). By 1925, the full list of open recreational spaces within municipal limits numbered only fourteen places, of which a large proportion like the Padang, Raffles Reclamation Ground, the Old Gaol site in Stamford Road, Hong Lim Green, and the Race Course were under the control of recreational or sporting clubs or 'alienated' for specialized activities such as cricket, golf, and horse-racing. There were hence few public spaces which were accessible to the ordinary public (SIT, 191/25, 13 May 1925, 'Provision of Parks and Open Spaces': 'Existing Open Spaces').

148. *Housing Difficulties Report*, 1918, Vol. 1, pp. A38–A39.

149. ARSM, 1925, 'Appendix B: Singapore Improvement Trust [1925]', p. 28.

150. 'Singapore Improvement Trust [1925]', p. 28.

151. *Straits Times*, 21 November 1921.

152. Between 1920 and 1923, eight open spaces were reserved and in 1924, another four were added (ARSM, 1923, 'Appendix C: Improvement Trust for Singapore, Report for 1923 by the Deputy Chairman', p. 6; 'Improvement Trust for Singapore, Report for 1924', pp. 9–10).

153. 'Singapore Improvement Trust [1925]', p. 28.

154. 'Singapore Improvement Trust [1925]', p. 29; SIT 600/24, 26 September 1924: 'Proposed Layout for the Block Bounded by Church Street, Philip Street, Chuliah Street, South Canal Road and Synagogue'.

155. ARSM, 1926, 'Appendix: [Singapore Improvement Trust, 1926]', p. 3.

156. King, 'Exporting Planning', p. 210.

5
Municipal Versus Asian Utilities Systems: Urban Water Supply and Sewage Disposal

Wealthy cities like Singapore were expected to find their own water supply.[1]

The perfection of cleanliness would be that all refuse matters should from the very beginning, pass away inoffensively and continuously.[2]

> Our little system cannot fail
> We'll have other but the Shone,
> We'll keep the pail—the model pail—
> And pay the piper with a Loan.[3]

Water Supply, Sewage Disposal, and 'Filth' Diseases in Nineteenth-century Cities

BY the beginning of the twentieth century, the municipal agenda of measures to render the urban environment clean and orderly had become increasingly broad and complex. The preceding chapter concentrated on municipal attempts to enforce the spatial rearrangement of urban built form as a strategy of improvement to combat urban malaise and disease. Another strategy was that of displacing Asian 'utilities' systems of supply and disposal developed over an era of *laissez-faire* with municipally controlled, public utilities. This chapter examines two urban utilities of particular importance to the health of the city—the supply of water and the disposal of human waste—to illustrate some of the debates and difficulties underlying the transition from a privately managed to a public system of utilities. Such a transition, although successful in the final analysis, was far from smooth or inexorable, but was instead continuously challenged by Asian interest.

Whilst an adequate and accessible supply of water was crucial to the sustenance of burgeoning towns and cities, it became clear in the nineteenth century that it could also become a highly efficient vehicle for the delivery of pathogenic elements. By the mid-nineteenth century, it had become axiomatic among sanitarians and physicians in Britain and elsewhere in Europe that an adequate supply of *pure* water was central to sanitary reform and essential to the health of towns. It was increasingly recognized in medical circles that impure water, whether standing or used for washing, cooking, or drinking, bore heavy responsibility for the transmission of diseases including the so-called 'fevers', dysentery, diarrhoea, hepatitis, and cholera. In Britain, the discovery of a connection between impure water supply and choleraic disease was attributed to Dr John Snow, who from his investigations at the Broad Street pump in 1854, 'dramatically and conclusively traced cholera deaths to houses supplied by the suspect water of the Southwark and Vauxhall water company'.[4] By the 1860s, the declaration of John Simon, the well-known medical officer to the Local Government Board, that good health was contingent upon 'cleanliness, ventilation, and drainage', but more importantly depended on 'the use of perfectly pure drinking water' was tantamount to an official dictum.[5]

Among the various contaminants of potable water, sewage was regarded as the most dangerous as a harbinger of disease.[6] The accumulation of excrement was an unavoidable and ever-increasing by-product of urban growth; its removal was a vital challenge to urban societies. As A. S. Wohl, public health historian, observed, though an unpleasant and indeed scatological subject, amassed excrementitious filth formed part of the urban topography of Victorian cities which was 'immediately evident to nose and eye', and 'excrement removal deserves far greater attention than historians have given it'.[7] In Britain, the campaign for an efficient and systematic means of sewage disposal was spearheaded by health officers during the second half of the nineteenth century. Surveys and reports containing abundantly graphic accounts of cesspools in cities with their 'fetid pollution and poisons', the 'unspeakable abomination of volatile contents', and their association with 'filth' diseases provided the evidence for the campaign to eradicate cesspools and to adopt in their place a more efficient and sanitary system of sewage disposal. By the second half of the nineteenth century, the dry conservancy method gradually replaced the cesspool as the predominant form of excrement removal throughout British cities, and by the 1890s, water-borne sewerage systems and modern filtration and treatment plants began to become increasingly widespread although the dry pail system continued to persist in some cities well into the twentieth century.[8]

The corpus of knowledge established by several lines of enquiry which associated an impure water supply and sewage pollution of the environment with the spread of diseases contributed critically to the medical transition away from Hippocratic and climatological perspectives, and an emphasis on homoeopathy, towards preventive, public medicine. This medical transition originated in the middle of the eighteenth century when the Hippocratic tradition underwent a fundamental change, shifting from a fatalistic or passive view of the environment as the cause of disease to embrace 'a medicine of avoidance and prevention ... which sought to show mankind which disease-conducive circumstances to evade, and to determine what aspects of the environment might be modified to weaken or eliminate their capacity to cause disease'.[9] Eighteenth-century efforts to remove or modify pathogenic environments included the drainage of fetid swamps, lavation of stench-laden refuse accumulated on streets, ventilation of closed areas to remove noxious air, and reinterment of corpses outside the city. In the nineteenth century, environmentalism continued to underline 'the new ecological relations that big cities create among large groups of people and between them and their natural environment' and the importance of cleaning up breeding grounds of disease within cities.[10] The nineteenth-century search for a pure water supply and an efficient means of sewage disposal continued the environmentalist campaign to remove pathogenic elements from the urban environment. This movement not only gathered momentum in European towns and cities throughout the nineteenth century but the importance of an unpolluted environment characterized by pure water, free circulation of air, abundant sunlight, and the efficient removal of waste also became entrenched in public health debates in outlying parts of the British Empire. Turn-of-the-century Singapore provides an example where the dictates of Western sanitary science and in particular the relation between water, sewage, sanitation, and the people's health were translated and worked out in an Asian environment.

The first half of this chapter focuses on the development of a municipal supply of water for the town of Singapore. It first examines the origins of Singapore's first waterworks and its expansion under municipal control during a period of rapid demographic growth and increasingly pressing urban demands on water. It then analyses the centrality of a clean and abundant water supply to public health arguments and attempts to weigh the degree of success of the municipalization of water in terms of the acceptance of municipal water among a predominantly Asian population. The second half of the chapter turns to debates surrounding the question of replacing a privately run system of waste disposal in the hands of Chinese syndicates by a municipally organized system. The debates not only revolved around the issue of private versus public management, but also pivoted on the question of what constituted an 'appropriate' system of disposal for an Asian population.

The Origins and Expansion of a Municipal Water Supply

In the early years after the founding of Singapore, the problem of supplying fresh water to ships which called at the port gained pre-eminence over the needs of the local population. In 1823, for example, the Resident, John Crawfurd, proposed to spend $1,000 on a new reservoir and waterworks to cater to the needs of the port but nothing was done for the local urban population, which continued to depend on wells for their daily supply of water.[11] It was not until 1852 that the government surveyor, J. T. Thomson, proposed a scheme for the supply of 546 million gallons of water to the town from the headwaters of the 'Singapore Creek'.[12] Given the lack of government and public interest, the scheme failed to materialize. Five years later, Tan Kim Seng, a wealthy Straits Chinese merchant and public benefactor, offered to finance Singapore's first waterworks but the project of bringing 'an efficient supply of wholesome water to the town' soon foundered in the mire of government bureaucracy and parsimony. Tan Kim Seng offered the government $13,000 for the construction of the waterworks on condition that the works would 'always [be] maintained in an efficient state' and that 'the water be available to the inhabitants free of all charge'.[13] Both the offer and its terms were accepted by the Straits government in January 1858 but three years later, when the actual estimates sent in proved to be far above $13,000, the Government of India refused to sanction the work on the grounds that 'they had been misled' and that the work could not 'justly fall upon Imperial resources'.[14] However, to repay Tan Kim Seng his contribution and give up the work would not reflect creditably on the government, which 'deem[ed] itself bound to carry out the work' on the condition that 'no superfluous expense should be incurred in so doing'.[15]

The waterworks was eventually built and comprised an impounding reservoir in the vicinity of Thomson Road, and a masonry conduit to convey the water to within 200 feet of the Singapore River.[16] The Impounding Reservoir, as it was then known,[17] had a catchment area of about 1,890 acres and was completed in 1868 although pumps and distributing works were only finished in 1877. Singapore's first waterworks was finally opened in 1878, twenty years after Tan Kim Seng's philanthropic gesture and fourteen years after its benefactor's death.[18] On completion, responsibility for the waterworks and the distribution of piped water throughout the town was transferred to the municipality which levied a two-part tariff in order to sustain the work, the first part being a public water rate of 3 per cent levied on all property within the water area, and the second a domestic rate measured by taps and meters which fell only on domestic property which took a water supply.

The completion of the waterworks, however, did not immediately guarantee an efficient and plentiful supply to the town. The town service was only laid on for twelve hours a day and had to be further curtailed during the dry season every year.[19] Water filtration was only introduced in 1889 and up to the first decade of the twentieth century, only about a third of the total amount of water delivered was filtered. There were frequent public complaints that municipal water was characterized by a brown coloration and a turbid appearance and contained offensive sediments, although the municipal authorities maintained that these did not detract from the wholesomeness of the water.[20] By far the most pressing problem confronting the water authorities, however, was ensuring that supply kept pace with the rapidly growing demand for water. In 1894, the reservoir was enlarged by the construction of an embankment at a point lower down the valley and a second line of cast iron pipes was laid along Thomson Road and brought into use in 1898.[21] Towards the closing years of the nineteenth century, pressure on existing water resources escalated, largely as a result of increasing demand for water from 'the Docks and Wharves at Telok Blanga[h] and New Harbour and from suburban districts like Mount Elizabeth and Orchard Road and Chinese streets like Havelock Road, Chin Swee Road and Kelang Road'.[22] By the turn of the century, the average daily supply of water was about 4 million gallons, and given an estimated 5 per cent increase annually, at least 6.5 million gallons a day would be required by 1910.[23] It was essential to enlarge the supply of water immediately and a plan was proposed to channel water from the upper portion of the Kallang River into the Thomson Road Reservoir[24] and also to open a second high service reservoir at Pearl's Hill.[25] The urgent need for a more elastic supply of water was underscored by water famine in the town during drought periods in 1901, 1902, and 1904. It was evident that during the drought, the Thomson Road Reservoir could not cope with the combined effects of falling levels and increased demand on the municipal supply when wells in the town dried up. As a result, a number of domestic services had to be curtailed, and the hours of supply strictly limited.[26]

In order to accommodate additional water drawn from the Kallang River, the Thomson Road Reservoir was extended in 1905 by raising the embankment an additional 5 feet, thus making the maximum depth 26½ feet and the total capacity 1,004 million gallons. When the Kallang Tunnel Works were completed in 1907, the town was ensured a minimum supply of 5.5 million gallons per day.[27] However, even before this work was completed, the necessity of securing additional sources of water in order to meet the projected demand, revised to 9 million gallons per day by 1911, was urgently sounded. In 1902, the municipal engineer, Robert Peirce, outlined a scheme to construct a new reservoir with a storage capacity of 845 million gallons, known as the Kallang River Reservoir,[28] by building an embankment across the valley of the Kallang River.[29] This reservoir, completed in 1910, was capable of yielding about 3.5 million gallons a day. Hence, by the beginning of the second decade, the town was potentially ensured of a total of about 9 million gallons a day through its two reservoirs[30] (Figure 5.1).

In the early 1920s, however, it was recognized that the margin of safety in the amount of water supplied was so slim that 'disaster' could no longer be averted unless 'large and bold measures [were] taken to improve the water supply without delay'.[31] The municipal water area already covered all the main populated areas within the municipality and was rapidly expanding (Figure 5.2). In 1920, the average quantity of water pumped each day had risen beyond 9.5 million gallons, in excess of the maintainable yield of the existing two reservoirs.[32] As early as in 1904, Peirce had predicted that even if all the potential water resources on the island of Singapore were to be developed and utilized, the supply of water would not be able to keep up with the rapidly burgeoning urban population by the 1920s.[33] The search for

FIGURE 5.1
The Supply of Municipal Water, 1911

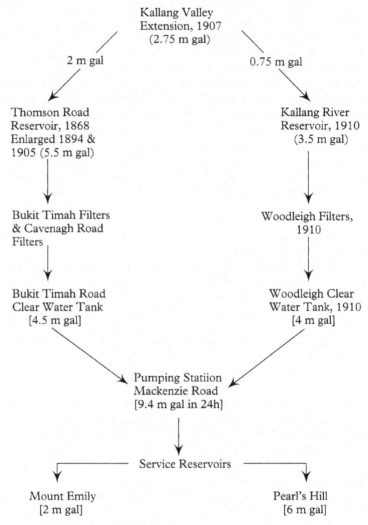

Figures in ( ) represent 'safe minimum discharge per day'.
Figures in [ ] represent 'total capacity'.

Source: Singapore Municipality Waterworks.

water had to be extended further afield. In 1912, Peirce made preliminary investigations in the Pulai district in Johore and in 1920, the municipality commissioned S. G. Williams, the water engineer, to make a thorough investigation of water resources in the southern portion of Johore State. Williams's findings were submitted to Sir Alexander Binnie, Son & Deacon, a firm of consultant engineers in

FIGURE 5.2
Municipal Water Limits as at 31 December 1924

Britain, which reported in favour of the Gunong Pulai Scheme primarily because it was the only scheme among several[34] which could be obtained by gravity and was hence the most economical. Work on the Gunong Pulai scheme was completed in December 1929 and the reservoir named the 'Sultan Ibrahim Reservoir' after the sovereign of Johore.[35] In 1931, work on a second reservoir, the Pontian Kechil Reservoir about 5 miles beyond Gunong Pulai (Figure 5.3), was also completed.[36]. The two Johore reservoirs had a combined daily yield of 15.5 million gallons of water,[37] which, added to existing supplies on Singapore Island itself, was estimated to be able to meet all of Singapore's water requirements for another twenty years.[38] To cope with the Johore supplies, a new service reservoir with a capacity of 30 million gallons was constructed on top of Fort Canning Hill and put into service at the end of 1928.[39]

FIGURE 5.3
Water Supply from Johore, 1920s and 1930s

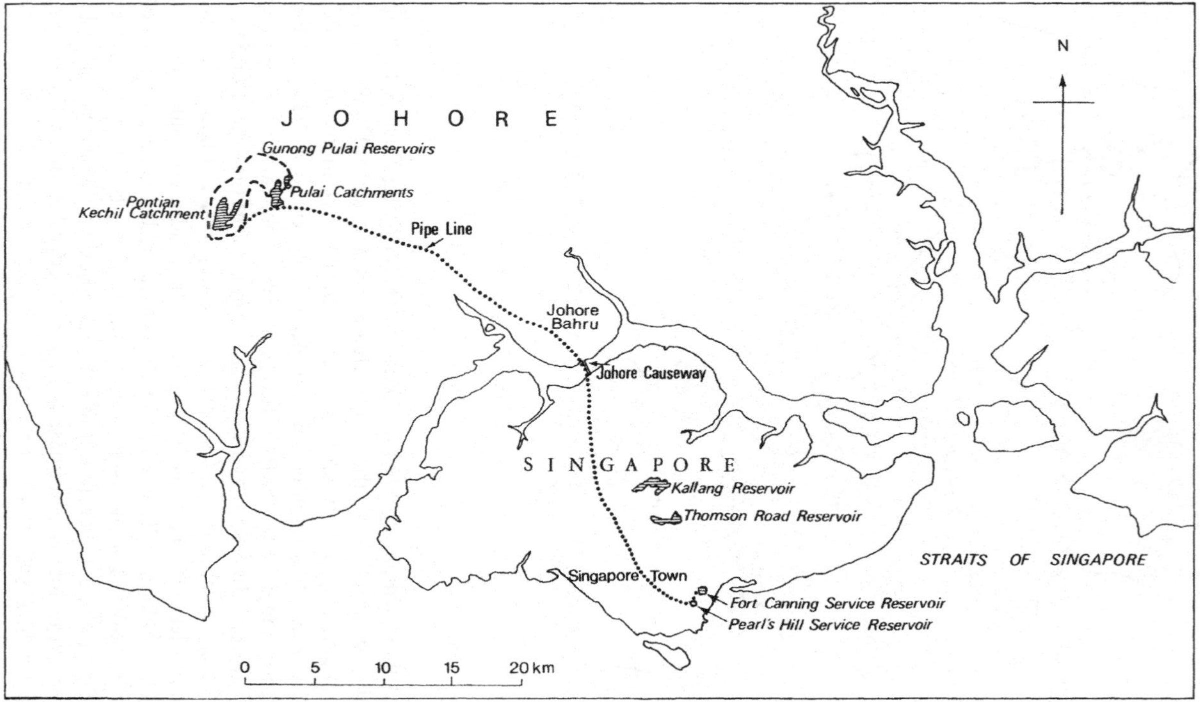

Water Supply and Public Health:
Well versus Municipal Water

As already discussed in Chapter 3, according to contemporary Western medical opinion in late nineteenth-century Singapore, diseases were mainly caused by noxious effluvia emanating from damp, contaminated soils and fetid waters. The miasmatic theory was far from moribund and was frequently invoked as a general cause of the unhealthiness of the city. Medical men in Singapore, however, also acknowledged the primacy of polluted water, particularly that contaminated by sewage, in transmitting diseases such as cholera, enteric fever, dysentery, and diarrhoea. The local press alleged that there was no medium 'so sure, so insidious, and so fatal in the propagation of the germs of epidemic disease as water' and that these 'poisons [were] of so subtle a character as not to be weighed in the balance or measured in the glass'.[40] The 'lack of a proper and efficient water supply' left Singapore 'open not to the chance, but to the absolute ultimate certainty of epidemics which invariably attend the neglect or absence of well-recognised sanitary measures'.[41] According to the municipal health officer, Dr W. R. C. Middleton, if the city were to achieve any significant reduction in mortality rates, two of the most pressing problems which had to be first solved were that of providing a 'supply of pure water to every dwelling within the water area and the subsequent closing of polluted wells' and 'the systematic collection of night soil and its conversion to some harmless form of manure'.[42]

Dr Middleton considered the condition of surface wells in slum areas in Singapore 'the chief factor in maintaining the prevalence of cholera, enteric fever, diarrhoea and dysentery and a correspondingly high death rate'.[43] The typical house well was a shallow well with a broken brick lining and no other provision against surface soakage from the contaminated ground around it.[44] According to a 1896 report, nearly every house in the town proper possessed a well which was situated close to and 'often in actual contact' with one and 'occasionally three or four' cesspools in the same compound.[45] An analysis of the water derived from some of these wells, in particular those found in the densely populated Asian parts of the town, revealed distinct signs of contamination (Table 5.1).[46] A survey of existing wells within municipal limits conducted in 1902 confirmed the view that contamination of well water was rife: of the 3,877 wells returned, 3,265 or nearly 85 per cent were considered 'obviously bad' and unfit for domestic use.[47] Municipal health reports constantly reiterated the observation that the conditions of surface wells in the Asian parts of Singapore were 'such as would not be tolerated for a moment in any European City'[48] and were in fact, 'more or less diluted cesspools'.[49] It was stressed that the most dangerous wells were those which were agreeable in colour and taste and which had no offensive odour because Asians preferred these for drinking, particularly during the heat of day, as the well water was cooler than water obtainable from standpipes.[50]

TABLE 5.1

Analyses from Surface Wells in Various Parts of the Municipality, 1896

	Location of Well	Chlorine (grains/gallon)	Nitrites	Nitrates	Free Ammonia (parts/million)	Albuminoid Ammonia (parts/million)
1.	Lorong at 48 Rochor Road	16.00	Distinct trace	Abundant	0.02	0.22
2.	47 Havelock Road (*rikisha* coolie lodging house)	42.00	Distinct trace	Distinct trace	0.32	0.80
3.	Armenian Church compound	3.00	Absent	Absent	0.00	0.05
4.	128 Victoria Street (lately a *rikisha* coolie lodging house)	21.05	Distinct trace	Considerable trace	0.26	0.22
5.	12 Sago Street (eating and baker's shop)	14.00	Distinct trace	Distinct trace	0.42	0.34
6.	16 China Street (lodging house)	18.00	Distinct trace	Distinct trace	0.64	0.25
7.	62 Clyde Street (Hadji lodging house)	21.00	Abundant traces	Abundant traces	0.02	0.20
8.	66 Teluk Ayer Street (Baba's private house)	2.00	Trace	Trace	0.08	0.18
9.	Private well in Grange Road	0.03	Absent	Absent	0.00	0.06

Source: 'Report on the Special Investigation into the System of Death Registration', p. 22.

From the perspective of contemporary Western medical opinion, the threat of polluted well water to the health of the city was not only engendered by the chemical and bacteriological impurity of the water itself, but amplified and perpetuated by the insanitary 'habits' and 'prejudices' of the Asian people. Dr Middleton claimed that even if contaminated well water were not used for drinking purposes, the dangers were by no means diminished:

It is well-known that it is the universal practice among the Chinese, and probably among other Asiatics to wash out their mouths at the same time as they bathe and the danger from such a pratice [sic] is little less, if any, than if the same water were drunk. The water, however, if not drunk is used for 'domestic purposes', which may include the cleaning of cooking and eating utensils, of clothes and of houses. The idea of 'cleansing' cooking and eating utensils with dilute sewage is sufficiently disgusting, but the danger does not end here. It has been stated to be extremely probable that the sputa of phthisical patients, when expectorated on the floors of houses and undergoing drying is apt to be inhaled by others and so give rise to the disease in them. Is it not probable that such polluted water, containing swarms of microorganisms when thrown on the floors of houses and drying should afford a medium for the propagation of disease, even without the necessity for drinking it?[51]

In accordance with the strong antipathy against these practices, Dr Middleton argued that the serious threat to public health posed by the continued existence of polluted wells could only be obviated by 'the compulsory instalment of a constant water service in each house and the subsequent closing of wells'.[52] If a constant municipal water service could be secured for every dwelling, then, it was argued, 'all excuse for the maintenance of the foul wells which abound all over the town would disappear by the closure of these wells' and as a result, 'one of the most prolific sources of fever and cholera would be removed'.[53] He felt that the multiplication of standpipes in public places alone would only constitute a partial solution because 'the ordinary Asiatic [was] content to use the polluted water [from their own wells] rather than avail himself of the purer supply provided at the neighbouring stand pipe'.[54] This was not difficult to understand because whilst well water was free and unlimited as long as a drought did not occur, householders had to hire water carriers at a rate of between 0.5 and 1 cent for a picul or two buckets (approximately 16 gallons) of water, depending on the distance from the standpipe.[55]

In 1897, the Municipal Health Department attempted to enforce 'the systematic closing of polluted wells' within the Rochor district, part of the northern congested area.[56] This, however, encountered a certain amount of 'active' opposition and widespread non-compliance on the part of Asian inhabitants who were unconvinced that wells were closed on sanitary grounds alone.[57] Instead, it was a widely held view among the Asian population that closing of wells was a municipal ploy to deny them access to a free and conveniently located source of water in order to promote the taking up of municipal water services and the augmentation of municipal revenue.[58] Typical of Asian attempts to evade or inflect municipal sanitary control, the strategies resorted to by the Asian labouring classes usually took the form of unobtrusive non-compliance rather than outright opposition. As an illustration of this, whilst active opposition to well closing was somewhat sporadic and unsustained, the widespread and unobtrusive reopening of closed wells continued unabated throughout the early decades of the twentieth century. Complaints that the closure of wells was 'not taken seriously by the Asiatics', that 'closed wells [were] often re-opened and all the [municipality's] good work undone', and that the provisions of the Municipal Ordinance were incapable of dealing with such rampant offences were frequently ventilated in municipal reports.[59] In 1910, despite a more vigorous campaign against polluted wells and the issue of a larger than usual number of notices to close wells, the municipal health officer had to admit at the end of the year that polluted wells were 'still with us' as 60 per cent of well notices were not complied with.[60] Furthermore, as there were no laws forbidding the digging of new wells, the municipal authorities were unable to eliminate wells altogether for while 'the Health Department [was] engaged in closing wells, ... the people generally [might] be equally busy digging wells and making excavations'.[61]

The difficulty of preventing the reopening of wells was further exacerbated by the frequent occurrence of dry seasons when the municipal supply ran short.[62] As seen earlier, up to the second decade of the twentieth century, the margin of safety in the quantity of municipal piped water which could be supplied by the combined resources of existing reservoirs was critically slim.[63] In the early twentieth century, the chronic insufficiency of municipal water to meet growing demands was a consequence not only of rapid population growth and the extension of water services, but was also exacerbated by the extension of a water-borne sewerage system approved in 1911 (see later discussion), the adoption of sanitary fittings such as wash-basins and baths, the establishment of rubber factories requiring heavy water consumption, and increased demands from shipping.[64] The daily rate of consumption per head rose from 21.7 gallons in 1901, 22.8 gallons in 1911, to 30.01 gallons in 1921.[65] In cases of drought and falling water levels in the reservoirs, certain municipal services such as the supply to shipping and for street watering and drain flushing had to be curtailed, the hours of domestic supply restricted, and certain private services cut off.[66] In the water famine which ensued, wells previously condemned and closed were reopened for use, a measure which the Municipal Health Department had little choice but to tolerate.[67] It was tacitly acknowledged by the municipal authorities that without the assurance of a regular and abundant supply of municipal water for every eventuality, a large proportion of polluted wells must be allowed to remain open.[68]

The Cost and Conservation of Water: Taps versus Meters

From the municipal perspective, replacing contaminated well water with a pure, piped supply as the main source of domestic water was an important but partial step towards improving health conditions in the city. Even when the Asian had converted to the use of piped water, municipal vigilance was necessary to ensure that the purity of the water was not contaminated on the premises or wasted. It was argued that 'free water mean[t] waste of water'[69] and that computation and payment by meter as opposed to the number and size of taps must prevail in order that 'those who indulge in a so-called "domestic supply" really pa[id] for what they use'.[70] In January 1899, meters were attached to a large number of private services taken by coolie lodging houses, brothels, and other houses with large numbers of occupants in order to check water wastage.[71] The metering of water and the charge of 40 cents per thousand gallons consumed aroused strong objections from lodging housekeepers. Whilst only 22 out of 500 of these stopped taking a municipal supply in consequence, the majority of lodging housekeepers reacted by passing on the incidence of the water tax to their lodgers. On average, the monthly lodging rent was increased from 35 to 45 cents per coolie, whilst a *jinrikisha* puller had to pay about $1.00 per month for lodgings with free water for bathing and washing his vehicle.[72]

On 19 December 1900, a special committee was appointed comprising both Chinese and European municipal commissioners to enquire into any alleged grievances of the labouring classes if the metering of municipal water were made compulsory for every household.[73] Alex Gentle, a former president of the Municipal Commission, in giving evidence before the committee, considered the average 'Asiatic' householder incapable of judicious water management as long as municipal water remained unmetered:

The Asiatic consumer is no economist of what costs him nothing; or rather, that once his domestic water payment is fixed at $1.50 or $3 a month, as the case may be, he proceeds to build a bathing tank and use or waste water as he can, by way of getting the value of his money out of the Municipality.... The average householder lets his tap run all day into his tank or barrel, leading the water from one vessel to another by means of bamboos or pipes, and even his neighbour's house if he is not detected, and often supplying the water of 30 to 40 coolie lodgers.[74]

Metering was hence the only safeguard against 'the wasteful consumer, the manufacturer who surreptitiously use[d] water in his trade, the cattle keeper who bathe[d] his oxen and buffaloes, and the coolie lodging-house keeper who ha[d] a bath for all comers'.[75] Despite the initial outlay as well as running cost involved in using meters, Gentle contended that 'it would be better in the long run to place a meter in every house and to maintain a small army of meter readers than to let [the] water revenue escape [municipal] grasp'.[76]

Gentle justified metering not only on economic grounds but bolstered his claims by citing health and sanitary considerations. He contended that if municipal water supplied to Asian households was not properly metered, the outcome would be 'the prodigal misuse [of water] which ... [was] both wasteful and insanitary. The centre of the [coolie] house [was] never free from pools of water and the broken floor [was] trodden into unwholesome mire.'[77] From the municipal perspective, given the inherently 'wasteful tendencies' of the Asian consumer, even attempts at cleanliness could become detrimental to health. It was argued, for example, that in the absence of metering, the 'Asiatics' habit of flushing and cleansing floors and houses' with a plentiful supply of water could 'act deleteriously [sic] by adding to the moisture of an already humid atmosphere in houses, and thus increase the dampness of dwellings erected on a saturated soil with insufficient access to light and air'.[78] The municipal president, J. O. Anthonisz, further contended that the experience of Indian towns such as Calcutta, Bombay, and Madras showed that the introduction of an abundant supply of piped water without safeguards against waste led to a rise in malarial fevers in the towns because waste water tended to gather in 'pools and places where the air [was] cool and moist during the hot dry season', hence providing 'conditions eminently favourable to the existence of that pestiferous insect [the mosquito]'.[79] It was also argued that an unmetered domestic water supply 'indirectly conduce[d] to overcrowding' because lodging house keepers who paid a fixed monthly charge for an unlimited supply of water but levied a water charge of 5–8 cents per head were tempted to 'squeeze' an excessive number of coolies into their houses in order to maximize profits.[80]

The committee on water supply invited leaders of the Asian communities to channel grievances and represent the views of the labouring classes on the question of metering. In general, Asian leaders either argued to keep the tap rate or for a very low domestic water rate should the service be metered. Tan Kiong Siak, for example, felt that water should be supplied at cost price to the poorer classes and that the 'Commissioners should make up their minds not to ... profit from the water supply [but] to raise other rates instead'.[81] Tan Jiak Kim expressed concern that given the typical household structure whereby a large number of coolies subrented from a head coolie who paid the house rent and taxes, a metered water supply would inevitably result in head coolies restricting the supply to tenants, a measure which would be detrimental to health.[82] Similarly, Tan Chun Fook argued that any increase in the cost of water supply would ultimately be passed on to tenants in the form of increased rents.[83]

In their findings, the water committee endorsed the municipal view that an unmetered municipal supply of water not only resulted in heavy losses on the part of the municipality, but was 'very undesirable from a sanitary point of view'.[84] It concluded that 'the most efficient way of preventing waste [was] the use of the meter'.[85] However, on account of some of the views expressed by the leaders of the Asian communities, the committee felt that it was unable to recommend compulsory metering, primarily because 'metering tended to lead to a restricted and insufficient use of water among Asiatics'.[86] Instead, it advised that consumers should be induced to recognize the advantages of meters by reducing the charge for metered water from 40 to 20 cents per thousand gallons and increasing the rates for unmeasured supply by tap.[87]

The chief recommendations of the water committee were implemented in 1902 and the municipal president reported that Asian opposition to the introduction of meters was much less than expected, chiefly because of the reduced price at which metered supply was charged.[88] From a total of 2,425 meters in use in 1902,[89] the number of metered services steadily grew to 12,021 in 1917,[90] by which time it accounted for 88 per cent of municipal water services. Towards the end of the Great War, difficulties in obtaining meters and meter parts resulted in a fall in the number of metered services and a partial reversion to the tap rate[91] but by the mid-1920s, almost all water services within the town of Singapore were metered.[92] The introduction of meters signalled another step towards the realization of the municipal vision of a healthy, sanitized city equipped with a pure, abundant, and carefully conserved supply of water. The path towards fulfilment, however, was not without conflict nor negotiation, and concessions had to be continually made to accommodate Asian priorities and perceptions. It was also unclear to what extent a municipal water supply helped improve the health of the city. When the waterworks were first undertaken, the municipal authorities confidently anticipated that when this 'great engineering work' was completed, 'there would be an appreciable diminution in the death rate, and that the general health of the community would be vastly improved'.[93] The difficulty of testing such an assumption lies partly in the unreliability of morbidity statistics in colonial Singapore, but more generally in the difficulty of disentangling the effect of improved water from the impact of other public health measures as well as socio-economic, nutritional, and institutional factors. In reality, death rates from bowel complaints did not begin to fall consistently until after the Great War (Figure 5.4), that is, four decades after the opening of the waterworks. Mortality rate from bowel complaints appeared to have peaked when municipal water consumption levels fell. This could be explained by the fact that given the narrow margins of safety in the quantity of municipal water available, the occurrence of drought inevitably resulted in a shortfall in municipal water which in turn led to the reopening of wells and the increased consumption of polluted disease-carrying well water. It was several decades before the municipal water supply was of sufficient quality and quantity to have an impact on the city's health problems. It has been suggested that in Britain, 'the direct environmental benefits of increased water deliveries for sanitary purposes were probably limited before the whole range of water services, including sewage treatment and river conservancy, were modernized'.[94] Similarly, in Singapore, it was observed that 'the filling of wells and the supply of filtered water will

FIGURE 5.4

Municipal Water Consumption and Death Rate from Bowel Complaints,
1895–1925

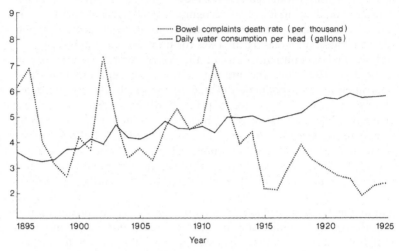

Source: ARSM, 1895–1925.

effect only a partial removal of the conditions productive of dysentery and other bowel complaints [and] leave untouched the existing method of sewage removal and its dangers'.[95] It was unlikely that the installation of a municipal supply of piped water alone would eliminate water-borne diseases without a complementary sewage disposal system which would remove excrementitious filth from the urban environment efficiently. As will be seen in the second half of this chapter, the municipalization of sewage disposal was accomplished at a much later date than that of water.

The Chinese System of Night-soil Disposal

Up to the 1880s, the removal and disposal of night-soil from the town of Singapore was entirely in the hands of Chinese syndicates who organized the collection of night-soil in buckets and their transfer to market gardens and plantations on the outskirts of the town. Such a system appeared to have served the town well in the earlier days, when market gardeners and small planters were able to dispose of all that the town produced, and were in fact willing to pay for the privilege of removing night-soil from every house once in every three days.[96] However, the extent of cultivated area had not grown apace with the rapid growth of the urban population in the last quarter of the nineteenth century, and as a result, the supply of night-soil generated by the town quickly outstripped the demand. Vegetable gardeners could no longer absorb the amount of night-soil produced and instead of paying for what used to be the privilege of removing night-soil, collectors were only willing to perform the service if allowed to do so for nothing. Even this arrangement did not survive for long and in the course of time, the value of night-soil had depreciated to such an extent that the positions of householder and collector were reversed: by the 1880s, the former had to pay the latter for the removal of night-soil.

From the municipal perspective, the problem of night-soil disposal could no longer be left entirely in the hands of private enterprise. An 1890 survey of the conditions of latrines found in two streets—one in the central district inhabited principally by the Chinese and the other in Rochor, a largely Malay enclave—attempted to 'quantify' repugnance in terms of the number of persons per latrine, the size and condition of latrine receptacles, and the depth of accumulated faeces. It became an oft-cited piece of evidence indicative of the disgusting state of affairs of four-fifths of the privies in the municipality, 'which had no catchment apparatus whatever, other than the bare earth (with or without a hole dug in it), or a large cesspool built of brick, the bottom being frequently unpaved and the mortar quite porous'.[97] The sixty houses surveyed along the Chinese street had a total population of 1,137 and 48 latrines in the form of tubs, jars, and cesspools, averaging 18.95 persons to each house and 23.69 persons to each latrine. Quantitatively, the Malay street appeared more respectable, with 76 houses, 489 persons and 72 latrines among them, yielding averages of 6.43 persons per house and 6.79 persons per latrine. However, the large majority of latrines in both streets contained large accumulations of excrement ranging in depth from a few inches to a few feet and often 'one mass of writhing maggots'.[98] These were removed only when required for agricultural purposes or when overflowing, which was infrequent, given the continual loss by soakage or leakage into the unpaved floor of the cesspool.

In the typical Chinese dwelling house, privies on each floor were connected with each other and with the privy on the ground floor by means of brick shafts or downshoots which were often in a defective state of repair, allowing 'the escape of noxious gases from their filth encrusted interiors into the dwelling rooms of the house'.[99] The lower privy was provided with a bucket or a cesspool from which the contents were ladled out into buckets and removed by the *toti* or night-soil coolie through the front of the house, a process which distributed 'fumes and gases' all over the house.[100] From a series of observations and calculations based on the daily production of night-soil per head of population supplied by the gaol, James MacRitchie, the municipal engineer in the 1880s and 1890s, concluded that the 'startling' amount of 50 tons of semi-liquid sewage a day was not removed from domestic latrines and cesspools in the town proper and absorbed into the ground as a consequence.[101]

The press reported that such inefficient removal of sewage 'defile[d] the soil, contaminate[d] the air around, breeding diseases such as typhoid, enteric or "cesspool" fever which had hitherto been almost unknown in Singapore but were now rife amongst all classes'.[102] Contemporary medical opinion held that unremoved, human excrement left putrefying in the soil became a source of danger causing 'filth' diseases either by means of direct infection or by 'tainting' the atmosphere, food, or water with the products of organic decomposition. The authority of Sir John Simon himself was quoted by Dr Gilmore Ellis, the acting municipal health officer, in substantiating his claim that the 'septic unwholesomeness' of excremental nuisances affected householders not only when excremental filth was ingested but also when the 'poisoned air' was breathed in:

Privies and privy drainage, with their respective stinkings and soakings and the pollutions of air and water which are thus produced have in innumerable instances been the apparent causes of outbreaks of enteric fever.... Filth diseases are apparently enabled to run their course, with successive inoculations from man to man, by the instrumentality of molecules of excrement, which man's carelessness lets mingle in his air and food and in drink.[103]

From the municipal perspective, a private system of night-soil removal in Asian hands was 'an extremely dangerous one'.[104] As MacRitchie warned,

we are completely in the hands of a set of men who might, from some real or fancied injustice, strike at any moment, and leave the Town and the Municipal Commissioners who are responsible for its sanitation in an extremely serious position.... A strike [would leave] the excrement of a population of about hundred thousand people ... festering in the cesspools and privies of the town.[105]

Furthermore, not only did the Chinese 'cesspool and pail' system of night-soil removal result in the pollution of the domestic environment, it also constituted a public nuisance in its passage through the streets of the town as well as at its final place of deposition as manure in vegetable gardens. The Chinese practice of applying night-soil to vegetable gardens aroused strong condemnation from Western medical men who considered it 'unscientific from [a] health point of view, as it [helped] to increase, if not actually cause, the incidence of Diarrhoea, Dysentery, Typhoid and diseases of that group'.[106] An observer who had studied night-soil disposal systems in several tropical cities described the daily scene in Singapore:

Strings of Chinamen may be seen and smelt every morning pursuing their way into the country with [night-soil] pails upon their shoulders. Bullock carts are also used, carrying a number of pails at one time standing on the cart-floor. The Chinese market-gardeners, ... place great value on the pail-contents; they store the material in open pits in their gardens. They also store in the same place all the urine they can get hold of. The mixed product they distribute with scrupulous care on individual plants. At night (this time is selected because the sun in the day time would quickly dry up the manure) they may be found going along the rows of vegetables and, lighted on their way by a lantern, pouring small quantities of liquid manure from long-necked cans upon each individual plant. The results, from the gardener's point of view, are excellent ... [but] there is no doubt that the direct transfer of pail contents to gardens is to be deprecated, [and using] the water in which the pails were washed ... for the plants would be as bad as [using] crude faeces.[107]

Frequent complaints from the European public were made to municipal officers of 'nuisances affecting the health and comfort of Europeans, generally caused by their Chinese neighbours' who store night-soil on their premises for purposes of manuring their fields'.[108] It was alleged that 'improperly and offensively manured' vegetable gardens in close proximity to public roads and dwelling places, if not directly injurious to health, were 'breeding grounds of every imaginable abomination'.[109] The market gardens of Singapore, averred *The Straits Times*, were

vegetable horrors, reeking with decomposed excreta and totally unregulated by the Municipality.... There are many roads in Singapore that, in themselves, are picturesque and convenient, and very suitable for the erection of good European houses, but they are rendered undesirable, if not positively unhealthy, by the fact that market gardens, plentifully manured with extremely evil-smelling manure, abut upon those roads.[110]

The Asian system of night-soil removal was hence undesirable in every way: not only did it result in the pollution of the domestic environment, it also infused the public urban environment with the pungent smells of disease.

Establishing a Municipal System of Night-soil Removal: The 'Pail' versus 'Sewer' Debate and the MacRitchie Reports

By the late nineteenth century, the replacement of the Asian system of sewage disposal with a municipally controlled one become increasingly prominent on the municipal agenda. Report after report of the municipal health officer and other medical men warned of the 'gravest dangers' associated with the insanitary state of latrines and cesspools, the sewage pollution of well water, the practice of manuring vegetables with crude night-soil, the 'disgusting pollution' of the atmosphere, and the discharge of sewage into the Singapore River from untrapped house drains connected to dwelling places on the river banks.[111] In 1890, the municipal engineer, James MacRitchie, submitted a report on the disposal of night-soil in which he recommended a 'complete pail system' under municipal supervision as the best method of sewage disposal most suited to the town. The system was based on MacRitchie's study of pail systems in use in towns in the north of England such as Manchester and Birmingham. He advocated dividing the town into five collection districts of about 30,000 houses each and for house-to-house collections to be carried out by municipal servants using pails which were then emptied into covered bullock carts for transfer to night-soil depôts. From the depôts, Shone ejectors would force the night-soil by pneumatic pressure through cast-iron mains, in part to iron tanks placed at points near principal roads where cultivators could remove night-soil to their gardens and plantations, with the balance to be discharged into the sea.[112]

In his report, MacRitchie affirmed his conviction that the pail system was best suited to local conditions on several grounds. First, given the flat topography of the town and suburbs, the soft nature of the underlying soil, and the heavy rainfall, MacRitchie contended that constructing sewage drains and maintaining powerful pumping engines to raise the sewage to a level to allow it to gravitate to the sea or collecting tanks for treatment would require a 'practically prohibitive outlay'.[113] Second, a pail system would obviate the dangers associated with sewage gas which could turn a water-borne system of drainage into 'the most admirable machinery for distributing death and disease found in one locality as widely as possible into the very houses of the people'.[114] Third, unlike a water-carriage system of sewage disposal, the pail system would not be dependent on a plentiful supply of water. Not only was the municipality unready to guarantee a continuous water service necessary for a water-carriage system, it wanted to retain the prerogative of cutting off water supply in cases of arrears in payment of bills without making themselves 'the probable active agents in introducing sickness into the house from which supply was cut off'.[115] Most crucially, however, MacRitchie considered it inadvisable to introduce a water-closet system amongst a population 'so varied in nationality and habits' as the Asian communities in Singapore for 'unless carefully looked after and put to their legitimate purpose only', water closets were 'apt to become a source of danger'.[116] In contrast, a pail system was already familiar to the Asian population, and would require minimal structural alterations to closets already in use in Asian dwellings.[117] MacRitchie's view that a water-carriage system would be 'dangerous' if introduced among Asians was firmly endorsed by Dr Gilmore Ellis:

Water-closets have been recognised by those best fitted to judge, as being the most advantageous form of privy possible for an educated and civilized community, such a community as would be the first to call in skilled aid should anything become deranged so as to affect their proper working, but not only do our local physical features make their adoption next to an impossibility, but above all, our populace is neither fitted, nor even likely to be fitted, to be trusted with such a system, especially considering the grave dangers of an inefficient water-closet, dangers greater than those of our non-system.[118]

In a language highly derogatory of Asian custom and civilization and at the same time curiously oblique and reticent in referring to actual sanitary practices, what these 'grave dangers' involved were never officially explained, let alone investigated. It appeared that part of the mistrust of a water-carriage system stemmed from fears that the Asian use of sand, shell, and other materials in connection with their domestic cleaning, ablutions, and sewage disposal habits would lead to a misuse of water closets.[119] Both officers advocated the pail system as 'the simplest, most serviceable, and least expensive of all systems, and one eminently adapted to such a population as is gathered together in [the] town of Singapore'.[120]

The municipal engineer's scheme met with the approval of both medical men and the municipal commissioners but further investigations quickly revealed a major flaw in the scheme. Experimentations with tidal currents off the coast near Pasir Panjang demonstrated that there was a strong possibility that foul matter ejected into the sea would be carried shorewards.[121] The project had to be modified and in a second report on the disposal of sewage, MacRitchie suggested two alternative plans: the first was to send the whole supply of town sewage along with street sweepings far out to sea in hopper barges, and the second was to dry the sewage in hot air cylinders and convert it into poudrette which could be sold to agriculturists as manure.[122] MacRitchie preferred the latter as a valuable and hygienic manure could be secured by such a means. Whilst the municipal commissioners deliberated on his report, MacRitchie was sent on a special mission in early 1893 to the principal cities of India and Burma to study the working of various schemes for the disposal of sewage used there and to examine their suitability to local conditions.[123] On his return, MacRitchie reported that the water-closet system was not favourably received by the inhabitants of Indian towns and that Indian local authorities did not encourage their adoption as a result of 'doubts as to its successful management by the natives'.[124] He reaffirmed his conviction that 'the pail system for excreta [was] undoubtedly the best system to be adopted'.[125]

Improving the 'Asiatic Pail System'

Up to the end of the nineteenth century, despite pressure from the municipal engineer and the health officer that immediate measures should be taken to deal with night-soil removal in the town, little was accomplished in establishing a municipally controlled system. Instead, certain 'improvements' in the 'Asiatic system' were enforced to abate the so-called 'night-soil nuisance'. In 1889, a by-law was passed restricting the hours of operation of night-soil coolies except those who converted from using wooden pails to municipally supplied galvanized iron buckets imported from England.[126] Municipal officers, however, complained of the 'apathy and indifference' of *toti* and the difficulty of winning their co-operation and assistance in putting a stop to the 'wooden bucket nuisance'.[127] In 1891, the by-law regulating the removal of night-soil was amended to entirely prohibit the use of wooden buckets and to enforce the adoption of regulation buckets of the pattern and construction approved by the municipality.[128]

Efforts were also concentrated on abolishing cesspits under the floors of houses and their attendant downshoots from latrines on the upper floors, and to substitute privies with pails or glazed earthenware jars placed on impervious floors.[129] In 1895 alone, at the start of the campaign against cesspools and downshoots, a total of 3,848 notices were served on houseowners, of which about 70 per cent were complied with.[130] The scheme met with 'a good deal of passive resistance' and 'fanciful and frivolous objections' from Chinese houseowners and occupiers, who, unaccustomed to the daily removal of excreta from their houses, found the 'innovation' irksome.[131]

Another proposal to help alleviate the amount of 'nuisance' created within the home was the establishment of municipal latrines. In the 1880s, the Municipal Board embarked upon a programme of providing public latrines in congested areas such as near markets and the junctions of busy thoroughfares and streets despite strong objections of property owners in the immediate vicinity of such latrines. Amongst the first were those established in Teo Chew Street, the Cecil Street–Market Street junction, Telok Ayer Market, Tanjong Pagar, and the Merchant Street–Hong Lim Quay junction in an attempt to abate 'the misuse of drains and the accumulation of foul matter in houses'.[132] The municipal engineer was confident that by increasing the number of public conveniences in the streets, a 'large part of the night-soil would be therein deposited and much less nuisance and danger caused by privies inside the houses'.[133] It appeared that in houses occupied solely by male adults, where public latrines were easily accessible, house privies were abolished as residents preferred to resort to public ones.[134] By 1896, there were eight public pail latrines in the division of town north of the Singapore River and eleven in the southern division, producing a total of about 205 pails of night-soil per day.[135] Within five years, the number of latrines in both divisions had more than doubled[136] and in the first two decades of the twentieth century, between forty and forty-five public pail latrines were maintained at municipal expense.[137] In 1896, the municipal president, Alex Gentle, reported that 'popular prejudice' against municipal conveniences was subsiding as their usefulness became apparent.[138] A survey taken in March, April, and May 1897 showed that at fifteen latrines, 15,000 visits were paid in twenty-four hours.[139] There were, however, difficulties in cleansing public latrines under a 'makeshift system of sewage removal'[140] for although the latrines were municipally owned, the task of removing sewage from them was only partly carried out by night-soil coolies under the superintendence of municipal officers, and partly farmed out by contract to Chinese syndicates which reportedly 'worked very badly'.[141] As a result, some of these public conveniences were in themselves malodorous nuisances and contributed to rather than helped solve the night-soil problem. The municipal engineer complained that it was necessary to upgrade these latrines continually as they were 'situated generally in slum areas and [were] used by the lowest class of people who ha[d] to be educated in sanitary methods'.[142]

The municipal commissioners also attempted to put a stop to the 'abominable habit' of storing night-soil in pits and tubs in vegetable gardens close to the public road. In 1895, summonses were served under section 152 against some thirty market gardeners living in the

Serangoon Road area for keeping night-soil on their premises for more than twenty-four hours.[143] The prosecutions were, however, unsuccessful as the magistrate dismissed the case on the ground that the section did not apply to filth stored on the premises for the purpose of manuring fields.[144] The municipal commissioners were also unable to proceed against the 'nightsoil nuisance' on the basis of complaints from European residents as in such cases, the nuisance was considered a private rather than public one.[145] Under the new Municipal Ordinance passed in 1896, the municipal commissioners were empowered under section 211 to designate roads where the use of the 'Asiatics' favourite manure' could be prohibited within 100 yards. In 1897, the municipal commissioners designated the vicinity of Serangoon Road as a prohibited area and instituted prosecutions against uncompliant vegetable gardeners in the area.[146] However, without 'a larger staff of detectives', it was difficult to obtain the evidence needed to secure the imposition of prohibitive penalties.[147] As it was, the imposition of fines was an insufficient deterrent in preventing market gardeners from persisting in the use of night-soil as manure.[148]

Up to the end of the nineteenth century, except for the enforcement of regulation buckets, the abolition of downshoots, the introduction of movable pails in domestic latrines, and the multiplication of public conveniences, little progress was made to inaugurate a municipally organized scheme for the collection of town sewage. In 1889, T. Shelford, one of the unofficial members of the Legislative Council, disparagingly summarized the efforts of the municipality in dealing with night-soil in one sentence: '[T]hey have substituted zinc for wooden pails.'[149] Inadvertently, the abolition of the cesspool and the substitution of the pail exacerbated the problem of night-soil removal because the daily amount of night-soil to be removed had increased as less material was allowed to percolate into the ground under the house.[150] While contamination of the soil was reduced as a result of municipal efforts, little was done to counteract the problem of an oversupply of night-soil which was causing further dislocations in the Chinese manure market. In the late nineteenth century, several Chinese syndicates attempted organizing a scheme to try to obtain a monopoly of household latrine cleansing and to cut a corner in the manure trade. Members of a *kongsi* under one Chang Ah Ng reportedly forced themselves upon householders for the purpose of night-soil removal, claiming that they had been authorized by the municipal commissioners to do so.[151] In another instance, the chop Chew Seng Kongsi petitioned the municipal commissioners to license their members as the only persons authorized to remove night-soil from private houses in the town.[152] These tactics were symptomatic of falling profitability in an oversupplied manure market.

In the municipality's defence, the president reported that it had not been possible to organize a system of removal for private houses or even public buildings such as police stations and government buildings because of intractable difficulties connected to cleansing vessels on a large scale, as well as the availability and control of coolies, cartage, and disposal of night-soil.[153] Experimental poudrette works had been installed in Trafalgar Street in July 1898 but this could only deal with a proportion of the collection from public latrines. The process of converting sewage to poudrette was both tedious and expensive and its commercial success only possible if it could command a high price in the manure market or if the cost of manufacture were thrown onto the householder in addition to the cost of removing the raw material.[154] The attempt to induce vegetable gardeners in the Serangoon Road area to use poudrette instead of crude night-soil failed and the processed product had to be sold under contract for shipment to estates out of Singapore such as those in Java and Province Wellesley at a loss.[155] Having proved an unprofitable method of night-soil disposal, the municipal commissioners resolved to discontinue the process of poudrette conversion at the end of 1903.[156] Well into the first decade of the twentieth century, the town of Singapore was still without any signs of the much advocated municipal night-soil collection system.

The 'Pail' versus 'Sewer' Debate Revisited: The Peirce and Simpson Reports

Fifteen years after MacRitchie submitted his first report on the organization of a municipal system of night-soil collection, Robert Peirce, the municipal engineer, was directed to prepare a new scheme for the systematic disposal of sewage. Peirce's scheme was essentially similar to MacRitchie's: it advocated that night-soil be brought in pail vans to receiving stations where it would be pumped by Shone ejectors to an outlying station to be loaded on to barges which would carry the night-soil out to sea. As there were differing views on the Municipal Board whether such a method of disposal implied a waste of valuable manure, Dr Middleton and G. A. Finlayson, the municipal bacteriologist, were asked to criticize the scheme.[157] Marshalling together the opinions of experts in Britain on the viability of 'infective organisms' in ordinary soil and in sewage, Middleton and Finlayson argued that the evidence suggested that certain virulent bacteria could persist for long periods or even indefinitely in excrement.[158] Furthermore, even if the view held by certain sanitarians in Britain that returning human excrement to the soil was innocuous as infective organisms were rapidly killed by other organisms found in the soil was acceded to, they were of the opinion that there were crucial differences between the Chinese method of utilizing night-soil and the practice in Britain. In the latter, night-soil was buried to render it harmless; in Singapore however, 'the faeces were not buried but poured in solution over the vegetables'.[159] The danger to health was multiplied by the fact that among the Chinese, 'filth in pits and jars [was accumulated] near dwellings and accessible to flies' and also because 'the vegetables grown by the Chinese [were] not confined to plants of the cabbage order which would be boiled before consumption but include[d] also radishes and lettuce which [were] eaten raw'.[160]

The municipal bacteriologist reported that samples of vegetables bought in the open market showed degrees of sewage contamination although neither the B. *typhosus* or the B. *dysenteriae* could be isolated in the laboratory.[161] Middleton and Finlayson concluded that the contention that 'rotten nightsoil' was an innocuous manure could not be upheld given the Chinese method of night-soil management and gave their support to Peirce's proposal of dumping night-soil at sea as the most satisfactory method of disposal.[162]

Peirce's scheme for the disposal of night-soil under municipal supervision was approved by the Board and in 1905, despite the protests of the Chinese commissioners Choa Giang Thye and Dr Lim Boon Keng and that of Dr Murray Robertson against the introduction of an important motion involving increased taxation without adequate notice to members of the Board, provisions were made to put the municipal engineer's scheme on trial in a district bounded by Rochor Canal Road, Kallang Road, Crawfurd Road, Arab Street, and the sea.[163] At this point, just as the long awaited municipal night-soil collection scheme appeared to be finally materializing, the project was unexpectedly nipped in the bud by the arrival of Professor W. J. Simpson in May 1906 on his mission of enquiry into the sanitary conditions of the city (see Chapter 3). The municipal commissioners were instructed by the government to suspend operations in connection with indents for apparatus required to give effect to the municipal engineer's scheme in order to await Professor Simpson's expert report.[164] In evaluating the town's sewage disposal scheme, Simpson concentrated his attention on the incidence of diseases such as dysentery, diarrhoea, cholera, and enteric fever which were generally accepted to be highly prevalent in the town. Whilst the unreliability of information about such diseases and the 'grave difficulty in tracing case after case of sickness' militated against a rigorous scientific demonstration of their association with a defective system of sewage removal, Simpson claimed that it was possible to remedy this defect by extrapolating conclusions derived from a careful inquiry into conditions which occurred in microcosm in institutions such as the public jail and the Alexandra army barracks.[165] It was evident, claimed Simpson, that 'though food and water were good, yet arrangements for the collection and removal of the excreta from the jail [and barracks] were so defective that it was impossible for infected dejecta to be at times prevented from infecting the drinking water'.[166] There was 'no difficulty whatever', continued Simpson, 'in showing that the same conditions which have given rise to dysentery in the Alexandra barracks and the jail ... also exist very abundantly and generally [among the civil population] in Singapore' and as proof of his point, Simpson enumerated a catalogue of possible 'facilities' through which the 'gross pollution' of food and water could occur such as the pouring of pail washings down surface drains, the Chinese custom of storing their drinking water in barrels and ladling it out when needed, and the agency of flies and dust.[167] The dangers inherent in the Asian system of sewage removal was further accentuated by the fact that as it was a matter of private arrangement between the householder and the *kongsi*, and not a municipal affair, no special arrangements for the thorough disinfection of houses visited by typhoid or dysentery could be secured, thereby increasing the risk of the infection being carried from latrine to latrine by the night-soil coolie.[168] Such an account of 'dangers' to the health of the town had been rehearsed many times; however, Simpson's solution was radically different from those proposed before. The Professor rejected Peirce's scheme for municipal removal based on the pail system as unsatisfactory.[169] Instead, Simpson swept aside all prejudice against the introduction of underground sewers and water-closets among the Asian population and strongly advocated a water-carriage system of sewage disposal based on the Shone system of evacuation by automatic ejectors worked by compressed air.[170] He cited the example of Rangoon's Shone system which had been successfully operating for over twenty years despite initial objections that the system would be unsuited to Oriental habits.[171] The only difficulty Simpson foresaw in introducing a sewerage system in Asian areas in Singapore was the absence of backlanes between houses but nevertheless, he argued that a large and rapidly growing town like Singapore 'must face the cost of the introduction of a sewerage system if it [were] to retain and improve in health and bring itself in line with the comforts of a modern city'.[172]

The municipal committee on health and disposal of sewage set up to evaluate Professor Simpson's recommendations, however, preferred the pail system to a water-carriage system.[173] Dr Middleton tenaciously maintained that a water-carriage scheme could not be applied to 'native quarters of the town' and that 'at best a water-carriage system [might] be introduced for certain districts along with a pail system in others'.[174] Resolution of the night-soil question was once again delayed as the municipal commissioners sought information from Rangoon, Hong Kong, Calcutta, Madras, Karachi, and Bombay, where sewers had been established for some time, 'in order to ascertain the experiences and views of these towns as to the suitability of a sewerage system in an Eastern city'.[175]

In the meantime, the 'abominable public nuisances' associated with an uncontrolled system of night-soil collection continued to plague the town.[176] Each year, the municipal health officer chalked up numerous offences against the conservancy by-laws, including the non-removal of night-soil from privies, throwing pail washings into house-drains, driving night-soil carts through the streets without providing against the escape of offensive gases, and carrying insufficiently covered offensive matter through the streets.[177] However, a large proportion of offenders were not detected, or if detected, not summoned as the municipal officers feared that too vigorous a campaign against night-soil coolies and carters might catalyse a strike which would paralyse the entire city.[178] *The Straits Times* also alleged that the private system of night-soil collection was intolerable as it placed householders at the mercy of Chinese *toti* who made extortionate demands for payment at the instigation of triad societies.[179] It cited a 'typical instance' in which Hylam night-soil scavengers boycotted the home of an European resident who had dismissed his Hylam servants and replaced them with Cantonese ones. An appeal to the Municipal Health Department for assistance turned out to be futile as the night-soil coolies involved were private servants over whom the Department had no control. Such 'gruesome grievances' not only 'penalise the householder in [a] disgraceful fashion', but also 'practically force the householder who refuse[d] to submit to extortion to endanger the health of himself and his neighbours'.[180]

The Move towards Sewering the Town

In October 1909, in order to break the deadlock in deciding on a sewage disposal most suited to Singapore, the municipal commissioners decided to engage the services of G. Midgeley Taylor, a sanitary engineer from the British firm Taylor, Sons & Santo Crimp, to report on the question of drainage and the disposal of sewage in Singapore.[181] Taylor recommended a system of underground sewers to conduct sewage to a pumping station, from which it could be pumped to purification works consisting of tanks and bacteria beds before gravitating through an outfall main into the sea at Pasir Panjang.[182] The proposals were estimated to cost $4.5 million, exclusive of the cost of land, easements, legal, and engineering charges.[183] In the same year as when Taylor's report was received, a municipal enquiry commission set up to investigate the working of the Singapore municipality heavily criticized its failure to introduce an organized system of sewage disposal. The 1910 report condemned the manner of night-soil collection in vogue in the town in no uncertain terms as 'the greatest blot on the administration of the Singapore Municipality', particularly in comparison with the well-organized system of night-soil collection already established in one of her sister Settlements, Penang.[184] The addition of an official voice to the growing censure of the European public, the press, and their own health officers for 'neglecting to provide better means for the disposal of night-soil' finally spurred the commissioners into action.

Taylor's scheme was, however, considered too ambitious and instead, the commissioners directed the municipal engineer to prepare a modified scheme capable of being established in sections.[185] The modified scheme was adopted in 1911 and an application made to the government for authority to borrow $2 million to be expended on the first phase of the scheme which provided for the sewering of public institutions, clubs, and public latrines, the erection of three pumping stations, and the construction of a rising main and outfall works.[186] The proposed sewerage scheme was not a complete one for the entire municipal area, but a partial one dealing only with the built-up part of the town. It divided the built-up area into three sewerage divisions— north, central, and south—each with its own set of sewers and pumping station from which sewage was pumped to the treatment and disposal works at Alexandra Road. After allowing solids to settle, the liquid waste was spread by revolving sprinklers over bacterial filter beds composed of coral. The sewage liquid was then collected in humus tanks where the final silt was allowed to settle and the clear effluent passed into the Singapore River.[187] It was also proposed to combine removal by sewers with a system of collection by pails and carts to central depôts in order to reduce the initial number of main sewers required and to allow cart collection to be gradually superseded as new sewers were constructed.[188] In the next two decades, sewers were gradually constructed in various parts of the town, generally where backlanes were available. Certain sections of the English Public Health Amendment Act of 1907 relating to water-closets were adapted to local requirements. It was also proposed to make compulsory the installation of water-closets on premises situated within 100 feet of a sewer, provided in the case of tenement houses that a common water-closet be erected within the compound.[189] Every new house was also to have water-closet accommodation and a municipal water supply laid on.

In reality, however, the extension of sewers was a piecemeal and protracted business. Its execution was retarded by a number of factors, the chief of which was the insufficiency of water.[190] Up to the third decade of the twentieth century, the supply of municipal water was barely adequate to meet normal demand without the additional burden of a sewerage system.[191] Financing for the sewerage scheme was also constrained by the fact that the bulk of municipal loans and revenue were already committed to the extension of the town's water supply. Water expenditure must come first, claimed the municipal president, as 'water [was] the first necessity of life'.[192] It was hence impossible to spend 'anything but dribble on matters other than those connected with the supply of water'.[193] Given the financial circumstances, sewer construction in the north division had to be 'purposely retarded'.[194] The onset of the war in 1914 and the practice of war stringency further delayed the construction of sewers in various parts of town.[195]

The utility of sewers was also greatly diminished by the difficulty of constructing backlanes, without which it was in the opinion of the municipal president, 'inconvenient if not dangerous' to connect most of the houses up to sewers.[196] In 1921, he forecasted that given the large number of back-to-back houses, conversion to backlanes and sewers would be protracted.[197] In the same year, a committee appointed to enquire into the financial position of the municipality urged that in order that funds might be concentrated on the expansion of the town's water supply, the extension of sewers should be suspended as the construction of backlanes, without which the utility of sewers was suspect, was far from completed.[198] The work of getting individual houses connected to existing sewers was also 'slow and laborious': the high cost of installing water-closets and connections causing owners to delay action 'by every means in their power'.[199] By the early 1920s, only about 2 per cent of houses in the municipal area had been connected up to the water-borne sewerage system and as such, the sewers were carrying 'a mere tithe' of the load for which they were designed.[200]

The 'Two-pail System': Compulsory Municipal Night-soil Collection as an Interim Solution

By the 1920s, it had to be acknowledged that 'many years [would] elapse before the majority of houses within Municipal Limits [would] be connected to the water-carriage system' and that in the interim period, it would be necessary to make the existing pail system as efficient as possible by placing it under stricter municipal supervision.[201] In the early 1920s, the municipal engineer reported that enormous quantities of night-soil had been found in public drains in the early mornings, chiefly because of exorbitant fees demanded by private night-soil carriers.[202] The monthly fee for the service had increased from 50 cents or $1.00 to over $5.00 per house within a short space of time. As a result, the poorer classes who could not afford to pay had resorted to surreptitious dumping of night-soil in the night. The establishment of a municipal service was seen as the only solution to abate such nuisances. In 1913, a municipally administered scheme for the collection of night-soil from designated districts using the so-called 'two-pail system' was inaugurated to replace private collection by Chinese syndicates in an attempt to abate long-standing 'night-soil nuisances'. The scheme was so-named as two pails were involved in each collection: the soiled pail and contents removed from the privy by the municipal coolie and a clean pail containing disinfectants replaced.[203]

Priority was first given to the predominantly European suburban areas. In April 1913, Tanjong Katong, one of the suburbs where residents had been most voluble in their complaints against the private *toti*, was designated a municipal area followed by the Tanglin district in July 1915.[204] The first Asian urban district selected as a compulsory municipal collection area in October 1915 was Sanitary District No. 4, a predominantly Chinese area with about 750 houses bounded by Havelock Road, Magazine Road, Boat Quay, and New Bridge Road.[205] The municipal scheme was gradually introduced into several other urban areas in the following years.[206]

The extension of the municipal night-soil collection scheme soon collided with the vested interest of Chinese syndicates who had paid considerable sums of goodwill money to maintain their share of the manure market.[207] It was initially feared that trouble might arise amongst private night-soil coolies who would be dispossessed of their business as a result of the introduction of the compulsory municipal system. A threatened strike was, however, avoided with the aid of the protector of Chinese who explained that the municipal scheme would only be gradually introduced and that *toti* would have time to find other employment.[208] As an interim measure, the compulsory municipal collection service only had a limited impact as it affected only a proportion of latrines within the municipal area. By 1923, a decade after the scheme was first inaugurated, the municipal service only covered seven areas totalling 5,961 latrines in 4,406 houses, roughly 15 per cent of the total number of houses within municipal limits.[209]

Although hailed as a major improvement in the collection of night-soil by the municipal engineer, the municipal scheme did not necessarily solve some of the long-standing problems associated with the Chinese *toti*. Under both the private and the municipal system, the contents of the pail remained open to visits by flies and insects for any period of time up to forty-eight hours, thus becoming one of the contributory causes of 'the high and rising death rate from dysentery, diarrhoea and enteric in the town'.[210] There were also frequent complaints to the municipal commissioners regarding nuisances arising from the conveyance of night-soil, whether by private or municipal means, through major thoroughfares such as Orchard Road, the transfer of night-soil from pails to carts, and noxious smells due to the tipping of night-soil at the Peoples' Park sewage collection station.[211]

Ironically, when the much-feared night-soil coolies' strike actually occurred, leaving unremoved filth festering in the town, it was the municipal coolies rather than privately hired ones who downed their tools. In February 1920, the municipal conservancy and night-soil coolies 'struck' in a body in order to articulate their demands for higher wages.[212] The municipal commissioners attempted to 'bring the latter class to their senses' by means of prosecutions in court but this was unsuccessful. Faced with the prospect of an unscavenged town, the municipal officers had little choice but to give way to the demands of the labour force.[213] The increase in pay, however, did little to improve labour discipline, the average coolie working only twenty-five out of thirty days a month.[214] The acute labour situated eased towards the end of the year as a result of a local trade depression but the problem of retaining 'suitable men' for night-soil collection remained problematic throughout the 1920s.[215]

Negotiating Control over Urban Utilities

In condemning 'Asiatic systems' (or 'non-systems', in the words of Dr Gilmore Ellis) of water supply and sewage disposal and replacing them with 'improved' municipally controlled systems of greater 'purity' and 'efficiency', the authorities legitimized their actions by drawing on the superior claims of Western medical theory and the findings of medical and engineering experts 'imported' from Britain. What was pursued in the name of environmental improvement, however, rested not so much on abstract medical theory or the application of scientific criteria, but on the value judgements of those in positions of power and what they considered 'appropriate' for an 'Asiatic' population. This was evident in the insistence of water meters for coolie classes with 'insanitary' and 'wasteful' tendencies, and in the debates surrounding the choice of a system of sewage disposal for the town. The long delay in establishing a modern sewerage system was in part due to the tenacity with which municipal officers held on to the unsubstantiated assumption that a water-carriage system was 'dangerous' and inappropriate for an 'Asiatic' population. Confusing cause with effect, the sanitary habits of Asians were perceived not as an adaptation to the want of proper facilities, but a failure of their civilization, a view which only served to confirm racial prejudices and to heighten the sense that the Asians were indeed a race apart, as much removed from the rudiments of modern civilization as they were from the respectability of water-closets. Not only was such an assumption held without proof for several decades, it was finally shown to be untenable when water-closets were fitted into Asian houses adjoining the Cross Street–Chin Chew Street backlanes in 1916. It was reported that the working of the water-closets was 'entirely satisfactory', no nuisances were created, and that the change from the old pail system was 'much appreciated' by the inhabitants of the houses.[216] By 1923, David B. McLay, the acting sewerage engineer, reported that there were numerous enquiries for sewers and that tenants were 'anxious to have water-carriage fittings installed in their houses, and in most instances [were] simply waiting for the Commissioners to go ahead and complete the sewerage of the town'.[217]

The transition from privately organized systems of water supply and sewage disposal to municipally controlled public utilities was a protracted one punctuated with many difficulties not only because of financial and technical problems but also because the municipal vision of a sanitized city and the means to achieve it was one which was imposed upon rather than shared by the bulk of the Asian population. The difference in conception of disease and its causes between western and Chinese medical theories was most clearly illustrated by contrasting attitudes towards excrementitious filth. From the municipal perspective, night-soil harboured disease-causing germs, exuded noxious odours, and must be systematically removed from human habitation and destroyed. From the Chinese perspective, night-soil was a valuable source of manure to be accumulated in vessels and sold to farmers.[218] Even the *Straits Chinese Magazine*, a forum for English-educated Straits Chinese, defended night-soil as 'the best and cheapest manure', arguing that night-soil which had undergone a special process of fermentation contained no harmful organisms and did not expose the public to any danger.[219] Asian resistance to attempts to replace their own systems of water supply and sewage disposal took many forms, most generally through individual efforts to thwart municipal orders such as reopening wells and applying night-soil to vegetable gardens, but also occasionally through threatened strikes by night-soil coolies or by the collective channelling of their grievances through the leaders who represented them to the authorities.

The late nineteenth to early twentieth century was an era of uncritical faith in the ability of sanitary engineering to transform public health. The protracted attempt at municipalizing the town's water supply and sewage disposal, however, demonstrates that these were not neutral scientifically sanctified engineering works which could be arbitrarily imposed, but were subject to negotiation between those who imposed them and those who had to live with them.

1. PLCSS, 27 June 1870, p. 65.

2. ARSM, 1896, 'Supplement: Pollution of the Singapore River', p. 46.

3. Jingle quoted in a satirical account of a municipal meeting on the 'disposal of town sewage' (*Straits Produce*, 3 (July 1894), 5).

4. Anthony S. Wohl, *Endangered Lives: Public Health in Victorian Britain*, London: J. M. Dent & Sons, 1983, p. 125. Snow's study demonstrated a higher incidence of cholera among Londoners who consumed polluted water from the Southwark and Vauxhall company compared to that of similar social classes supplied by the Lambeth company which obtained its water from a less polluted source (John Snow, *On the Mode of Communication of Cholera*, London: J. Churchill, 1855). His work provided essential evidence for sanitarians such as Sedgwick, Chadwick, and Petinkoffer who were campaigning for the provision of clean water supplies as a means of preventing disease epidemics.

5. Quoted in Kerrie L. Macpherson, *A Wilderness of Marshes, The Origins of Public Health in Shanghai, 1843–1893*, Hong Kong: Oxford University Press, 1987, p. 83.

6. According to the science of bacteriology, it was clear that putrescent matter, however offensive and rich in emanations did not generate disease until a contagious factor was added to it and that this factor came from the bowel discharges of infected patients which contained the specific bacilli. These bacilli were thought to be capable of existing as 'pure and virulent saprophytes' in the soil under favourable conditions of temperature and moisture (Patrick Manson, *Tropical Diseases*, London: Cassell, 1898, p. 195; T. C. Allbutt (ed.), *A System of Medicine*, Vol. 1, London: Macmillan, 1896, p. 870). However, in colonial Singapore, the fear of excrement was far less discriminating and more deeply entrenched. Even with the advent of the germ theory in the late nineteenth century, the belief that disease arose spontaneously from miasma, effluvia, or noxious gases emanating from accumulated organic matter, the so-called miasmatic or pythogenic theory, still held its own within medical circles and in the public imagination. Excrementitious filth was considered 'the most dangerous of all [filths]' (ARSM, 1896, 'Supplement: Report on a Special Investigation into the System of Death Registration in Force in Singapore', p. 13).

7. Wohl, *Endangered Lives*, p. 80.

8. The sewering of London, Bristol, Cardiff, and a few other cities had occurred fairly early during the 1860s but progress was generally slower in the north of England such as in Manchester and Liverpool (Wohl, *Endangered Lives*, pp. 95–100 and 107–8).

9. James C. Riley, *The Eighteenth-Century Campaign to Avoid Disease*, Basingstoke: Macmillan, 1987, pp. x and 89–112.

10. Gerard Kearns, 'Zivilis or Hygaeia: Urban Public Health and the Epidemiologic Transition', in Richard Lawton, *The Rise and Fall of Great Cities*, London: Belhaven Press, 1989, pp. 107–8.

11. F. J. Hallifax, 'Municipal Government', in Walter Makepeace, Gilbert E. Brooke, and Ronald St. John Braddell, *One Hundred Years of Singapore*, Vol. 1, John Murray, p. 326.

12. Ibid., p. 327.

13. PLCSS, 27 June 1870, p. 66.

14. Ibid., p. 65, quoting a letter from Lieutenant Colonel H. Yule, Secretary to the Government of India, Public Works Department, Fort William, to the Governor of the Straits Settlements, 13 December 1861.

15. Ibid., p. 67.

16. PLCSS, 6 May 1868, p. ix.

17. This reservoir was later named the 'Thomson Road Reservoir' in 1907 and subsequently renamed 'MacRitchie Reservoir' in 1922 after James MacRitchie, who was the municipal engineer in charge of the waterworks from 1883 to 1895 (MMSC3, 10 November 1922).

18. Tan Kim Seng died on 14 March 1864 in Malacca at the age of fifty-nine (Song Ong Siang, *One Hundred Years' History of the Chinese*, Singapore: Oxford University Press, 1984 (first published in 1902), p. 49).

19. *Straits Times*, 18 June 1888. Only in 1913 did supply become continuous, night and day, throughout the year (ARSM, 1913, p. 24).

20. *Straits Times*, 27 August 1896; 29 October 1896. In the opinion of the municipal analyst, the brownish, turbid appearance of municipal water was an indication of the presence of dissolved vegetable matter which was in no way harmful to health (ARSM, 1909, 'Appendix K: Analyst Department', p. 69).

21. *Singapore Municipality Waterworks: Opening of New Works, 26 March 1912*, Glasgow: Aird & Coghill, 1912, p. 12.

22. ARSM, 1899, p. 7. Within a period of five years between 1897 and 1901, the daily average quantity of municipal water delivered to the town increased from 3 to over 4 million gallons.

23. ARSM, 1901, 'Appendix M: Minutes by the Municipal Engineer dated 28 November 1900 on the Extension of Private Services Throughout the Town', p. 6.

24. ARSM, 1900, p. 9.

25. The Pearl's Hill Reservoir was ideally situated to supply the southern side of the Singapore River, including the Tanjong Pagar wharves and the rapidly burgeoning coolie population around New Harbour (ARSM, 1897, 'Appendix M: New High Service Reservoir at Pearl's Hill', pp. 49–50). The first service reservoir was built in 1878 on Mount Emily on the northern side of the Singapore River (Hallifax, 'Municipal Government', p. 328).

26. ARSM, 1902, p. 11; ARSM, 1904, p. 11.

27. The Thomson Road catchment area yielded a daily average of 2,521,000 gallons whilst an average of 3,175,000 gallons were received into the reservoir through the Kallang Tunnel per day (ARSM, 1911, p. 23).

28. This reservoir was renamed 'Peirce Reservoir' in 1922 after the municipal engineer in charge of its construction (MMSC3, 10 November 1922).

29. ARSM, 1902, p. 15.

30. *Singapore Municipality Waterworks*, pp. 11–14.

31. ARSM, 1920, Municipal Engineer's Department, p. 63A.

32. ARSM, 1920, p. 8.

33. Singapore Municipal Comissioners, *Singapore Water Works, Water Supply from Johore*, Presented on the Occasion of the Completion of the Reservoir at Pontian Kechil, Singapore: Printers Ltd., 1932, p. 1.

34. Other schemes under consideration included the Pelapah scheme, the Lenggiu scheme, and the Scudai River scheme (ARSM, 1922, 'Appendix F: Singapore Water: Report by Messrs Sir Alexander Binnie, Son, and Deacon', p. 158).

35. ARSM, 1929, p. 15.

36. *Singapore Water Works, Water Supply from Johore*, p. 18.

37. Potentially, Gunong Pulai and Pontian Kechil could each yield 5.5 million gallons and 9 million gallons per day respectively.

38. *Singapore Water Works, Water Supply from Johore*, pp. 4–5 and 18.

39. ARSM, 1928, p. 11.

40. *Straits Times*, 29 December 1888.

41. *Straits Times*, 23 July 1888.

42. ARSM, 1896, Health Officer's Department, p. 86. A third 'pressing problem' cited was 'the destruction of Town refuse'. Although this is not discussed in this chapter, it should be noted that like the disposal of sewage, it represented yet another issue which called forth municipal measures to ameliorate pathogenic environments by removing urban detritus.

43. ARSM, 1896, Health Officer's Department, p. 88.

44. W. R. C. Middleton, 'The Sanitation of Singapore', *Journal of State Medicine*, 8 (1900): 698.

45. 'Report on a Special Investigation into the System of Death Registration', p. 7.

46. Ibid., pp. 7 and 22. Water from wells situated in Chinatown (Nos. 2, 5, 6, and 8 in Table 5.1) and those in the Rochor district (Nos. 1, 4, and 7) contained significant amounts of nitrites, nitrates, free ammonia and albuminoid ammonia which were indicators of water contamination in contemporary sanitary engineering tests of water quality. The presence of albuminoid ammonia revealed animal and vegetable decomposition in the water, a process in which free ammonia was produced. Nitrates and nitrites were tested for because sewage pollution brought nitrogen compounds into the water.

47. ARSM, 1902, Health Officer's Department, p. 154.

48. ARSM, 1897, Health Offier's Department, p. 89.

49. ARSM, 1905, Health Officer's Department, p. 32.

50. Ibid., p. 30.

51. 'Report on the Special Investigation into the System of Death Registration', p. 7.

52. ARSM, 1896, Health Officer's Department, p. 88.

53. ARSM, 1896, p. 9.

54. ARSM, 1896, Health Officer's Department, p. 88.

55. 'Progress Report for December 1900', para. 6, reproduced in ARSM, 1901, 'Appendix M: Notes of the Second Meeting of the Special Committee Appointed to Inquire into the Question of Water Supply, 15th March, 1901', pp. 26–7. Given that the average water consumed by a coolie was 15 gallons a day, the cost of having water carried from the standpipes would increase his monthly expenditure by 15 to 30 cents, which was a significant sum for water alone, given that the average charge for lodgings was only between 25 and 35 cents per month. As explained by one of the correspondents of the local press, 'the cost of carriage from the street hydrant on the main road being great, a large number of people naturally use[d] what they [had] at hand' (*Straits Times*, 9 May 1907).

56. ARSM, 1897, Health Officer's Department, p. 89.

57. Ibid.; ARSM, 1899, Health Officer's Department, p. 56; ARSM, 1901, p. 15; ARSM, 1910, Health Officer's Report, p. 46.

58. 'Evidence of Mr Tan Jiak Kim' cited in ARSM, 1901, 'Appendix M: Report of the Third Meeting of the Special Committee Appointed to Enquire into the Question of Water Supply, 1 May 1901', p. 2.

59. ARSM, 1896, Health Officer's Department, p. 88; MPMCOM, 5 May 1897; ARSM, 1910, p. 13, ARSM, 1910, Health Officer's Report, pp. 7 and 46.

60. ARSM, 1910, Health Officer's Report, p. 7.

61. W. J. Simpson, *Report on the Sanitary Conditions of Singapore*, London: Waterlow & Sons, 1907, p. 47.

62. 'Correspondence with Government on the Subject of High Death Rate in Singapore', p. 118.

63. ARSM, 1921, 'Appendix H: Forecast of Financial Requirements of the Singapore Municipality for the Period 1922–1931', p. 5; ARSM, 1921, Municipal Engineer's Department, p. 63A.

64. MPMCOM, 26 April 1918.

65. 'Singapore Water: Report by Messrs Sir Alexander Binnie, Son, and Deacon', p. 159.

66. ARSM, 1902, p. 11; PLCSS, 22 December 1905, p. C280; MPMCOM, 30 July 1920; MGCM, 13 August 1920.

67. For example, 663 polluted wells closed during previous years were re-opened by occupiers during a period of drought in the latter half of 1902 (ARSM, 1902, Health Officer's Department, p. 154).

68. ARSM, 1902, Health Officer's Department, p. 27.

69. 'Administrative Report for 1895', para. 12, reproduced in ARSM, 1901, 'Appendix M: Notes of the Second Meeting', p. 3.

70. ARSM, 1896, p. 8.

71. ARSM, 1901: 'Appendix M: Copy of Resolution Passed at Commissioners' Meeting of 23 November 1898', p. 12.

72. 'Progress Report for December 1900', p. 26.

73. ARSM, 1901, p. 11. The water committee that was appointed comprised J. O. Anthonisz (the municipal president), W. Evans, T. Sohst, A. Barker, Lee Choon Guan, and Choa Giang Thye.

74. 'Administrative Report for 1895', p. 2.

75. Ibid., p. 3.

76. Ibid.

77. Ibid.

78. 'Correspondence with Government on the Subject of High Death Rate in Singapore', p. 118.

79. ARSM, 1902, p. 13.

80. 'Administrative Report for 1896', para. 13, reproduced in ARSM, 1901, 'Appendix M: Notes of the Second Meeting', p. 10.

81. 'Evidence of Mr Tan Kiong Siak' cited in ARSM, 1901, 'Appendix M: Report of the Fifth Meeting of the Special Committee Appointed to Enquire into the Question of Water Supply, 14 June 1901', pp. 5–6.

82. 'Evidence of Mr Tan Jiak Kim', p. 1.

83. 'Evidence of Mr Tan Chun Fook' cited in ARSM, 1901, 'Appendix M: Report of the Sixth Meeting of the Special Committee Appointed to Inquire into the Question of Water Supply, 26 June 1901', p. 8.

84. ARSM, 1901, 'Appendix M: Report of Special Committee Appointed to Inquire into the Question of Water Supply', p. 2.

85. Ibid.

86. Ibid., pp. 3–4.

87. Ibid. The committee recommended the increase of the domestic unmetered rate as follows:

Type of Service Pipe	Old Rate	Recommended Rate
Half inch	$1.50 per month	$2.00 per month
Three-quarter inch	$3.00 per month	$4.00 per month
One inch	$5.00 per month	$6.00 per month
For Each Additional Tap:		
Half inch	$0.50 per month	$1.00 per month
Three-quarter inch	$0.75 per month	$1.50 per month
Half inch	$1.00 per month	$2.00 per month

88. ARSM, 1902, p. 13.

89. Ibid.

90. ARSM, 1917, p. 25.

91. ARSM, 1918, p. 18; MGCM, 15 February 1918.

92. ARSM, 1924, Water Department, pp. 19I-20I.

93. *Straits Times*, 29 October 1896.

94. J. A. Hassan, 'The Growth and Impact of the British Water Industry in the Nineteenth Century', *Economic History Review*, 38 (1985): 543.

95. Simpson, *Report on the Sanitary Conditions*, p. 47.

96. ARSM, 1890, 'Report on and Estimates for the Disposal of Night-soil', p. 1.

97. ARSM, 1893, 'Report on and Estimates for, the Disposal of Night-soil at Singapore and for the Improvement of the Surface Drainage', p. 61.

98. Ibid.

99. ARSM, 1895, Health Officer's Department, p. 60.

100. 'Report on and Estimates for the Disposal of Night-soil', p. 3.

101. ARSM, 1894, p. 10; MPMCOM, 29 August 1894.

102. *Straits Times*, 17 February 1887.

103. 'Pollution of the Singapore River', p. 45.

104. 'Report on and Estimates for the Disposal of Night-soil', p. 1.

105. Ibid.

106. 'Chinese Gardeners and Disease', *Malaya Medical Journal and Estate Sanitation*, 1, 2 (1926): 33.

107. F. Smith, 'Municipal Sewerage: Part I', *The Journal of Tropical Medicine*, 6 (1903): 287.

108. ARSM, 1895, p. 14.

109. *Straits Times*, 2 September 1896.

110. Ibid.

111. 'Pollution of the Singapore River', pp. 44 and 47–8; 'Report on a Special Investigation into the System of Death Registration', pp. 6–7 and 11.

112. 'Report on and Estimates for the Disposal of Night-soil', pp. 4–6.

113. Ibid., pp. 1–2.

114. Ibid.

115. 'Extract from Health Officer's Monthly Report for October 1890', in ARSM, 1893, 'Report on, and Estimates for the Disposal of Night-soil at Singapore, and for the Improvement of the Surface Drainage', p. 52.

116. 'Report on and Estimates for the Disposal of Night-soil', p. 2.

117. Ibid.

118. 'Extract from Health Officer's Monthly Report for October 1890', p. 52.

119. Simpson, *Report on the Sanitary Condition*, p. 54. Neither MacRitchie nor Dr Gilmore Ellis was in favour of a dry privy or earth-closet system either, mainly because it was felt that Asian householders would find it too troublesome or expensive to keep a stock of dry earth, sand, or ashes for deodorising purposes and as a consequence, neglect its maintenance. The actual quantity of filth to be removed would also be doubled by the addition of dry earth and this would both increase cartage bills as well as reduce its value as manure ('Extract from Health Officer's Monthly Report for October 1890', p. 52).

120. 'Extract from Health Officer's Monthly Report for October 1890', p. 53.

121. MPMCOM, 13 October 1890; ARSM, 1891, p. 5.

122. ARSM, 1891, p. 5.

123. ARSM, 1892, p. 8.

124. ARSM, 1893, 'Report on Sanitary Arrangements of Towns Visited in India and Proposals for the Improvement of the Sanitation of Singapore', p. 50.

125. Ibid.

126. MPSM, 15 February 1889.

127. MPMCOM, 24 November 1890.

128. MPMCSM, 11 February 1891.

129. ARSM, 1895, p. 15.

130. ARSM, 1895, Health Officer's Department, p. 60.

131. Ibid.

132. MPMCOM, 27 April 1892; ARSM, 1894, p. 15; MPMCOM, 13 February 1895; ARSM, 1896, p. 16; MPMCOM, 15 July 1904.

133. ARSM, 1896, Municipal Engineer's Department, p. 32.

134. Ibid.

135. Ibid., p. 33.

136. ARSM, 1901, p. 14.

137. In the 1920s, these municipal pail latrines were gradually replaced by water-carriage latrines where sewer connections could be effected (ARSM, 1920, Municipal Engineer's Department, p. 21A).

138. ARSM, 1896, p. 16.

139. ARSM, 1897, Municipal Engineer's Department, p. 32.

140. ARSM, 1896, p. 16.

141. ARSM, 1896, p. 17, MPMCOM, 10 March 1897.

142. ARSM, 1920, Municipal Engineer's Department, p. 21A.

143. MPMCOM, 25 September 1895.

144. ARSM, 1895, Health Officer's Department, p. 61; MPMCOM, 9 October 1895.

145. ARSM, 1895, Health Officer's Department, p. 61. Recourse to the law against nuisances affecting private individuals was provided under the Summary

Criminal Jurisdiction Ordinance XII of 1872; however, it was 'impossible to induce even the loudest complainants to avail themselves of [its provisions]' (ARSM, 1895, p. 14).

146. ARSM, 1897, p. 16.

147. ARSM, 1897, Health Officer's Department, p. 88.

148. ARSM, 1898, Health Officer's Department, p. 96.

149. PLCSS, 14 November 1889, p. B135.

150. ARSM, 1895, Health Officer's Department, p. 60. In 1901, Dr Middleton's meticulously calculated statistics comparing estimates of the amount of solid excreta produced and the volume evacuated from the town area per day indicated that only a comparatively 'trifling' amount of solids was not regularly removed and allowed to seep into the soil under houses (ARSM, 1901, 'Appendix J: Disposal of Night-soil', p. 59).

151. MPMCOM, 12 August 1896.

152. MPMCOM, 10 March 1897.

153. ARSM, 1898, p. 21.

154. Ibid.

155. ARSM, 1899, p. 15; ARSM, 1899, Health Officer's Department, p. 57; ARSM, 1901, p. 14.

156. ARSM, 1903, p. 10.

157. ARSM, 1904, 'Appendix: Middleton and Finlayson to President, Municipal Commissioners', p. 1; MGCM, 12 October 1904.

158. 'Appendix: Middleton and Finlayson to President', p. 6.

159. Ibid., pp. 2–3.

160. Ibid.

161. ARSM, 1895, Bacteriologist's Department, p. 4.

162. 'Appendix: Middleton and Finlayson to President', p. 1.

163. MPMCOM, 18 March 1906.

164. MPMCOM, 1 June 1906; 27 July 1906.

165. Simpson, *Report on the Sanitary Conditions*, p. 46.

166. Ibid., pp. 44–5.

167. Ibid., pp. 46–9.

168. Ibid., p. 48.

169. ARSM, 1907, p. 14.

170. Ibid., pp. 53–4.

171. Ibid., p. 54.

172. Ibid., p. 53.

173. ARSM, 1907, Health Officer's Report, p. 42.

174. ARSM, 1908, Health Officer's Department, p. 39.

175. ARSM, 1907, Municipal Engineer's Department, p. 23.

176. ARSM, 1908, Health Officer's Department, p. 39.

177. ARSM, 1908, Health Officer's Department, p. 39; ARSM, 1909, Health Officer's Department, p. 41; ARSM, 1910, Health Officer's Department, p. 40.

178. ARSM, 1908, Health Officer's Department, p. 40.

179. *Straits Times*, 19 May 1913.

180. Ibid.

181. MPMCOM, 1 October 1909.

182. ARSM, 1910, p. 17.

183. Ibid.

184. 'Report of the Municipal Enquiry Commission', pp. C203–4.

185. ARSM, 1910, p. 17.

186. ARSM, 1911, p. 20; ARSM, 1911, Health Officer's Department, p. 42; MPMCOM, 28 April 1911.

187. ARSM, 1913, p. 21.

188. ARSM, 1910, p. 17.

189. MMFGPSYC, 14 April 1916; 9 May 1916.

190. MMSC2, 20 August 1920; MPMCOM, 26 January 1923.

191. When water-carriage fittings were ultimately installed in shophouses, the increased consumption of water per head was 3 gallons per day. Since the washing of

night-soil pails at the Peoples' Park Dumping Room necessitated the use of about $1\frac{1}{2}$ gallons per head, the nett increase resulting from the introduction of a water-carriage system amounted to about $1\frac{1}{2}$ gallons per head per day for the population served (ARSM, 1923, Scavenging and Conservancy Department, p. 3H).

192. MMSC2, 20 August 1920.

193. Ibid.

194. ARSM, 1915, p. 20.

195. ARSM, 1916, pp. 21–2.

196. MPMCOM, 26 January 1923.

197. ARSM, 1921, 'Appendix H: Forecast of Financial Requirements of the Singapore Municipality for the Period 1922–1931', p. 5.

198. PLCSS, 1922, 'Appendix A: Report of the Commissioners Appointed by His Excellency the Governor to Inquire into the Financial Position of the Municipality, 1921', p. C6.

199. ARSM, 1920, p. 8.

200. ARSM, 1921, 'Appendix H: Forecast of Financial Requirements', p. 8.

201. ARSM, 1920, 'Pearson to President, Municipal Commissioners, 24 June 1921', p. 64A.

202. ARSM, 1920, Municipal Engineer's Department, p. 21A.

203. ARSM, 1915, p. 21.

204. *Straits Times*, 8 May 1910; ARSM, 1913, p. 20; ARSM, 1915, p. 21.

205. ARSM, 1915, p. 21.

206. These were Sultan Ali Estate and Kampong Kapur in 1917; Rochor in 1921; Conservancy District No. 8 in September 1922; District No. 3A in June 1924; District No. 10 in September 1926; Conservancy District No. 4 in August 1920, the Lavender Street area, Kampong Kapur Extension, and Sultan Ali Extension in January 1930; the Tanjong Pagar area in February 1930; and the Race Course area in March 1930.

207. ARSM, 1915, p. 21.

208. Ibid.

209. ARSM, 1923, Scavenging and Conservancy Department, p. 22H.

210. ARSM, 1921, 'Appendix H: Forecast of Financial Requirements', p. 5.

211. MMFGPSYC, 24 March 1916; MPMCOM, 25 August 1916, and 23 February 1917; MGCM, 9 August 1918, 9 January 1920, and 27 February 1920.

212. ARSM, 1920, p. 10; MPMCOM, 26 March 1920.

213. The daily wage of municipal night-soil coolies was increased from the 1919 rate of 52–65 cents to 75–83 cents in 1920 (ARSM, 1920, p. 10; ARSM 1919, Municipal Engineer's Department, p. 16A; ARSM, 1920, Municipal Engineer's Department, p. 16A).

214. ARSM, 1920, Municipal Engineer's Department, p. 20A.

215. ARSM, 1923, Scavenging and Conservancy Department, p. 15H.

216. ARSM, 1916, p. 22.

217. ARSM, 1923, Scavenging and Conservancy Department, p. 3H.

218. Night-soil, properly treated, also featured significantly in the Chinese pharmacopoeia. According to De Groot, no other excretions were made use of by the Chinese more extensively than faeces and urine for medical purposes. For example, people living close to urban markets buried empty jars in dung for a number of years in order to filter out a black and bitter fluid called 'yellow dragon-soup' used for curing pestilential disease. Ashes of faeces, and faeces and urine fermented with boiled rice for a long time, were prescribed for phthisis whilst dried excrements were highly recommended for stomach complaints and all sorts of eruptions including smallpox (J. M. M. de Groot, *The Religious System of China*, Vol. 4, Taipei: Ch'eng-wen Publishing Co., 1969, pp. 398–9).

219. 'Some Points in the Municipal Administration of Singapore', *The Straits Chinese Magazine*, 8, 4 (1904): 165.

PART II

PART II

Ordering the Public Environment

THE earlier chapters (3–5) have largely, though not exclusively, focused on the private, domestic environment as the object of medical discourse and the arena of conflict between the municipal authorities and the Asian communities. The cubicle, the latrine, other interior aspects of the Asian house as well as the people who inhabited and used these domestic spaces had been the objects of municipal measurement and surveillance, spatial rearrangement and manipulation (Chapter 3). The municipal vision of a healthy city had necessitated, in the first place, an invasion of domestic privacy in order to expose what lay hidden—filth, darkness, disease, and overcrowding—to the sanitizing gaze of the authorities. That this gaze also extended to encompass the wider, public environment was evident in the authorities' concern with town planning and the provision of public spaces to 'open up' the closely textured city (Chapter 4) as well as in their concern with establishing municipally controlled, public utilities as systems which would not only revolutionize domestic practices of water storage and sewage disposal but also ensure that only 'pure' water and 'neutralized' waste traversed the wider, public environment (Chapter 5).

While the colonial project of sanitizing the city remained vital to the manner in which the authorities attempted to structure and control the public built environment, it was joined by the equally pressing aims of producing a public landscape which was orderly, disciplined, easily policed, and amenable to the demands of urban development and efficiency. To the vision of a sanitized city was blended the image of a progressive, civilized city. Singapore was to exemplify modern principles of order and efficiency as befitted its status as a great commercial entrepôt in the British Empire. Imposing a semblance of order and discipline on the urban environment was particularly crucial to the colonial enterprise. A well-ordered city was not only an object which engendered colonial and civic pride but it also facilitated mobility and surveillance and promoted good health. Timothy Mitchell describes the process of ordering as 'enframing', a method of 'dividing up and containing' space so as to render colonial society 'picture-like and legible' and hence 'available to political and economic calculation'.[1]

The next three chapters (6–8) focus on municipal endeavours to create in the city's public spaces a landscape of clarity, order and progress. From the second half of the nineteenth century, in its role as chief architect of the public landscape, the Municipal Board of Singapore progressively enlarged its powers to control the making, naming, and use of public space. The imposition of order on the urban built environment, however, did not proceed unchallenged by those whose livelihood, daily routines, and communal rituals depended on their command and use of the city's public spaces. The public built environment provided a framework for different forms of human action and behaviour, contained and communicated conflicting meanings for diverse groups, and was often used to fulfil different ends.[2] It provided an arena, or what some have called a 'stage' or a 'theatre',[3] on which conflicts between the authorities and the Asian communities were enacted. The public environment was, however, not simply an immutable setting for conflict but was actively involved as the object of claims and counter-claims, and as a resource for pressing further claims. As such, the public landscape was constantly transformed by the process of conflict and negotiation through everyday strategies as well as articulated discourse.

Chapter 6 examines the question of effecting order in the public landscape by focusing on municipal endeavours to signify public space and render it more 'legible' through enforcing a system of colonial street- and place-names. The emergence of alternative systems of Asian names for urban places illustrated an incipient contest for the meaning of the city's public spaces. The juxtaposition of different systems of street-names testified to the syncretic character of the city and belied the municipal vision of one city well-integrated by a single, overarching system of signification. Chapter 7 turns to conflicts between the municipal authorities and the Asian communities over the meaning of public order and what constituted the legitimate functions of public space. The focus here is on the verandah, an element of the urban built environment which became the arena not only for daily conflicts and negotiations, but also for the outbreak of an urban riot of crisis proportions. Chapter 8 turns to yet another landscape of conflict—the Chinese burial grounds around and within the city. Pre-eminent here was the question whether the traditionally 'sacred' landscapes of the Asian communities were immune from the changes demanded by urban development and progress championed by the municipal authorities.

---

1. Timothy Mitchell, *Colonising Egypt*, Cairo: American University of Cairo Press, 1989, pp. 33–45.

2. The built environment as an arena of conflict and crisis is a theme often ignored in urban studies. Jamel Akbar's recent work which examines the intersecting claims of ownership, control, and use of major publicly or collectively owned elements of the trad-itional Muslim city provides a stimulating exception (Jamel Akbar, *Crisis in the Built*

Environment: The Case of the Muslim City, Singapore: Concept Media, 1988, pp. 107–28).

3. R. Trexler, *Public Life in Renaissance Florence*, New York: Academic Press, 1980; Charles Tilly, 'Notes on Urban Images of Historians', in Lloyd Rodwin and Robert M. Hollister (eds.), *Cities of the Mind: Images and Themes of the City in the Social Sciences*, New York: Plenum Press, 1984, pp. 130–1; Denis Cosgrove and Steve Daniels, 'Fieldwork as Theatre: A Week's Performance in Venice and Its Region', *Journal of Geography in Higher Education*, 13 (1989): 170–1.

6
The Naming and Signification of Urban Space: Municipal versus Asian Street-names and Place-names

World maps prepared for use within a given country misrepresent in some degree the place names of countries of other speech. When different alphabets are involved, the misrepresentation is magnified; and when a different system of language is used, ... the distortion puts the cultural and geographic realities completely beyond the grasp of the layman.[1]

Difficult as it may be to distinguish clearly between so-called 'spontaneous' naming of places by the man on the ground, so to speak, and the 'self-conscious' work of the man at the desk, the distinction must be kept in mind if place names are to tell us how space has been structured cognitively.[2]

Illustrating what might be called, after its well-known advocate, the 'Humpy Dumpty position', ... names are rooted neither in reality nor custom, but express instead the power of the namer over the thing named.[3]

The Significance of Street-names and Place-names in Singapore

IN colonial Singapore, the municipal authorities were empowered to establish a network of street- and place-names to facilitate the identification, demarcation, and differentiation of the urban built environment for the purposes of colonial rule. From the early days of the Settlement, Stamford Raffles decreed that 'each street should receive some appropriate name' and that it was 'the duty of the police to see [that they were] regularly numbered'.[4] Under section 28 of the Indian Act XIV (Conservancy) of 1856, the commissioners were empowered to affix in a 'conspicuous ... place at each end, corner, or entrance of every street' in the town of Singapore a board on which was 'the name by which such street is to be known'.[5] A clear and well-ordered system of street- and place-names was essential to the colonial and municipal authorities for a number of practical purposes. Accurate addresses and clearly signposted streets were necessary for levying house assessments and public utilities rates as well as for efficient postal, fire-fighting, and transport services. Portions of streets in colonial Singapore were occasionally renumbered, reclassified or renamed to accommodate the requirements of the municipal assessment office.[6] Street-names which were phonetically similar were often changed in order to avoid confusion and delay in summoning the fire brigade to the correct location in the event of a fire. In 1858, it was noted in the municipal meetings that much confusion reigned among the streets of Singapore because not only were certain streets, canals, and squares nameless, there were others where the same name had been given to two or even three streets.[7] The municipal commissioners embarked on the process of removing some of the confusion by naming and renaming some of the streets[8] but the duplication of names was not entirely remedied. Half a century later, during a fire in early 1908, the fire brigade was delayed as a result of the confusion of D'Almeida Street (a street in the European business quarter off Raffles Place) with Almeida Street (a street in Chinatown off South Bridge Road). Consequently, the municipal commissioners resolved that Almeida Street be renamed 'Temple Street' and also that a suburban road known as Almeida Road (Stevens Road to Bukit Timah Road) rechristened 'Balmoral Road'.[9]

The legibility of the urban environment was also crucial to the surveillance functions of the state, functions which ranged from the taking of a population census, orthodox police work such as inspecting houses, instituting arrests, posting notices, and serving summons on occupiers, to public health concerns such as tracing the source and spread of 'dangerous infectious diseases'. A well-organized system of street-names was necessary if clandestine activities, dangerous diseases, and the Asian population in general were to be rendered less amorphous, more visible to the observation and 'gaze' of the authorities and hence more accessible to control.

Besides containing shorthand descriptive content which helped to systematically differentiate one place from another, street- and place-names also possess cultural and symbolic meaning. If the urban landscape is treated as a 'text' or as 'cultural form' which upon interrogation 'reveals a human drama of ideas and ideologies, interest groups and power blocs',[10] then one of the most obvious cultural products to which an informed 'reading' may be brought are the names of places. Consequently, place-names have been critically 'read' for a variety of purposes, as 'signifiers' of wider societal trends. The toponymic significance of place-names honouring public figures has been invoked as indicators of nationalistic fervour[11] or community identity,[12] whilst the renaming of streets in post-colonial societies interpreted as an ideological tool to divest the landscape of its colonial associations and achieve political legitimation.[13] Street-names bearing allusions to particular types of activity have been used to recover the localization of trades and tradespeople.[14] The analysis of place-names, their elaboration and mutation, has also been instrumental to reconstructing the 'sequent occupance' of colonizing or immigrant groups,[15] the 'ecological invasion and succession' of operational territories by urban gangs,[16] as well as the chronology and geography of social and cultural change.[17] David Robinson, for example, demonstrates that in Latin America, 'when regimes fall, empires collapse, or one local élite replaces another, very often a name-changing process is initiated in particular places' and in urban areas in particular, 'a veritable historical geography can often be read from the street-names'.[18] The establishment of a network of official street- and place-names hence not only introduces order and differentiation into an originally amorphous landscape but also reflects the mental images of the dominant culture. The power of 'nomination' ... is often the first step in taking possession',[19] and street-names are 'among the first to undergo a refurbishing to commemorate new regimes'.[20]

The Naming Process

The naming of place, whether as a conscious, deliberate event or a more informal process of evolution, is in varying degrees a social activity. As Gordon Pirie has argued, this is so 'either by virtue of it involving joint decision-making and/or in respect of it occurring within a given social milieu in which there are formal or informal conventions of name selection, assignment and adoption'.[21] The naming process is hence not only of toponymic significance but also embodies some of the social struggle for control over the means of symbolic production within the urban built environment.

In colonial Singapore, official street- and place-names were assigned at municipal meetings on the approval of the commissioners.[22] Names for consideration were normally proposed by the municipal assessor, although suggestions sometimes originated with the municipal president, one of the commissioners, or the requests of property owners. Once decided, notice of new or changed street-names was advertised in the press and the schedule circulated amongst various heads of government departments, including the chief police officer, the commissioner of lands, and the secretary of the Fire Insurance Association. As such, the official naming process was strongly dominated by the opinions of municipal and government officers, and occasionally, those of influential property owners. It was relatively impervious to the views of those who actually lived or used the streets themselves and was hence generally detached from the social milieu of the plebeian classes.[23] Street nomenclature became a means by which the authorities were able to project on to the urban landscape their perceptions of what different areas within the city represented.

The process by which streets and places became named was, however, far more complicated than the institutionalized procedure of naming suggested. The 'unanimity of name selection which seems encrusted in the singularity of a[n] [official] place-name' is misleading for it detracts from the problematic nature of the naming process.[24] In colonial Singapore, although the authorities controlled the naming process in so far as the selection and application of names were concerned, it was the Asian populace who determined whether a particular assignment was adopted, ignored, or substituted. Official street nomenclature and their representations of meaning did not automatically pass into local currency but instead encountered barriers posed by alternative systems of street-naming and place-signification pertaining to the various Asian communities. This chapter examines debates over the content of municipal street nomenclature and what they signified. This is then contrasted with alternative systems of street-naming, and in particular that adopted by the Chinese communities. The final section examines some implications of the disjunctions between representations of the meaning of place imposed by the authorities and that used by the Asian plebeian classes in their daily routines.

Municipal Street-names and Place-names

An analysis of municipal street-names in Singapore suggests that certain themes tended to predominate in the christening of the city's streets.[25] The street-name was often used as a medium to commemorate prominent figures, particularly those who were considered to have contributed significantly to public work and urban development. In 1905, among 225 municipal street-names with Chinese (Hokkien and Cantonese) equivalents listed by H. W. Firmstone, an official with the Chinese Protectorate,[26] about 45 per cent served a commemorative function (Figure 6.1). Of these, two-thirds were given over to honouring European personages, whilst the remaining one-third paid tribute to Asian characters (Figure 6.2). European candidates for this form of secular 'canonization' included Resident Councillors and Governors (as found in Raffles Place, Crawfurd Street, Prince Street, Fullerton Road, Bonham Street, Jervois Road, and Cavenagh Road); royalty (Alexandra Road, Victoria Street, Albert Street, Coronation Road, and Empress Place); war heroes (Havelock Road, Neil Road, Outram Road, Jellicoe Road, and Nelson Road); distinguished visitors (Dalhousie Lane, Edinburgh Road, and Connaught Drive), and prominent citizens and civil servants (Still Road and Makepeace Road after journalists; Paterson Road and Palmer Road after leading members of the mercantile community; Pickering Street after the first Chinese protector;[27] Cook Street and Oldham Lane after missionaries; Hullet Road after a schoolmaster; and Everitt Road after a lawyer). Municipal commissioners and officers were also not reluctant to have their names attached to streets, and between 1880 and 1930 over forty streets in the municipality were dedicated to their memory.[28] In 1927, for example, municipal commissioner Dr K. Kiramathypathy moved that 'Jalan Tambah'[29] in the Kampong Kapor district be renamed 'Veerasamy Road' in honour of a fellow leader of the Indian community, Dr N. Veerasamy, on the grounds of his association with various public activities in the city of Singapore for the last thirty-five years, his eleven-year service on the

FIGURE 6.1
Comparison of Municipal and Chinese Names of 225 Streets

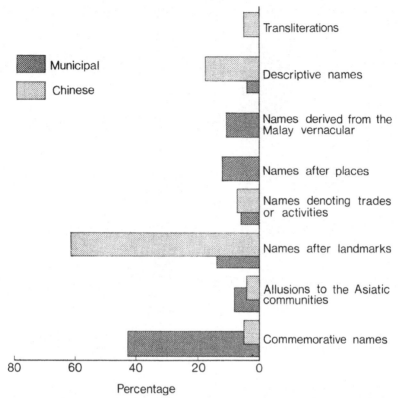

Source: Calculated from H.W. Firmstone, 'Chinese Names of Streets and Places in Singapore and the Malay Peninsula' [hereafter cited as 'Chinese Names of Streets'], *Journal of the Straits Branch of the Royal Asiatic Society*, 42 (1905): 53–208.

Municipal Board and the 'high esteem' accorded to him by 'the humble folk of the city, especially those living in Kampong Kapor'.[30] The selection of suitable streets to honour particular individuals was a carefully calculated process aimed at ensuring that the length, location, and importance of the chosen street were commensurate with the esteem due to the person commemorated. For example, in choosing road names for the Tiong Bahru area opened up by the Singapore Improvement Trust in the early 1930s, it was suggested that 'Mr Walter H. Collyer [Manager of the Improvement Trust] would not mind if we associate[d] his name with one of the streets—although he would be quite entitled to object to a street named after him unless it [was] a really long one'.[31]

The requests of land and property owners who had contributed land towards the making of streets or defrayed the cost of making, metalling, or draining them to name these streets after themselves or

FIGURE 6.2
Commemorative Street-names: European versus Asian Personalities

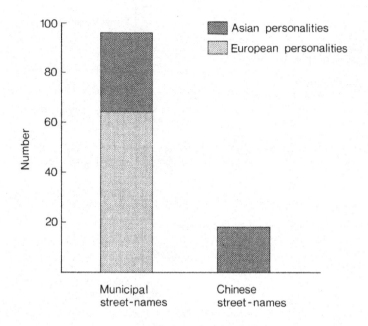

Source: Calculated from Firmstone, 'Chinese Names of Streets', pp. 53–208.

their family members were generally acceded to. It was through such a means that streets in Singapore acquired names such as 'Norris Road',[32] 'Chin Nam Street',[33] 'Cheng Yan Place',[34] 'Teck Guan Street',[35] 'Yow Ngan Pan Street',[36] and 'Eu Tong Sen Road'.[37] By perpetuating the names of 'deserving citizens' and 'eminent public servants', municipal street nomenclature represented the city as one dominated by enthusiastic public service and a whole gamut of civic talent. Commemorative street-names inscribed into the city's built environment a tangible record of all those who had contributed either materially or in the form of services to the making of a public-oriented city.

As a city which prided itself as being the capital of the Straits Settlements and the Malay States and 'the Clapham Junction of the Eastern Seas',[38] a certain proportion of Singapore's street-names also recalled linkages with important places in the Malay Peninsula and sur-rounding countries. These accounted for just over 10 per cent of the street-names listed by Firmstone (Figure 6.1) and included Malacca Street,[39] Johore Road, Muar Road, Ophir Road, Pahang Street, Trengganu Street, Enggor Street, Tras Street, Bernam Street, Raub Street (all commemorating Malayan places),[40] Bencoolen Street,

Manila Street, Japan Street, Pulau Saigon, and Rangoon Road. Municipal commissioners often favoured the orderly progression of an onomastic pattern, which once established, tended to perpetuate itself. Thus, when a road in close proximity to Rangoon Road, Mandalay Road, and Moulmein Road came up for naming in 1929, a 'Burmese name' was considered the most suitable. 'Martaban Road' was chosen followed in later years by the addition of 'Pegu Road' and 'Bhamo Road'.[41] This resulted in the establishment of 'colonies' of related place-names based on a common theme.

Although residential segregation along racial lines was not legalized in the city, colonial street-names often indicated an unofficial dichotomy between European and Asian residential areas. Roads in certain parts of the city, in particular the European residential suburbs, were given names which denoted the symbolic transfer of sentiment and the imagery of colonial hopes. In Tanglin and Claymore, the 'aristocratic' European suburbs, roads such as Orchard Road and River Valley Road 'present[ed] the appearance of well-shaded avenue[s] to English mansion[s]', comparable in their quiet but effective beauty to 'Devonshire lanes' (Plate 14).[42] Road-names such as 'Devonshire Road' and 'Chatsworth Road'[43] which conjure up the idyllic imagery of the English countryside were considered appropriate complements to the suburban landscape. European suburban roads also often originated as plantation carriageways or estate boundaries and were thus named after the estates and residences of well-known European inhabitants[44] (for example, Dalvey Road, Emerald Hill Road, Nassim Road, Irwell Bank Road, Leonie Hill Road, Orange Grove Road, Killiney Road, Cairnhill Road, Oxley Road, and Scotts Road in the Tanglin and Claymore districts).[45]

In the late 1920s, the choice of non-local road-names did not pass uncommented upon by some of the municipal commissioners. In October 1927, commissioner Syed Mohammed bin Omar Alsagoff raised the question why a new road from Orchard Road to Cairnhill Road had been christened 'Bideford Road'. The municipal president's explanation that 'Bideford was the town from which the owner of the land on one side of the road came' failed to satisfy Alsagoff who remarked that 'he had never heard of Bideford'.[46] In January 1929, Commissioner Laycock objected to the naming of new roads on the Lavender Road reclamation area after Cawnpore, Lucknow, Simla, Lahore, Benares, and Karachi on the grounds that 'there was no natural connection between this Colony and these places'.[47] He asserted that there was 'no need to go outside Malaya for names [for] there were plenty of people deserving recognition', a view supported by Dr H. S. Moonshi, an Indian Muslim member of the Board.[48] Despite these reservations about the use of non-local names, names derived from English places in particular continued to abound. When Tanjong Katong was developed as a European residential area and seaside resort in the 1920s, the roads in the area were named after English seaside resorts such as Bournemouth Road, Branksome Road, Boscombe

Road, Clacton Road, Margate Road, Parkstone Road, Poole Road, Ramsgate Road, Swanage Road, and Wareham Road.[49] The names of English counties and towns also provided the inspiration for a number of road names such as Dorset Road, Hertford Road, Norfolk Road, Derbyshire Road, Essex Road, Truro Road, Carlisle Road, Bristol Road, and Shrewsbury Road.[50] The abundance of Anglicized road-names in premier European suburbs allowed Europeans driving home to their bungalows sequestered in park-like expanses to escape the impress of the tropics and 'native' culture and symbolically return home to English settings.

Areas associated with different Asian communities were also clearly reflected in street-names assigned by Singapore's first town committee appointed by Raffles in 1822 to 'appropriate and mark out the quarters or departments of the several classes of the native population' in order to 'prevent confusion and disputes'.[51] Following Raffles's instructions, the committee designated separate divisions for different racial groups: the European division was dignified by the term 'town' while the Asian immigrants were relegated to separate *kampung*. Street-names in each of these divisions were clearly identified with the intended inhabitants, as evidenced by 'Arab Street' in the Arab *kampung*, 'Chuliah Street', 'Nellore Street', 'Nagapatnam Street', and 'Patna Street' in the Chuliah *kampung* and 'Church Street' in European Town (see Figure 2.7). In the Chinese *kampung*, street-names such as 'Canton Street', 'Hokien Street', 'Chin Chew Street', 'Macao Street', and 'Nanking Street' testified to the care taken by the town committee 'to advert to the provincial and other distinctions among this peculiar people'.[52] Despite the fact that some of the proposed street-names did not materialize and the principle of residential separation did not survive in its entirety beyond the nineteenth century, the plan provided the basis of a system of street-names which associated racial identity with specific places. Although the rapid influx of Chinese immigrants soon extended the perimeters of the original Chinese *kampung*, overflowed its bounds, and spread throughout the town, street-names associated with the Chinese race (for example, 'Amoy Street', 'Teo Chew Street', 'Pekin Street', 'Hong Kong Street', 'Pagoda Street', and 'Wayang Street') continued to abound in the traditional heart of 'Chinatown'. Similarly, in the vicinity of Arab Street, street-names such as 'Haji Lane', 'Bussorah Street', and 'Shaik Madersah Lane' (the latter two names commemorating prominent members of the Arab community) evoke the historical presence of the Arab *kampung*. Street-names on the site of the original Bugis *kampung* such as 'Java Road', 'Sumbawa Road', and 'Palembang Road' also retained their association with the Indonesian islands. It appears that the Chuliah *kampung* marked on Lieutenant Jackson's plan never materialized, but an early concentration of south Indians established on the western fringe of Commercial Square left its impress on the name of one of the streets leading off the Square. 'Kling Street' was already extant on the earliest comprehensive plan of the town drawn from an actual survey,[53] and was named after southern Indians known in Singapore as 'Klings', a Malay/Javanese corruption of 'Kalinga', the ancient empire of southern India which had trading connections of great antiquity with the Malay Archipelago.[54] By 1905, according to Firmstone's list, about 10 per cent of municipal street-names alluded to associations with the various Asian communities (Figure 6.1).

In later years, similar attempts were made to associate place with race, albeit with increasing complications. In 1925, for example, it was proposed that a new road off Serangoon Road in the Indian district be named 'Bombay Road'.[55] Further investigations, however, revealed that such a name was actually offensive to the Indian community as it kindled the association of the site, which was a former convict burial ground, with the phrase *orang kena buang Bombay*,[56] a pejorative term for Indian convicts transported to Singapore in the days when the Settlement served as a station for transmarine convicts. The racial association was ultimately abandoned and the road named 'St George's Road' by virtue of its proximity to the already existing 'St Michael's Road'.[57] To take another example, in selecting a name for a new road leading from Kreta Ayer to Anson Road over the site of a former Chinese (Cantonese–Kheh) burial ground, the collector of land revenue, J. Lornie, preferred one which would reflect the original association of the place with the Chinese community. One of the earlier suggestions—'Aljunied Road'—was hence rejected on the grounds that it 'might more appropriately be given to an area in which Mohammedans had some interest'.[58] 'Choon Guan Road', in memory of Lee Choon Guan, a well-known Straits-born Hokkien shipping and real-estate tycoon who had died recently, [59] was strongly favoured. However, some of the municipal commissioners feared that as the former burial ground belonged to the Cantonese and Kheh communities, objections might be raised against a street-name commemorating a member of another dialect group. 'Man Sau Road', after Leong Man Sau, a prominent member of the Cantonese community who died in 1916, [60] was suggested instead, but turned down by the municipal assessor on the grounds that it was phonetically similar to 'Mansoor Street', an existing road off North Bridge Road. It was ultimately decided to abandon the search for a Chinese name and to christen the road 'Maxwell Road' in honour of 'the eminent services to Malaya of three generations of public servants [who were] all members of one family'.[61] Less controversial was the naming of streets newly laid in Sultan Ali's Estate[62] in 1909 when the estate was improved. In order to reflect its traditional association with the Muslim population as well as to complement existing street-names such as 'Jalan Sultan', 'Sultan Gate', and 'Arab Street', names derived from Islamic cities such as 'Baghdad Street', 'Kandahar Street', and 'Muscat Street' were chosen.[63] By the end of the nineteenth century, although the basis of community life was no longer rigidly tied to a territorial dimension, the endurance of street- and place-names invested with racial and cultural connotations testified to the tendency manifest in the colonial consciousness to order society by separating the colonized into different, recognizable racial containers.

Another category of municipal street-names (about 15 per cent of Firmstone's list; see Figure 6.1) owed their origins to adjacent or nearby physical features, landmarks, and other material symbols. Of these, only about a quarter testified to the presence of non-European landmarks such as religious buildings (Pagoda Street, Mosque Street, and Synagogue Street) and places of entertainment (Club Street and Wayang Street); the bulk referred to symbols pertaining to transport, the military, or natural topography. These included bridges (North, South, and New Bridge Roads), quays (Boat Quay), canals (North and South Canal Roads), transcommunications (Telegraph Street and Cable Road), military installations (Battery Road, Cantonment Road, Fort Road, and Magazine Road) and topographical features (Hill Street, River Valley Road, and Beach Road).

A small number of municipal place-names (about 6 per cent of Firmstone's list; see Figure 6.1) offered information on the localization of early trades or agricultural activities. Most prominent amongst these was 'Commercial Square',[64] the heart of European business on the south bank of the Singapore River. Certain streets had an occupational term as its first element (Carpenter Street and Merchant Street), whilst others either indicated a particular product being manufactured or for sale in the street (Fish Street and Sago Lane) or the location of a particular form of agricultural pursuit (Orchard Road, Garden Street, and Buffalo Road). These street-names had tended to persist despite the fact that the particular trade or agricultural activity which they recalled had been dispersed or relocated elsewhere.

Although the municipal authorities attempted to choose street-names which alluded to the presence of the Asian communities in parts of the city which they deemed appropriate, and despite the fact that Asian landowners succeeded in toponymically claiming several streets for themselves, most municipal street-names honoured the perceptions of powerful European namers rather than those of the people living in the places so named. Even when original names derived from the Malay vernacular (of which some might have predated British settlement) were preserved (equivalent to 11 per cent of street-names listed by Firmstone; see Figure 6.1), they often had to be transmogrified to fit English-speaking tongues.[65] In 1889, H. T. Haughton observed that in Singapore, 'as the old Malay inhabitants ... died out or migrated, it [was] probable that, before very long, the [Malay] names of places [might] have become corrupted (as some already [had] been) almost beyond recognition'.[66] Thus, 'Sa-ranggong', an area named after a long-legged water-bird called the *ranggung* was gradually transmuted into 'Serangoon', 'Kálang Púding' (the latter word referring to the garden croton, a shrub with variegated leaves) evolved into 'Kalang Pudding', and 'T'mpenis' (referring to the name of the Riau ironwood tree) became at first 'Tempenis' and later 'Tampines'.

Chinese Street-names and Place-names

Unlike the official municipal street-names assigned and decreed by law, Asian names for streets and places developed informally within certain conventions and parameters set by each community. Practices which give places in the city names and meanings were likely to be partly idiosyncratic and partly socialized.[67] Asian place-names often differed not only from those assigned by the authorities, but also between ethnic and dialect groups, and occasionally within each group there could be more than one name for a particular place.[68] As names which had evolved spontaneously through usage rather than deliberate assignment, Asian street- and place-names tended to retain a certain plebeian character typical of the local patois. The vast majority of Chinese street- and place-names (65 per cent of those listed by Firmstone; see Figure 6.1) utilized material symbols and landmarks which formed an integral part of daily lived experience as their terms of reference. In contrast to symbols of military significance (Cantonment Road) or which encapsulated the capitalist business ethos of the city (Commercial Square) favoured in municipal street-names, Chinese street-names possessed a more prosaic character. For the Chinese, temples, theatres, markets, gambling dens, bridges, and wells lent their names to streets and places as exemplified by 'Kwong Fuk Min' (Cantonese name for Lavender Street) meaning 'Kwong Fuk Temple street', 'Ngau Chhe Shui Hei Yun Kai' (Cantonese name for Smith Street) meaning 'theatre street in Kreta Ayer',[69] 'Sin Pa Sat Pin' (Ellenborough Street) meaning 'beside the new market [Ellenborough Market]', 'Kiau Keng Khau' (Church Street) meaning 'the mouth of the gambling houses',[70] 'Ang Kio Thau' (one of several Hokkien names for Thomson Road meaning 'head of the red bridge'), and 'Tai Cheng Keak' (Cantonese name for Kampong Glam beach meaning 'foot of the big well').[71] Inscribed into Chinese street-names were a wide variety of material symbols and edifices of significance to the daily practice of urban life of the Chinese communities. Certain Chinese street-names also provided clues to the 'hidden' dimensions of everyday life for they signified the territorial boundaries of different groups within Chinese society. Streets such as 'Ghi Hok Koi' (Carpenter Street), 'Ghi Hin Koi' (China Street), 'Hai San Koi' (Upper Cross Street), and 'Siong Pek Koi' (Nankin Street) were named after the headquarters of major secret societies, which might well have implied that these streets were at one time the operational territories of their eponymous secret societies.[72]

The close association between the signification of places and the local, daily life of the plebeian classes is also evident in another group of Chinese street-names which comprised those denoting particular trading, artisanal, and agricultural activities. These formed about 7 per cent of Chinese street-names listed by Firmstone and included certain streets well-known for particular trade specializations such as 'Tau Hu Koi' (Upper Chin Chew Street) meaning 'bean curd street', so-named

because of the number of bean curd manufacturers located along this
street; 'I Sion Koi' (Pekin Street) literally 'clothing-box street' on
account of the large number of cabinet-makers living here; 'Sio Po
Phah Thit Koi' (Sultan Gate) meaning 'blacksmiths' street in "small
town"';[73] and 'Bih Lang Koi' (Lorong Teluk) meaning 'bamboo basket
street'. More macabre though equally essential preoccupations in
street-naming were to be found in the Chinese aliases for Macao Street
('Kuan Chha Tiam Koi' translated as 'coffinshop street') and Sago
Lane ('Sey Yun Kai', a Cantonese name meaning 'street of the dead'
on account of the preponderance of 'death houses' where the chronic
sick and dying waited out their days as well as the number of shops
selling funeral paraphernalia existing side by side in this lane).
Agricultural pursuits also featured in names such as 'Chhai Hng Lai'
(Lavender Street) meaning 'within the vegetable gardens', and 'Eng
Chhai Ti' (Jalan Penang), translated as 'ground where a vegetable
called "eng chhai" is planted'.

The plebeian character of Chinese street-names is also evident in
another category of names which were purely descriptive or directional
in content (15 per cent of Firmstone's list, the second largest category;
see Figure 6.1). These included names which indicated the location of
the street by specifying a general area or proximity to another street
such as 'Go Cho Lut Bo Bue Hang' (meaning 'blind alley off Rochor
Road', a name referring to the Bernard Street–Farquhar Street–Carnie
Street area), 'Kam Kong Ka Poh Huen (or Toa) Koi' (meaning
Kampong Kapur cross (or big) street, the Chinese equivalent of
Dunlop Street) and 'Kam Kong Ma Lak Kah Bue Tiau Koi' (meaning
'the end street in Kampong Malacca', the Chinese name for Solomon
Street). Other names in this category which developed in the late nine-
teenth century identified streets on the basis of the number of buildings
originally erected along a particular street. Examples included 'Ji Chap
Keng' (Jalan Sultan) meaning 'twenty buildings'; 'Chap San Kang' and
'Chap Peh Keng' (both names referring to Boat Quay) meaning 'thir-
teen shops' and 'eighteen houses' respectively; and 'Pelt Keng A'
(Cheang Hong Lim Lane) meaning 'eight small buildings', a not altogether inappropriate name for a very short street. The numerical mode
of description was also encountered in the names of main thoroughfares on both sides of the Singapore River: 'Tai Ma Lo' ('great horse[-
carriage] way') and 'Ji Ma Lo' ('second horse[-carriage] way') were
Cantonese names for South Bridge Road (Plate 15) and New Bridge
Road on the south side as well as North Bridge Road and Victoria
Street on the north side. Other descriptive street-names owed their
origins to the physical structure of the streets: among these were names
like 'Gu Kak Hang' (Cheng Cheok Street) meaning 'ox-horn lane', an
apt name for a crescent-shaped street leading off from and back to
Tanjong Pagar Road, and 'Tan Pin Kai' (Cantonese name for North
Canal Road as well as Upper Macao Street) literally meaning 'one-
sided street', so-called because houses were only built on one side of
the street.

In contrast to municipal street-names which mainly served as memorials to important personages, the naming of places within the Chinese communities seldom performed commemorative functions. Only 5 per cent of Chinese street-names (as opposed to 65 per cent of municipal street-names; see Figure 6.1) listed by Firmstone honoured particular eminent persons. Of these, none were Europeans and where a particular street was called after an Asian personage (or his *chop*), it was normally because of the proximity of the street to a material feature (such as a house or a shop) belonging to that personage rather than as a memorial to his contributions to Singapore society. Examples included 'Seng Po Toa Chhu Au' (Armenian Street) meaning 'behind Seng Poh's big house', 'Chin Seng Chhu Pin' (Coleman Street) meaning 'beside Chin Seng's house' and 'Heng Long Kai' (Robinson Road) meaning 'street where *chop* "Heng Long" is located'.

In general, there was little direct correspondence between the content of municipal and Asian street-names. A street dedicated to British royalty such as Albert Street, for example, had several much less august aliases: to the Chinese, it was known either as 'Bo Moan Koi' ('the street where sesamum oil is pressed') or 'Mang Ku Lu Seng Ong Kong' (Bencoolen street district joss house) whilst to Tamil-speaking Indians, it became 'Thimiri Thirdal' ('place where people tread fire', referring to the fire-walking ceremony held during the Thaipusam festival in this street).[74] Neither was there significant correspondence between the English and Chinese phonetic pronunciation of place-names. According to Firmstone's list, less than 5 per cent of municipal street-names were known among the Chinese in their transliterated form (Figure 6.1). Among the few where transliterations had become part of the Chinese vocabulary of street-names such as Cecil Street (known to the Cantonese as 'Si Shü Kai'), Bencoolen Street ('Mang Ku Lu'), Craig Road ('Ka Lek Lut'), Duxton Road ('Tok Sun Lut'), Robinson Road (called 'Lo Man San Kai' by the Cantonese) and Short Street ('So Si Tek Hang'), distortions in pronunciation when transposed into a different tongue were inevitable.[75] Furthermore, transliterated names were usually supplemented by other alternative means of referring to particular streets. Firmstone himself noted that whilst 'Lo Man San' (Robinson) was 'quite Chinese in sound', there was no guarantee that 'the name [would be] intelligible to the ordinary Cantonese-speaking Chinaman' and that 'a very long rigmarole' would be necessary to interpret 'Robinson Road' clearly.[76]

The contrast between municipal and Asian place-names, however, went beyond differences in etymological content and phonetics. They also represented different ways of signifying the built environment. In contradistinction to municipal street-names whose primary significance was to identify the urban landscape with the authorities' notions of civic progress and racial ordering in the city, Asian street-names were devoid of such ambitions. Instead, Chinese street nomenclature was strongly anchored to local features, symbols, and activities which formed a significant part of quotidian experience. This was evident in the preponderance of Chinese street-names which embody daily activities and local landmarks, and was also indicated by the fact that whereas over 10 per cent of municipal street names were inspired by the names of non-local (often British) places, Chinese names for Singapore streets reflected purely domestic influences (Figure 6.1). This tendency appeared to be characteristic not only of Chinese place-names in Singapore: in his study of place-names in China, J. E. Spencer notes that 'although Chinese names indicate both domestic cultural and geographical influences, they almost never indicate cultural influence from other parts of the world.[77] In contrast, municipal place-names aimed at transcending parochial influences and projecting a representation of the landscape which accorded with the municipal vision of a vital, progressive outpost of the British Empire. In summary, whilst Chinese place-names tended to match the 'use' values of the streets to which they were attached, municipal street-names attained a level of signification which conveyed meaning over and above the immediate material functions of the streets themselves.

Another major difference in the way in which the municipal authorities and the Asian people signified place lay in the precision with which the urban built environment was differentiated and defined according to a system of names. Municipal street-names identified and defined exact, clearly bounded streets whilst Asian street-names tended to specify general locations relative to particular landmarks or distinguished by the presence of certain activities. For the authorities, clearly labelled and defined streets were crucial to the colonial enterprise of governing and policing the city. The naming of streets and places, coupled with numbering houses, was one of the means of organizing information about the built environment for such purposes. For the Asian communities, however, the naming of places was to serve as signposts to daily activities: as such, the signification of places was inseparable from the substance of everyday social practice but did not necessarily require the delineation of precise streets.

The Contest for the Meaning of the
Urban Built Environment

In Singapore, whilst colonialism established a network of official street- and place-names reflecting the mental images of the dominant culture, names given by indigenous or other cultures continued to persist beneath the surface. The two systems of place-names had little common ground: the full English meaning and nuances of municipal street-names was not comprehended in the Chinese cultural repertory and vice versa. The language of place not only facilitates the interpretation of the mental images of various cultural groups, but where the naming process differs between groups, it also provides a valuable tool in understanding contrasting representations and uses of the colonial landscape and its physical artefacts. The capacity of Asian communities to develop and utilize their own names and signifiers to denote and differentiate various parts of the built environment also implied that each community had a certain latitude in bringing a diversity of socio-cultural influences to bear on the urban landscape. As seen, practices which conferred names and meaning to the urban built environment were not entirely idiosyncratic, but dependent on Asian perceptions and uses of particular places in terms of their associations with particular types of socio-economic activities, religious or symbolic sites, and territorial gangs. The persistence of different systems of signification within the city implied that municipal representations of the landscape did not command an unchallenged hegemony. Instead, it testified to the syncretic character of the city and belied the municipal vision of a unitary city divided into racial containers but well-integrated by a single overarching system of authority and its signification.

In colonial Singapore, municipal street- and place-names were seldom comprehended, and frequently ignored by the Asian communities. Haughton observed in 1891 that 'the names given by the municipality to the various streets [were] only used by the European portion of the population, and the Chinese, Tamils and Malays [had] names for the streets very different from their Municipal titles'.[78] Although successive town committees had selected what had been perceived as suitably 'ethnic' names such as 'Hongkong Street' and 'Macao Street' on the one hand and 'Jalan Sultan' on the other for Chinese and Malay areas respectively, 'the fact remain[ed] that Municipal names [were] ignored by the natives, with the exception of the police, who [were], of course, compelled to learn them'.[79] Colonial and municipal authorities despaired at what was seen as the haphazard and imprecise manner in which Asians identified places and furnished addresses. Firmstone remarked:

It is characteristic of the Chinese that in identifying streets, accuracy is the last thing that strikes them as essential. If you ask a Chinaman—or better still a Chinese woman—newly arrived and resident on Singapore, where he lives, the invariable answer will be 'Singapore'. A second query will perhaps elicit information as to the district of the town or island, but it will take many questions before the actual address can be ascertained, though it might have been given directly, if the person questioned had thought that it was of any importance.[80]

Not only do the Chinese have their own sets of street- and place-names, their 'happy-go-lucky way of using one expression to describe any one of perhaps a dozen streets' was from the European perspective a frustrating tendency which defied accurate identification of addresses.[81] The name 'Tek Kah' or 'Foot of the Bamboos',[82] for example, described an ambiguously defined territory at the town end of Bukit Timah Road and included Albert Street, Selegie Road, Short Street, and any of the numerous lanes in that neighbourhood. According to Haughton, the Malays carried this tendency of ambiguous identification even further, for they '[took] but little notice of streets, and as a rule, only describe[d] places by *kampungs*'.[83] Asian street-names also tended to be incomplete and uneven in their coverage of the town, abounding in areas occupied by or associated with their own communities and sparse in others. Unlike municipal street-names which were literally invented at committee meetings and officially assigned to particular streets from specified dates, Asian place-names developed through an informal process of evolution as initially name-less streets and thoroughfares grew in importance and function. As such, their multiplication was both unsystematic and uncharted, becoming known to members of the community who lived in or used the streets but often impervious to the authorities, including govern-ment interpreters who remained 'lamentably ignorant' of Asian desig-nations of place.[84] Asian disregard for municipal names for places and the corresponding ignorance on the part of the authorities of Asian place-names were of practical significance for those involved in govern-ing and policing the city. Often, addresses could not be ascertained with any degree of accuracy, whether in the case of instituting arrests, serving court summons, or tracing the spread of infectious diseases. The latter, for example, was often frustrated by either the deliberate falsification of addresses where victims of infectious diseases had stayed on the part of relatives and friends, or the inability or reluctance of Asians to furnish accurate addresses.[85] The non-comprehension and non-acceptance of municipally assigned street- and place-names and the use of their own alternative systems of place signification thus ren-dered the Asian population less open to the surveillance strategies of the colonial state.

Municipal attempts to enhance the acceptability and usage of muni-cipal street- and place-names amongst the various Asian communities were limited by various difficulties. In 1912, municipal commissioner Dr Yin Suat Chuan proposed that Chinese and Malay characters cor-responding phonetically to the existing English names should be added to the name-plates of the roadways in town as a means of pop-ularizing municipal place-names.[86] Although the proposal was unani-mously agreed to by the Municipal Board, it was left in abeyance till 1921 when the issue was again raised by another Chinese commis-sioner, Wong Siew Qui, who urged that immediate steps be taken to put the 1912 resolution into effect.[87] The municipal president, R. J. Farrer, explained that though the resolution had not been formally rescinded, the matter had been dropped because of the high cost involved in the project[88] and also the difficulty of expressing the names of many of the roads phonetically.[89] When pressed further by Asian commissioners and the Mohammedan Advisory Board to add Asian characters to street-name plates, Farrer expressed his scepti-cism as to the usefulness of Asian name-plates, arguing in the case of Malay names that the addition would only benefit 'an infinitesimal section of the population' as there were 'very few inhabitants of the town who [could] read Malay (Arabic) characters but [were] unable to read Roman characters'.[90] This was refuted by one of the Malay commissioners, Che Yunus bin Abdullah, who argued that Malay characters would be extremely useful and would be of especially 'great assistance to Malay policemen who [were] unable to read the Roman characters'.[91]

What prompted Asian community leaders to press for the addition of Asian names to the streets of the city appears to have been a desire to stake their communities' claims over the landscape. Dr H. S. Moonshi, for example, prefaced his arguments for the addition of Malay characters by reminding the Board of what he saw as the basic Malay character of Singapore. He argued that 'as Singapore is a Malay country and the prevalent language is Malay, Municipal Commissioners should add Malay characters in the new street-name plates to be put up'.[92] In the same way, for the Chinese municipal commissioners, the addition of Chinese names represented a means of impressing on the built environment the Chinese character of the city. Although the requests for Asian street-names to be put up was ultimately turned down on grounds of the need for financial stringency, they indicated that the contest for the signification of the urban built environment was also expressed at the level of discourse.

As a process, the naming of places is hence not the simple prerogative of the municipal authorities but is contingent on 'the social relations of deference and defiance'[93] inherent in society. Whilst the authorities had the power of selecting what were considered appropriate names and formally assigning them to the streets of the city, the Asian communities comprised the social milieu which retained the power over whether these names would be adopted. The existence of alternative Asian systems of place signification in daily use implied that there were competing representations of the urban landscape rather than a single municipally imposed image. The naming of places was also an important element of the colonial enterprise of government and surveillance; the failure to impose and enforce the adoption of one uniform system of naming places was hence in part a reflection of the fact that colonial power was not absolute and was often countered in the creative spaces of everyday life.

1. Joseph E. Spencer, 'Chinese Place Names and the Appreciation of Geographic Realities', *Geographical Review*, 31 (1941): 71.

2. David E. Sopher, 'The Structuring of Space in Place Names and Words for Place', in David Ley and Marwyn S. Samuels (eds.), *Humanistic Geography: Prospects and Problems*, London: Croom Helm, 1978, p. 252.

3. Donald K. Emmerson, '"Southeast Asia": What's in a Name?', *Journal of Southeast Asian Studies*, 15 (1984): 4. This observation is based on Alice's encounter with Humpty Dumpty who believed that words are not rooted in the nature of things but mean just what he chooses them to mean (Lewis Carroll, *The Philosopher's Alice: Alice's Adventures in Wonderland and Through the Looking-Glass*, annotated by Peter Heath, London: Academy Editions, 1974, pp. 192–4).

4. Raffles to Town Committee, 4 Nov 1822 (reproduced in C. B. Buckley, *An Anecdotal History of Old Times in Singapore*, Singapore: Oxford University Press, 1984 (first published in 1902), p. 84.

5. J. A. Harwood, *The Acts and Ordinances of the Legislative Council of the Straits Settlements from 1st April 1867 to 1st June 1886*, Vol. 2, London: Eyre and Spottiswoode, 1886, pp. 1263–4.

6. MPMCOM, 13 December 1907 and 17 January 1908; ARSM 1921, Assessment Department, p. 3A; SIT, 651/26, 14 June 1926: 'Name Given to a Private Road and also to a Kampong off Grange Road'; SIT 796/26, 14 January 1926: 'Names Given to Three Private Roads and One Kampong off Bukit Timah Road'; SIT 893/29, 11 June 1929: 'Area Served by a Small Lane Near Newton Road to Be Named as a Kampong'.

7. 'Municipal Minutes, 8 March 1858' (reproduced in Buckley, *An Anecdotal History*, p. 667.

8. Ibid. Places named at the municipal meeting included 'Dalhousie Canal', 'Havelock Road', 'Neil Road', 'Outram Road', 'Trafalgar Square', 'Albert Street' and 'Lavender Street'. 'Tavern Street' and 'Commercial Square' were renamed 'Bonham Street' and 'Raffles Square'. Names such as 'Church Street', 'Flint Street' and 'Market Street' on the north side of the Singapore River which were duplicated on the south side were changed to 'Waterloo Street', 'Prinsep Street' and 'Crawfurd Street'. The Indian Mutiny of 1857 probably triggered off the preponderance of names associated with British war victories—Havelock, Neil, and Outram were heroes of the Mutiny while Trafalgar and Waterloo were recollections of famous battles.

9. MPMCOM, 28 February 1908. Only D'Almeida Street off Raffles Place was retained in commemoration of Jozé d'Almeida, a Portuguese surgeon and one of the most respected inhabitants of early nineteenth-century Singapore (S. Durai Raja-Singam, *Malayan Street Names: What They Mean and Whom They Commemorate*, Ipoh: Mercantile Press, 1939, pp. 96–7). In another case in 1914, the municipal assessor pointed out the confusion and liability to delay in cases of fire due to the similarity of street-names commemorating the well-known opium and spirit farmer and property owner, Cheang Hong Lim and members of his family. Hare Street, Covent Garden, Covent Row, Covent Alley, Covent Street, Taipeng Road, and Calcutta Street were eventually substituted for Cheang Hong Lim Lane, Cheang Hong Lim Market, Cheng Hong Lim Lane, Cheang Jim Kheng Street, Cheang Jim Chuan Street, Cheang Wan Seng Road, and Cheang Jim Hean Street. Hong Lim Quay was also renumbered as part of Boat Quay (MMFGPSYC, 24 July 1914; MPMCOM, 31 July 1914). In the same year, Syed Ali Road was renamed Newton Road (after Howard Newton, the assistant municipal engineer in the late nineteenth century) to avoid confusion with Syed Alwi Road (MMFGPSYC, 16 October 1914).

10. David Ley, 'Styles of the Times: Liberal and Neo-conservative Landscapes in Inner Vancouver, 1968–1986', *Journal of Historical Geography*, 13 (1987): 41; James S. Duncan, *The City as Text: The Politics of Landscape in the Kandyan Kingdom*, Cambridge: Cambridge University Press, 1990.

11. Wilbur Zelinsky, 'Nationalism in the American Place-name Cover', *Names*, 31 (1983): 1–28; R. W. Stump, 'Toponymic Commemoration of National Figures: The Case of Kennedy and King', *Names*, 36 (1988): 203–16.

12. Alisdair Rogers, for example, interprets the dedication of a boulevard, a school and a vest-pocket park in a mixed ethnic neighbourhood in Los Angeles after the Black Civil Rights leader, Martin Luther King Jr., as the assertion of Black identity in the face of Latino in-migration into the area, redevelopment and gentrification (Alisdair Rogers, 'Cinco de Mayo and the 15th January: Contrasting Situations in a Mixed Ethnic Neighbourhood', in Alisdair Rogers and Steven Vertovec (eds.), *The Urban Context: Ethnicity, Social Networks and Situational Analysis*, London: Berg, 1995, pp. 117–40

13. Susan J. Lewandowski, 'The Built Environment and Cultural Symbolism in Post-colonial Madras', in John A. Agnew, John Mercer, and David E. Sopher (eds.), *The City in Cultural Context*, Boston: Allen and Unwin, 1984, pp. 237–54.

14. Eilert Ekwall, *Street-Names of the City of London*, Oxford: Clarendon Press, 1954, pp. 49–57.

15. Hsieh Chiao-Min, 'Sequent Occupance and Place Names', in Ronald G. Knapp (ed.), *China's Island Frontier: Studies in the Historical Geography of Taiwan*, Honolulu: University Press of Hawaii, 1980, pp. 107–14.

16. Mak Lau Fong, *The Sociology of Secret Societies: A Study of Chinese Secret Societies in Singapore and Peninsula Malaysia*, Kuala Lumpur: Oxford University Press, 1981, p. 81.

17. N. M. Holmer, 'Indian Placenames in South America and the Antilles', *Names*, 8 (1960): 133–49; Voon Phin-Keong, 'The Origins of Chinese Place Names', *Geographica*, 5 (1969): 34–47; David Robinson, 'The Language and Significance of Place in Latin America', in John A. Agnew and James S. Duncan (eds.), *The Power of Place: Bringing Together Geographical and Sociological Imagination*, Boston: Unwin Hyman, 1989, pp. 157–84.

18. Robinson, 'The Language and Significance of Place', p. 160.

19. Ibid.

20. Joel Richman, *Traffic Wardens: An Ethnography of Street Administration*, Manchester: Manchester University Press, 1983, p. 16.

21. Gordon H. Pirie, 'Letters, Words, Worlds: The Naming of Soweto', *African Studies*, 43 (1984): 43.

22. In the 1920s, with the increasing complexity of municipal affairs and the necessity for a division of labour amongst commissioners, the task of deliberating on street-names was assigned to a special committee. Minute papers, normally containing the municipal assessor's suggestions, were first circulated to the municipal president, committee members, and occasionally heads of various government departments for opinions to be registered before coming before the committee.

23. There were, however, a few instances where a name which had developed informally among the populace became formally recognized as the official name. In 1897, for example, on the question of naming a new street running from Cecil Street to Telok Ayer Street, the suggested name 'Jubilee Lane' was discarded in favour of 'Guthrie Lane', the name by which the street was already known among the Chinese (MPMCOM, 29 July 1897 and 12 August 1897).

24. Pirie, 'Letters, Words, Worlds', p. 51.

25. Similar themes prevailed in other parts of the British Empire (see A. J. Christopher, *The British Empire at Its Zenith*, London: Croom Helm, 1988, pp. 230–4).

26. H. W. Firmstone, 'Chinese Names of Streets and Places in Singapore and the Malay Peninsula', *Journal of the Straits Branch of the Royal Asiatic Society*, 42 (1905): 53–208. While not a complete listing of existing streets, Firmstone's list includes most of the major streets and thoroughfares in early twentieth-century Singapore. My own compilation of street-names from contemporary records and maps suggests that there were about 700 municipal street-names in existence by the 1920s and that Firmstone's list represents about one-third of the total.

27. From 1 January 1925, 'Macao Street' and 'Upper Macao Street' (where the original Chinese Protectorate was located) were renamed 'Pickering Street' and 'Upper Pickering Street' so that 'Pickering's name should not be forgotten' (SIT 598/24, 18 June 1924: 'Change of Names of Streets').

28. These include Broadrick Road, Gentle Road, Farrer Road, Hallifax Road, Peel Road, and Rowell Road named after municipal presidents; Carpmael Road, Hooper Road, and Peirce Road named after municipal officers; Anguilla Road, Boon Keng Road, Boon Tat Road, Cheng Tuan Road, Choon Guan Road, Croucher Road, Fowlie Road, Fraser Street, Jalan Eunos, Jiak Kim Road, Keng Chin Street, Keng Lee Road, Keong Saik Road, Kheam Hock Road, Kim Ching Road, Koek Road, Liang Seah Road, McKerrow Road, Manasseh Lane, Meyer Road, Nanson Road, Robertson Quay, Shelford Road, Sime Road, Sohst Road (changed to Mount Rosie Road in 1920 in commemoration of T. Sohst's wife), Teck Lim Road, Tessensohn Road, and Veerasamy Road named after municipal commissioners; and Cuscaden Road, Dunlop Street, Hare Street, McCallum Street, McNair Road, Pennefather Road, and Saunders Road named after government officers who served on the Board as official members. Some

commissioners were part of prominent Straits families whose names were already publicly recognized in street-names such as Braddell Road, Crane Road, Dunman Street, and Scotts Road.

29. This was the second time this street had been renamed. Formerly called 'Inche Lane' , it was renamed 'Jalan Tambah' when the Kampong Kapor roads were taken over as public streets in 1910. It was speculated that the street was so named because it stood above the level of the surrounding land (*tambah*, a Malay word, means 'increase' or 'addition' while 'jalan' means 'road' or 'way') (*SIT*, 451/27, 2 June 1927: '"Jalan Tambah" to be Re-named "Veerasamy Road"').

30. *Straits Times*, 29 January 1927.

31. SIT, 535/30, 4 June 1930: 'Road Names for Tiong Bahru'.

32. This was a new road formed in the 1890s and named after the Norris family (Richard Owen Norris and George Norris) who originally acquired the site from the East India Company in the 1830s (Raja-Singam, *Malayan Street Names*, p. 125; MPMCOM, 9 September 1897).

33. This was a new road opened by Cheang Chin Nam, a well-known dentist who practised in partnership with his brother Chin Heng as Cheong Brothers. They were also general merchants as well as land and estate owners (Raja-Singam, *Malayan Street Names*, p. 92; MPMCOM, 27 January 1905).

34. This was a private road off Victoria Street on which Lee Cheng Yan, a well-known commission agent and general trader with substantial dealings with Europeans, erected ten houses (MPMCOM, 12 January 1906).

35. Originally called 'Tampenis Road', this road which ran from Robertson Quay to River Valley Road was renamed at the request of Tan Chay Yan after his millionaire father Tan Teck Guan who once owned land in the area. The municipal commissioners initially demurred, but later agreed to the change in view of the fact that there were two roads in Singapore bearing the name 'Tampenis Road' (MPMCOM, 5 July 1907; 23 August 1907).

36. In 1910, Yow Ngan Pan, a Cantonese *towkay* and community leader gave up a strip of land for a new road to be built off Cantonment Road on the understanding that the road would be named after him (MMFGPC, 6 May 1910).

37. Originally part of Wayang Street, this road was rebuilt in 1919 by Eu Tong Sen, a wealthy tin and rubber magnate and property owner and renamed after him (MGCM, 1 August 1919).

38. *The New Atlas and Commercial Gazetteer of the Straits Settlements and the Federated Malay States*, Shanghai: Far Eastern Geographical Establishment, 1917, p. 24.

39. The third sister settlement, Penang, found its place amidst Singapore's street nomenclature in 1906 when a new loop road from Tank Road to the railway bridge at Fort Canning Road was named (MPMCOM, 4 May 1906).

40. In 1898, in assigning names to new streets laid out on either side of Anson Road near Tanjong Pagar, it was resolved by the municipal commissioners 'to use names of rivers and districts in the Malay Peninsula as being better adapted to the purpose than the names of persons or families' (ARSM, 1898, p. 14). The last four Malayan street-names listed here dated from this resolution.

41. SIT, 892/29, 8 June 1929: 'Naming of Road Running from Balestier Road to Mandalay Road'.

42. G. M. Reith, *1907 Handbook to Singapore*, revised by Walter Makepeace, Singapore: Oxford University Press, 1986 (first published in 1892), p. 35.

43. Chatsworth Road was originally named after the residence built by G. G. Nicol ('The Stranger's Guide to the Environs of Singapore, [Map of] Singapore Residency', J. Moniot, Surveyor-General, S.S., undated [c.1860s]). The name was obviously borrowed from that of the Duke of Devonshire's country house in Derbyshire, England. In 1929, when a private road off Chatsworth Road came up for naming, one of the commissioners E. A. Brown suggested using road-names associated with the Duke's country house such as 'Rowsley Road', 'Bakewell Road', 'Haddon Road', and 'Edensor Road'. The suggestion was turned down by the municipal president, R. J. Farrer who favoured 'simple names where possible as Singapore was not England' and 'the generality of

persons [in Singapore] [found] names like Rowsley difficult'. The road was eventually named 'Davie Road', but this was changed to Chatsworth Park two years later (*SIT* 876/29, 19 March 1929: 'Private Road Running from Bishopsgate to be Named "Chatsworth Park"').

44. Lee Kip Lin, *The Singapore House, 1819–1942*, Singapore: Times Editions, 1988, pp. 26–7.

45. In 'The Stranger's Guide to the Environs of Singapore', 'Dalvey', 'Emerald Hill' 'Nassim's Lodge', 'Irwell Bank', 'Leonie Hill', 'Orange Grove' and 'Killiney Bungalow' appear as private residences owned by A. L. Johnston & Co., W. Cuppage, Nassim, Boustead & Co., T. Campbell, Charles Carnie, and Thomas Oxley respectively. 'Cairnhill' was named after Charles Carnie, a businessman who built the first house and also owned a large nutmeg plantation on the hill in the 1840s (Buckley, *An Anecdotal History*, p. 406; Lee Kip Lin, *The Singapore House*, p. 26). 'Oxley Road' was named after Thomas Oxley, the senior surgeon of the Straits Settlements in the 1840s and 1850s, who owned a large estate called Oxley's Estate bounded by Orchard Road, Grange Road, Tank Road, and River Valley Road (Buckley, *An Anecdotal History*, p. 405). 'Scotts Road' was named after Captain William Scott, harbour master in the 1830s, who owned property in the locality (Raja-Singam, *Malayan Street Names*, p. 137).

46. MPMCOM, 7 October 1927 and 29 October 1927.

47. MPMCOM, 30 January 1929.

48. Amidst much laughter, Laycock also suggested that roads in the neighbourhood of Lavender Street 'might appropriately be named after flowers, such as "Rosemary Road" and "Thyme Road"'. The name 'Lavender Street' was already a standing joke for it was flanked on both sides by Chinese vegetable gardens and was reputedly one of the foulest smelling roads in the city (MPMCOM, 30 January 1929).

49. Raja-Singam, *Malayan Street Names*, pp. 85–6.

50. ARSM, 1921, Assessment Department, p. 3A.

51. 'Order signed by N. L. Hull, Acting Secretary, 1822' (reproduced in Buckley, *An Anecdotal History*, p. 81).

52. 'Raffles to Town Committee, 4 November 1822' (reproduced in Buckley, *An Anecdotal History*, p. 83).

53. 'Map of the Town and Environs of Singapore', drawn by J. B. Tassin, from an Actual Survey by G. D. Coleman, Calcutta, 1836.

54. John Crawfurd, *A Descriptive Dictionary of the Indian Islands and Adjacent Countries*, London: Bradbury and Evans, 1856, p. 198. In nineteenth-century Singapore, 'Orang Kling' was a bazaar Malay term which embraced southern Hindu Indians of all classes. After the Great War, however, the leaders of the Indian community, and its English-educated members in particular, felt that the street-name, 'Kling Street' had taken on a degrading connotation which it had not had in the past, since the only class of people referred to as 'Klings' after the war were the lowly Indian coolies (George L. Peet, *Rickshaw Reporter*, Singapore: Eastern University Press, 1985, p. 110). In 1918, Rev J. A. B. Coah wrote to the municipal commissioners suggesting that the name 'Kling Street' be changed as it was 'a cause of irritation to Indians' but the commissioners saw no reason to discard what they felt was an historical name (MGCM, 19 July 1918). The commissioners however yielded to increasing pressure to comply with the feelings of the Indian community three years later, and 'Kling Street' was rechristened 'Chuliah Street' on the motion of Dr H. S. Moonshi, one of the Indian members of the Board (MPMCOM, 24 Jun 1921). Under its new name, 'Chuliah Street' remained very much an Indian street, for it was the focus of a community of professional moneylenders from Madras known locally as the chetties (Peet, *Rickshaw Reporter*, p. 112).

55. SIT, 222/25, 9 April 1925: 'New Road off Serangoon Road Leading to a Government Rehousing Site to be Named'.

56. This bazaar Malay phrase literally means 'people who were thrown out of Bombay'.

57. 'New Road off Serangoon Road Leading to a Government Rehousing Site to be Named'.

58. SIT, 266/25, 28 March 1925: 'Selection of a Name for the New Road from Kreta Ayer to Anson Road'.

59. Lee Choon Guan (1868–1924) was highly active in the public life of the city. He was a municipal commissioner for five years, a Justice of Peace, president of the Straits Chinese Recreation Club for some years, a member of the Chinese Advisory Board and the Tan Tock Seng Hospital Management Committee, and a director of several commercial companies (Song Ong Siang, *One Hundred Years' History of the Chinese in Singapore*, Singapore: Oxford University Press, 1984 (first published in 1902), pp. 111–2).

60. Leong Man Sau was a native of Canton who came to Singapore at the age of twelve. Like Lee Choon Guan, he was a municipal commissioner and a member of the Chinese Advisory Board. He also served as the secretary of the Kwong Wai Sui Society, a Cantonese institution established for the purpose of controlling and managing their own schools and burial grounds (Song, *One Hundred Years' History*, pp. 432–3).

61. SIT, 266/25, 28 March 1925. The Maxwells were a well-known family which had seen three generations of service in Malaya. Sir Peter Benson Maxwell was Singapore's first Chief Justice after the 1867 Transfer. His second son, William Edward Maxwell, held several positions in the civil service, including that of Colonial Secretary and commissioner of land titles. William's brother, Robert Walter, joined the Straits Settlements Police Force and rose to the rank of Inspector-General of Police. A third-generation Maxwell, William George, also held multifarious government positions and was particularly well known in the Singapore context as the President of the Singapore Housing Commission (Walter Makepeace, 'Concerning Known Persons' in Walter Makepeace, G. E. Brooke, and R. S. J. Braddell (eds.), *One Hundred Years of Singapore*, Vol. 2, London: John Murray, 1921, pp. 431–42).

62. This estate, located in Kampong Glam, originated as land set apart in 1819 by Raffles for Sultan Hussein of Johore and his family. When Sultan Hussein died, he was succeeded by his elder son Sultan Ali who assumed legal right to the land. The estate became the subject of considerable litigation between descendants of Sultan Ali towards the close of the nineteenth century (PLCSS, 25 March 1902, pp. B70–B71).

63. MPMCOM, 1 October 1909.

64. Commercial Square was renamed 'Raffles Place' in 1858 in honour of the Settlement's founder (Buckley, *An Anecdotal History*, p. 667).

65. It goes without saying that where Chinese (Hokkien, Cantonese, or other dialects) personal names were used in street nomenclature, they were Anglicized transcriptions or romanizations of Chinese sounds. There were many irregularities in phonetic transcription, not least of all being the omission of all indication of phonetic aspirates.

66. H. T. Haughton, 'Notes of Names of Places in the Island of Singapore and its Vicinity', *Journal of the Straits Branch of the Royal Asiatic Society*, 20 (1889): 75.

67. Anthony P. Cohen, *The Symbolic Construction of Community*, London: Tavistock, 1985, p. 74.

68 The 225 municipal street-names listed by Firmstone had the equivalent of a total of 365 different Chinese (Hokkien and Cantonese) names. Using this list as a rough guide, there were an average of 1.6 Chinese names for each municipal name.

69. This street is so named by the Cantonese because of the location of the Lai Chun Yuen Theatre at 36 Smith Street. The theatre was particularly popular with the Cantonese community and exerted a strong influence in the naming of surrounding streets. Besides Smith Street itself, Trengganu Street became known as 'Hei Yun Wang Kai' ('street at the side of the theatre'), Temple Street as 'Hei Yun Hau Kai' ('street at the back of the theatre'), and Sago Street as 'Hei Yun Chhin Kai' ('street in front of the theatre'). As there were several 'theatre streets' in the city, these streets were differentiated from others by prefixing their names with the term for the district in which they were found, 'Ngau Chhe Shut' (meaning 'water cart'), the literal translation of the Malay name, Kreta Ayer (*Chinatown: An Album of a Singapore Community*, Singapore: Times Books International, 1983, p. 88).

70. The aptness of this name for Church Street was, according to Haughton, one which would 'strike those who [were] acquainted with police work in Singapore' (H. T. Haughton, 'Native Names of Streets in Singapore', in Mubin Sheppard (ed.), *Singapore: 150 Years*, Singapore: Times Books International, 1982 (article first published in 1891), p. 208).

71. Hokkien and Cantonese street-names given here and in the rest of this chapter are taken from Haughton, 'Native Names of Streets', pp. 209–17; and Firmstone, 'Chinese Names of Streets', pp. 53–208. Unless otherwise indicated, street-names quoted here refer to the Hokkien version.

72. Mak, *The Sociology of Secret Societies*, p. 77; *Chinatown: An Album*, p. 48.

73. 'Toa Po' ('big town') and 'Sio Po' ('small town') were frequently used general names for the two main districts on either side of the Singapore River. 'Toa Po' started from South Bridge Road and embraced the whole of Chinatown as far as Tanjong Pagar whilst 'Sio Po' described the area around North Bridge Road.

74. A list of 35 Tamil street-names is given in Haughton, 'Native Names of Streets', pp. 217–19.

75. Many residents of Chinatown continued to live in ignorance of the English name of the street in which they lived, largely because of the difficulties involved in phonetically translating English street-names into the vernacular. Examples of such translations as '"Nor Mee Chee" to mean North Bridge Road and "Sow Mee Chee" to mean South Bridge Road [were] typical of the inability of the Chinese language to give sound to an "r"' (Raja-Singam, *Malayan Street Names*, pp. 14–15).

76. Firmstone, 'Chinese Names of Streets', p. 125.

77. Spencer, 'Chinese Place Names', p. 81.

78. Haughton, 'Native Names of Streets', p. 208.

79. Ibid. Even the acquaintance of the average police constable with municipal street-names was suspect, for according to a leader of the Malay community, few Malay constables were able to read romanized characters (see p. 234.).

80. Firmstone, 'Chinese Names of Streets', p. 206.

81. Ibid.

82. According to A. Knight, an old resident, this area around Selegie Road was so named because part of the road used to be bordered by 'luxuriantly lofty bamboos' which were later removed to make room for houses (A. Knight, 'Chinese Names of Streets', *Journal of the Straits Branch of the Royal Asiatic Society*, 45 (1906): 287).

83. Haughton, 'Native Names', p. 208. This view is extended to the present day by Hans-Dieter Evers in his phenomenological study of differing conceptions of space and the image of the city between Malays and Chinese in Malaysian cities. In contrast to the attention that the Chinese pay to demarcating and differentiating urban space, Evers argues that the Malays have no clearly developed conception of bounded space and clear-cut boundaries in general. Malay areas are normally designated as *kampung*, within which there is little attempt at demarcation or orientation. Instead, there is usually no main street, plaza, or square, but 'only an apparently arbitrary system of winding foot-paths leading from house to house becoming narrower at times or ending in blind alleys' (Hans-Dieter Evers, 'The Culture of Malaysian Urbanization: Malay and Chinese Conceptions of Space', in Peter Chen and Hans-Dieter Evers (eds.), *Studies in Asean Sociology: Urban Society and Social Change*, Singapore: Chopmen Enterprises, 1978, p. 335).

84. Firmstone, 'Chinese Names of Streets', p. 206.

85. ARSM 1896, 'Supplement, Report of a Special Investigation into the System of Death Registration in Force in Singapore', p. 10; ARSM 1902, Health Officer's Department, p. 118; MPMCOM, 25 February 1927.

86. MPMCOM, 21 July 1912.

87. MPMCOM, 24 June 1921.

88. In 1913, when tenders were called from the Crown Agent for the supply of enamelled iron street-name plates, it was found that every set of six plates in English and one native language cost $15.40 and that a third language increased the cost by $12.00 to $27.40 (MMFGPC, 24 January 1913). In 1921, it was estimated that adding Chinese

characters alone to name-plates would amount to about $35,000 (MGCM, 24 June 1921).

89. MPMCOM, 24 June 1921; MMSC3, 8 July 1921.
90. MPMCOM, 29 December 1922.
91. Ibid.
92. Ibid.
93. Pirie, 'Letters, Words, Worlds', 51.

7

The Control of 'Public' Space: Conflicts over the Definition and Use of the Verandah

We believe we are representing European public opinion when we urge upon the authorities, be they municipal or otherwise, not to be turned aside from the measures they conceive necessary for the health and welfare of the community ... and to prevent nuisances on our public streets. British rule and British ideas must be paramount in Singapore—not Chinese mob rule and Chinese ideas.[1]

Bureaucratic methods, which appear to the people to be unjust and unfair, are likely to meet with at least a passive resistance, which will prevent success.[2]

Municipal Perceptions of Order in 'Public' Spaces: The Verandah

BY the end of the nineteenth century, the municipal authorities of Singapore expressed increasing concern that the ethos of *laissez-faire* which had hitherto prevailed in the use of public spaces within the city was threatening to undermine the city's public health, security and social order. The late nineteenth century saw the strengthening of municipal determination to impose standards of acceptable behaviour and permissible activities in public streets and spaces. The municipal commissioners increasingly appropriated the role of guardian of the public arena, taking upon themselves the rights and responsibilities of organizing, maintaining, and regulating public spaces such as streets, verandahs, open spaces, parks, halls, markets, and slaughterhouses in the interest of an abstract entity known as 'the public'. By maintaining an urban environment with uniform, well-demarcated public spaces, the municipal authorities hoped to influence the behaviour of the Asian communities in the public arena towards order and propriety. Those activities which failed to comply with municipal standards of good order and public appropriateness became circumscribed by a battery of legislative measures and subject to the discipline of police and magistrate's courts. In its attempt to impose order on the city's public spaces, the municipal authorities were confronted with two related

tasks: first, they had to identify the boundaries of 'public' space and
sharpen the distinction between what was perceived as 'public' and
'private' space[3] and second, they needed to define what constituted
appropriate forms of public activity as distinguished from the use of
public spaces for unsanctioned purposes.

The attempt to impose on the urban environment an unambiguous
definition of what constituted public space to be appropriated only for
municipally prescribed purposes in colonial Singapore was fore-
shadowed by major changes in the form and texture of nineteenth-
century British cities. The cellular, self-contained worlds of
eighteenth-century England, confined in courts and alleys and segmented by differences in class, trade, occupation, and the general
difficulty and danger of getting from one part of the town to the other,[4]
were gradually broken down in the course of the nineteenth century.
External space outside of private dwellings which had originally been
used in an 'ambiguous semi-public and semi-private' manner was
redefined as socially neutral, 'waste' space or 'connective tissue which
was to be traversed rather than used'.[5] The threshold between private
and public domains became 'less ambiguous and more definite, less
penetrable and more impermeable'.[6] In the new industrial order, with
the rise of suburbs in particular, the pavement, street, and road became
crucial spatial bridges linking home and workplace, as well as the
private and public sectors of society. The equilibrium which had
existed since medieval times between the spatial demands of communications and the social and communal functions of the streets[7] was
gradually destroyed by the increasing pressures of urban transit.[8] Street
life was eclipsed, and public space became 'dead',[9] socially sterile, and
ceased to be recognized as legitimate places for casual or purposeful
activities involving the community. Concomitant with the emergence of
the notion of a socially neutral, rigidly defined public space in British
cities was a movement away from a communal and customary to a
legal definition of acceptable standards of 'decent public behaviour'.
M. J. D. Roberts argues that 'standards, once the concern of "the community" to define and defend on a basis of informal consensus, [were] [in
the early nineteenth century] selectively appropriated by public authorities on grounds of public security and good order, and compliance ...
enforced either by official agents of the state or else by positive encouragement of private citizens to take their business to state tribunals of
adjudication'.[10] The appointment of professional policemen in the
1830s and 1840s further resulted in more stringent control of public
streets and places, and a high success rate in controlling public order
crimes.[11] What emerged was a modern sense of the term 'public
space', one which is 'designed, however minimally'; where 'everyone
has the rights of access'; where 'encounters in it between individual
users are unplanned and unexceptional'; and where 'their behaviour
towards each other is subjected to rules none other than those of
common norms of social civility'.[12]

The attempt to transplant emerging British notions of a legally defined standard of acceptable public behaviour and rigidly circumscribed public spaces into an Asian city such as late nineteenth-century Singapore was fraught with difficulties. This chapter examines in detail one particular element of the urban built environment of colonial Singapore—the verandah—to illustrate some of the tensions which developed over the definition and use of the city's 'public' spaces. The verandah, an open arcade or light-roofed gallery extending along the front of shophouses and tenement dwellings as a continuous walkway[13] became, by the late nineteenth century, the subject of contradictory definitions (municipal or Asian) and competing demands (public, private, or communal). Precisely because it occupied the ambiguous territory between the public street and the private house, it featured prominently in the conflicts between the municipal authorities and the Asian inhabitants in their attempts to assert their own conceptions and use of urban space. These conflicts were a reflection of the larger culture in a colonial context, for 'public space is the common ground where civility and [a] collective sense of what may be called "public-ness" are developed and expressed, ... serv[ing] as a reflection or mirror of individual behaviors, social processes, and ... often conflicting public values'.[14]

Verandah or 'five-foot-ways'[15] as they were popularly called in Singapore, constituted a distinct feature on the narrow street frontage of most terraced shophouses and tenement blocks in colonial Singapore. They were covered walkways, or what one police officer who patrolled the streets of Singapore in the 1920s called 'shadowed pavements, which stretched like cloisters under the jutting upper verandahs of shops and houses'[16]. It was Stamford Raffles, the founder of the town himself who decreed in 1822 that 'each house should have a verandah of a certain depth open at all times as a continued and covered passage on each side of the street'[17] for the accommodation of the public as part of his attempt to secure uniformity, regularity and the efficient use of space in his newly established town. Subsequent town committees and municipal authorities strove to maintain the verandah as an open, unobstructed public space for the convenience of foot passengers seeking protection from the tropical heat, torrential rains and the disorderly congestion of the street traffic;[18] and by means of building by-law controls,[19] ensured the continued existence of the verandah in the physical fabric of the town in the late nineteenth and early twentieth centuries.

Asian Perceptions and Use of the Verandah

While the authorities could ensure the supply of the verandah as a physical artefact, they were often unable to dictate that it be used by the Asian communities in a manner consistent with municipal expectations. Instead, the Asian people perceived the environment afforded by the five-foot-way through their own sets of expectations and priorities, and contrived their own ways of making this element of the environment effective according to their own purposes. What became strongly contested was the municipal definition of the verandah as 'public' circulation space to be strictly and continually kept open for free-flowing pedestrian traffic. The municipality championed public 'right of way' along verandahs as being faithful to the original intended function of the verandah and claimed that unobstructed 'public' space was also necessary for efficient movement in a congested town.[20] Furthermore, those who used the verandahs to store goods or set up stalls not only 'defile[d] the drains and deposit[ed] filth in every corner',[21] but by obstructing street drains, completely frustrated the efforts of the municipal scavenging staff to flush or sweep the drains.[22] Verandah encumbrance was hence objected to not only because it was an infringement of public right of way but also because it obstructed the municipal staff from carrying out their sanitary duties. An urban environment with open, clearly visible, public spaces was also essential to the policing and surveillance functions of the colonial state. In the words of Edwin Tongue, an assistant superintendent, 'good road and street discipline ... [made] good citizens and automatically facilitate[d] the detection of crime' and conversely, obstructed and obscured public spaces hampered police work and 'accentuate[d] the lack of citizen discipline of all classes'.[23] Cleanliness and clarity of the public environment were also crucial in representing, particularly to outsiders, an image of the city as one which engendered civic pride and municipal vigour. In 1888, for example, the press urged the newly constituted Municipal Commission to deal with verandah obstruction with the proverbial efficacy of 'new brooms' lest Singapore depreciated in value as a 'showcase' city. It cautioned that '[while] visitors to Singapore admire the width and fine appearance of the public streets and roads, ... the filth and obstructions in the native quarters of the town give them an unfavourable idea of municipal energy in the direction of hygienic improvement. A glance around show ample ground for disparaging criticisms of this kind.'[24]

The Asian communities, however, preferred a style of land and property use which emphasized diversity, and which was small in scale and mixed in function. It was one which sought to minimize travelling and to maximize the versatility of each individual locality. No economic opportunity, however small, was scorned and niches for survival were created wherever possible in a city where space commanded a high premium. The Asian landscape reflected a multiplicity of uses, coexisting either in mutual symbiosis or in direct competition, and this in itself was a mirror of the complexity of Asian life in an urban context. For the labouring classes, survival in the city required what Richard Sennett calls 'a multiplicity of contact points'[25] for none of the institutions in which they lived were capable of self support. Instead, individuals in the city required the support of an intricate network of personal relationships and penetration of a number of social regions in the course of daily activities. The Asian urban landscape was hence one which favoured flexibility and the juxtaposition of different activities in close proximity (Plates 16 and 17).

As such, contrary to municipal prescriptions, the Asian communities saw the verandah in a more ambivalent light, as space capable of accommodating more than one legitimate use at any one time, and often serving different functions at different times. It was neither rigidly 'public' nor 'private', but was sufficiently elastic to allow the co-existence of definitions. Owners of warehouses and shophouses considered the verandah as an integral adjunct to their houses, to be used for stowing boxes and bulky goods.[26] As the verandah was a structure open to public view, shopkeepers conveniently utilized the space to display merchandise, trade names and signboards (Plate 18) Verandah space could also be turned to economic use: just as the dwelling house was partitioned into cubicles rented out for profit, house owners and principal tenants often subdivided the verandah into plots rented out to hawkers and stallholders catering to the public and also to families living in overhead boxes in the passageway.[27] In the 1890s, the verandah space in front of each house was often subdivided to accommodate up to four petty stallholders, each paying a monthly rent of between $1.00 and $15.00.[28] On average, the house owner or chief occupier collected about $4.00–5.00 in verandah rents each month.[29] In the poorer quarters of Kampong Glam, verandahs were leased to petty traders from neighbouring Indonesian islands who used them as 'a covered place ... [to] sit and bargain, out of the sun and rain'.[30] Precisely because the five-foot-way was meant as a conduit for human traffic, hawkers and petty traders ranging from sellers of pork, vegetables, fruit, spices, and tea to tinsmiths, druggists, barbers, and money-changers found it economically attractive to set up stalls in the verandahs and sides of streets as they afforded immediate access to a large clientele (Plates 19 and 20). Itinerant hawkers who plied along the five-foot-ways offering their wares and services also found the verandah an ideal place for resting or as a place to lay down their baskets whilst conducting an economic transaction (Plates 21–25).

Apart from its economic value, the verandah was also perceived as social space and used as an open parlour where inhabitants of tenement blocks could escape the dim confines of their cubicles and meet for a chat or a smoke of opium. In most cases, tenements had no backyard, forecourt or back garden and the verandah was the sole place available for communal interaction. The encroachment on the verandah for social purposes was hence a necessary extension of cubicle living. It has been noted that though the notion of a boundary between the 'private' house and the 'public' street was always real and acknowledged, it tended to become less elaborate as wealth decreased.[31] Among the poor who inhabited tenements where private space was severely limited, the dominant locales for socializing were the public streets and verandahs. Private space hence tended to merge into communal usage, which extended not only within buildings (as in the communal hall or shared kitchen of tenements) but also included the

verandah and the street. Verandahs and streets were also converted
into arenas for communal celebration and public entertainment on
Chinese festive occasions when residents of particular streets subscribed towards staging *wayang*[32] as a means of placating the patron
deities of the streets or clans concerned.[33] Verandah space could also
serve ritual purposes for families hosting wedding or funeral parties
(Plate 26).

As five-foot-ways were semi-public, relatively neutral territory in
comparison to private dwellings, they provided those who inhabited
the city with easily accessible urban interstices for acts which required
the cover of anonymity such as the dumping of unwanted rubbish or
diseased corpses in such a way that responsibility could not be easily
traced to particular individuals. Whilst occupiers could be held responsible for accumulations of filth and rubbish within the confines of the
house, the cleanliness and upkeep of public verandahs were the responsibility of the municipal authorities.[34] Similarly, the disposal of corpses
on verandahs transferred the responsibility of burial to the police.[35]
The verandah was hence as much a *place*—the locus of economic,
communal, and occasionally clandestine activities—as a passage. As
social, economic, and highly mutable space, it was crucial to the social
reproduction of a casual labour force dependent on unreliable sources
of income and inadequate housing.

For the Asian labouring classes, the urban environment was also
'ordered' and 'structured' according to certain hidden dimensions
which undergirded Asian society in colonial Singapore. The city, and
in particular its business streets and thoroughfares, was segmented into
well-defined operational territories controlled by a number of secret
societies which lay claim to the 'right' to collect 'protection money'
from all who made use of the space bounded within their territories.[36]
Scholars who have studied the Chinese underclass in the late nine-
teenth- and early twentieth-century Singapore[37] have shown that
among coolies, *rikisha* pullers, and hawkers, there was a keen sense of
space and specialized knowledge as to which streets and places were
permissible to traverse or to work on and which implied running the
gauntlet of secret society *samseng*.[38] The concept of a 'public right of
way' along all public streets and verandahs was hence irrelevant to
those who made their living on the streets and who encountered daily
the might of the '*imperium in imperio* of the Chinese underworld'.[39] In
summary, rather than a fixed boundary dividing the city into 'public'
spaces accessible to all and 'private' spaces permissible to none except
those admitted by owners, there were degrees of accessibility and
exclusion to verandah space largely determined by the relations of the
persons involved, by time and by circumstance. As such, during certain
hours, the verandah might be permitted to one or more tenants who
had paid rent to the houseowner and 'protection money' to the secret
society controlling the street. At other times, it might be left completely
open to the pedestrian. On festive occasions, when the verandahs and
streets were turned into locales for Chinese *wayang*, they become

accessible to clan members who had subscribed towards the celebrations and sometimes to the wider public. The verandah and adjacent street were hence latticed with many timetables and competing claims, not just the explicit ones of the municipal scavenging crew or the routine police patrol, but also those of the itinerant hawker on his daily rounds, celebrations dictated by the cycle of Asian festivities as well as the covert activities of secret society gangs staking territorial claims on the urban environment.

Municipal Attempts at Verandah Reform in the Nineteenth Century

Municipal attempts to reform the verandah were based on rigidly demarcating the threshold between public and private spheres of activities and making the distinction between the two much less ambiguous. Verandah space which was kaleidoscopically multifunctional and sufficiently malleable in serving communal, interactive and economic purposes was redefined as socially neutral space, subservient to the public right of way and open to view and regulation. The complex, overlapping claims on the verandah were nullified, and superseded by one dominant purpose, namely, urban circulation. The campaign to liberate the verandah for pedestrian use dated from the earliest days of the Settlement. Though the majority of verandahs were built on private property, they were, by Raffles's decree, to be kept permanently free as the town's pedestrian conduits and could not be appropriated for any private or undesirable purposes. As early as 1822, the 'lower classes of Chulias' were prohibited from living in the verandahs of houses or anywhere else on the northern side of the town and were moved to a *kampung* marked out for them to the south of the Singapore River.[40] In the same decade, hawkers were prohibited from setting up stalls in the streets as this was considered 'a nuisance, prejudicial to the health and comfort of the inhabitants' and were allowed only within the confines of public government-owned markets.[41] Apart from the incursions of hawkers and the homeless, verandah space was often appropriated by proprietors and occupiers of houses for displaying or stowing goods. In February 1843, the press published a complaint that the natives had 'coolly appropriated the verandahs for their own special use' and that verandahs everywhere were impassable as they were 'choked out with the wares of Klings and Chinese'.[42] By June the following year, verandah obstruction was important enough to feature in the Grand Jury's presentment of 'nuisances' which had to be dealt with.[43]

In 1856, the town's first Conservancy Act XIV reasserted the right of the public to the free use of the verandahs and empowered the municipal commissioners to pull down or remove obstructions and encroachments of the verandah in cases of non-compliance. Three years later, the municipal commissioners first recorded their intention in the municipal minutes to 'assert the rights of the public to the free use of the verandahs'.[44] Verandah obstruction, however, remained prevalent and the whole question came to a head in the early 1860s in a celebrated case, *Syed Abdullah bin Omar Al Junied* v. *Municipal Commissioners*, arising from the former's attempt to enclose a verandah in Nankin Street for his own private use. On the evidence of John Crawfurd (a former Resident in the 1820s) and Sir George Bonham (a member of the town committee appointed by Raffles in 1822 and later, Governor of the Straits Settlements) as to the long-standing rights of the public to the free use of verandahs, judgement was delivered in favour of the municipal commissioners on 4 May 1863, whereupon the commissioners issued a notification declaring that proceedings would be instituted against all persons obstructing verandahs.[45] Municipal triumph, however, was short-lived and in practice, the obstruction of five-foot-ways persisted throughout the second half of the nineteenth century. In 1869, for example, the press observed that in Malacca Street, Telok Ayer Street, and Canal Road, Chinese shop-keepers and petty merchants allowed their coolies to sort out merchandise such as sugar and betel-nut on the verandahs in front of their shops 'to the great inconvenience of pedestrians'.[46]

In 1872, the law regarding verandahs was further complicated by section 18 of the newly passed Summary Criminal Jurisdiction Ordinance. Although the section defined verandahs as 'public thoroughfares for foot passengers', it also declared that verandahs were subject to 'all rights of property of owners of houses'.[47] Public and private claims on the verandah were hence so closely juxtaposed as to render the law difficult to interpret and apply. As a result of the legislative deadlock, it was reported that 'matters drifted from bad to worse, until the originally intended rights of the public had given way in a large degree to the claims of the owner or occupier'.[48]

The 'Verandah Riots'

In the late nineteenth century, tensions between the municipal authorities and Asian shopkeepers and traders over the use of the verandahs were heightened by the passing of the Municipal Ordinance IX of 1887 which, *inter alia*, empowered the municipal commissioners to remove any obstruction of verandahs, arcades, or streets which either hindered the work of scavenging or caused inconvenience to the passage of the public. When the ordinance came into effect at the beginning of 1888, the municipal engineer sent in a minute requesting to know whether the commissioners intended to take any action on the matter of verandah obstruction.[49] The commissioners resolved that one month's notice was to be given in the local English and vernacular journals as well as through placards placed throughout the town of the municipal intention to 'rigidly enforce the clearance of all open verandahs abutting on public streets in Town'.[50] The municipal president, Dr T. I. Rowell, stressed the importance of verandah clearance on sanitary grounds and anticipated that under his direction, the blocked-up state of verandahs in the town would be 'greatly improved' within twelve months. Typical of municipal manoeuvres, the new powers to order Asian society and the urban environment were couched in the language of improvement. Dr Rowell was, however, opposed by several of the non-official commissioners and in particular Thomas Scott, 'a gentleman of great influence with the Chinese community', who argued that enforced verandah clearance amounted to a confiscation of vested private rights and challenged the president's power to proceed without the consent of the commissioners.[51] The 'verandah question' was also strenuously debated in the press. *The Straits Times* applauded the municipal resolve to clear verandahs and anticipated that once the provisions of the Municipal Ordinance were 'strictly enforced without fear or favour', it would 'take effect beneficially on the street traffic, and be no inconsiderable boon to pedestrians'.[52] Like the municipal commissioners, it construed the 'verandah question' as a debate between 'public interest' and 'private ends'[53] and as such, public order could only be restored if the verandah were strictly demarcated as public space and its boundary with private property made less permeable. The rival daily, the *Singapore Free Press*, took the opposing stand that owners and occupiers of houses had a claim on the five-foot-way through 'long usage' and that the proposed campaign against verandah obstruction was 'harsh and coercive'.[54]

Whilst the press continued to debate the verandah issue, municipal inspectors under orders from Dr Rowell started to clear verandahs in the neighbourhood of Arab Street, Rochor Road, and Clyde Terrace Market (in the Kampong Glam district) on 21 February 1888.[55] At North Bridge Road, shopkeepers and traders reacted by closing their shops as a demonstration of protest against the municipal order to clear their verandahs.[56] News of this soon spread and shopkeepers in the vicinity eastwards of Bras Basah Canal followed suit by putting up their shutters.[57] The Inspector-General of Police, Colonel S. Dunlop, reported that from the 'quiet manner' in which the Chinese shopkeepers reacted, it was evident that the action had been preconceived in concert.[58] Contrary to the official stand that no unwarranted harshness or precipitancy was displayed by the inspectors in carrying out the law,[59] one of the local dailies reported that the verandahs were forcibly cleared, property damaged, sunshades pulled down, and shopkeepers roughly turned away.[60]

On the afternoon of the same day, the municipal commissioners convened an emergency meeting to evaluate the situation. The power of the president to order the clearance of verandahs without the consent of the commissioners was again called into question by some of the elected members of the Board and a motion stating that the requirement of the Municipal Ordinance would be considered satisfied as long as there was sufficient space for two persons to pass abreast along the verandah was passed under overwhelming pressure from the elected commissioners.[61] The motion was tantamount to a revocation of the president's original order and an acceptance of a more liberal definition of what constituted public right of way along the five-foot-ways. The

Governor, Sir Cecil Smith, was later to censure the municipal commissioners for their lack of support of their president and for 'practically yield[ing] to the opposition which the Chinese had shewn [sic] to even limited interference with their occupation of the ... verandahs'.[62] The municipal resolution failed to quell growing tensions and the next day (22 February), open rioting broke out in the streets, not only in Kampong Glam where verandah clearance had proceeded but elsewhere throughout the principal streets of the town.[63] All shops and markets remained shut and crowds of *samseng*[64] gathered in the streets, intimidating anyone who attempted to reopen for business. Stone and brickbat throwing was the order of the day, tramcars were set upon and damage done to public conveyances.[65] Members of the European and other communities were attacked and the police were called out in full force.[66] Trade and traffic were brought to a standstill. In the course of the afternoon, the municipal commissioners met again and ordered the circulation of the resolution passed the previous day intimating that the commissioners did not intend to act harshly and that only a clear passage of 3 feet was required.[67] On the morning of 24 February, inter-clan fighting broke out among the Cantonese and Hokkien coaling coolies at Tanjong Pagar and sporadic rioting continued throughout the town.[68] By noon, however, shops began to reopen for business and the rioting gradually petered out.[69]

The riots which originated in a confrontation between municipal inspectors and shopkeepers over the right to use one particular element of public space—the verandahs—rapidly escalated into a full-scale tussle for control over public streets and spaces in many parts of town. What was originally an act of disobedience on the part of shopkeepers spatially confined to a few streets became amplified by the involvement of the rowdier elements of Chinese society to become what the press described as 'an organized resistance to the law of the land'.[70] As with other instances of urban disturbance, friction or conflict between two relatively small groups quickly assumed an aspect of public violence, involving larger networks and loyalties which underlay Asian society. For a few days, the crowds were able to hold public places at ransom and to force the town into a standstill, albeit temporarily. The everyday routine was disrupted, and social spaces in the city momentarily reclaimed. The riots were also not quelled without the municipal commissioners being compelled to adopt a much less stringent definition of public right of way along verandahs compared to what was originally contemplated, thereby conceding much of the ground to Asian shopkeepers and traders who refused to relinquish their claim on the verandahs.

Whilst the 'verandah riots' represented by far the most violent aspect of the conflict over the use of 'public' spaces, it was not an isolated event but had instead been fomented in the wake of increasing ill feeling towards the authorities generated by a succession of unpopular measures implemented during the second half of 1887 and the early months of 1888.[71] In particular, the customary right conceded to the

Chinese to use the verandahs and the sides of public streets to burn sacrificial papers, display feasts for the dead, and hold street-*wayang* during the *sembahyang hantu* season[72] in August/September had been withdrawn by the government[73] in retaliation to a homicidal attempt on William Pickering, the Protector of Chinese, by Choa Ah Siok, a Teochew carpenter. The attack was interpreted by the government not only as a dastardly crime against the person of the Protector who 'ha[d] been for so long the friend and the benefactor of the Chinese community', but more crucially, as a 'great offence against the public order', which called for public punishment of the entire community and required public expiation on the part of the community.[74] The mode of punishment chosen, the withdrawal of the 'privilege' of using public space for communal celebrations, was calculated to reinforce the public nature of the crime.[75] The government further alleged that the Chinese could give information leading to the identification of those who had instigated the assault if they had a mind to do so and by withholding information, the responsibility for the perpetration of the crime lay upon the entire community. Until the Chinese co-operated to restore public justice and order, they forfeited all right to appropriate public spaces for their own use. The irony of the prohibition order following on the heels of the Queen's Golden Jubilee celebrations in June of the same year during which the municipal commissioners granted free permits to all the inhabitants of the town to erect poles for decorative purposes in the public thoroughfares[76] was not lost on some of the more articulate members of the Chinese community. During the Jubilee celebrations, the Chinese took 'a most prominent part', putting up 'lantern processions' with 'gigantic dragons' winding their way through the streets, 'gorgeous decorations and illuminations' and *wayang* scattered throughout the town.[77] The Chinese who wrote to the press remarked that neither 'the ink describing the grandeur of the displays in the streets and the loyalty of the people in celebrating Her Majesty's Jubilee'[78] nor 'the perspiration on the brow of [their] heads for having toiled so hard to make the occasion a marked event'[79] had dried before the right of the Chinese to use the same streets for their own celebrations was abruptly withdrawn within a space of two months. The *sembahyang hantu* passed without any major incident but as with the clearance of the verandahs six months later, the prohibition order impinged upon what the Chinese saw as their customary right to order and use the urban environment and in particular its public spaces, in a manner consistent with their own purposes. Against this background of growing resentment towards the authorities, and by the suspension of belief about the enormities of which 'the Other' was capable, a minor fracas quickly escalated into a full-blown riot. Although the 'verandah riots' were triggered off by what appeared to be conflict over a specific type of urban space, the fact that it rapidly assumed larger dimensions was an indication of widespread fears that the customary rights of the Chinese over the public environment were increasingly threatened.

The Verandah as an Arena of Daily Conflict:
Municipal versus Asian Strategies

The termination of the 'verandah riots' did not signify an unequivocal victory for either the municipal authorities or the Asian people; neither did the resumption of order signify a satisfactory conclusion to the 'verandah question'. The nature of public space and how it should be utilized continued to be contested in the daily arena over the next few decades, even though the contest never again assumed riotous proportions. Before the dust could settle after the 'verandah riots', the question of liberating the verandah from encumbrances was again ventilated. This time the initiative for the action originated with the government and the focus of attention was the numerous hawkers who set up pitches along verandahs and sides of streets.[80] In a letter to the municipal commissioners, the Colonial Secretary recommended registering and licensing hawkers, ostensibly as a check to the flagrantly irregular practices on the part of the native police in extorting money from hawkers.[81] The suggestion rested on the rather dubious premise that hawkers, once licensed, would readily complain to the authorities in cases of police extortion.[82] Clearly what was also contemplated was the possibility that by the enforcement of licensing quotas, hawkers who were unable to obtain licences would be compelled to retreat from verandahs and rent shop premises to conduct their businesses. The municipal commissioners, however, were of the opinion that licensing would unduly harass hawkers without actually checking police rapacity.[83] In the aftermath of the 'verandah riots', the commissioners were acutely aware that licensing could be easily interpreted as an additional squeeze on hawkers instead of a form of relief, and widespread discontent at this point could rapidly escalate into another major disturbance.[84] The attempt to regulate this one particular class of verandah tenants was hence put off until the twentieth century.

Meanwhile, there were no signs that shopkeepers, hawkers, and other verandah tenants had relinquished their claim to use verandah space. In the week following the riot, it was reported that 'verandahs [were] worse blocked than before, thus setting the authorities at open defiance'.[85] Two months after the riot, another correspondent who surveyed the verandahs on foot reported that whilst verandahs in Collyer Quay and certain neighbouring streets where European business houses were situated (Plates 27 and 28) appeared tolerably clear, several other streets where the Chinese operated such as Kling Street, Market Street, Boat Quay, and Circular Road were so obstructed that 'it [was] impossible to walk ten paces along the verandah before one ha[d] to turn out into the road'.[86] He urged the municipal commissioners to deal firmly with 'usurpers of verandah space' and complained particularly that there was no excuse to allow 'Asiatics' to encumber verandahs in areas such as North and South Bridge Roads which were 'equally parts of the European quarter of town'.[87] In the aftermath of the riots, however, the municipal commissioners were forced to reconsider their strategies. As enforced *en bloc* clearance of the verandahs had proved counter-productive, it was resolved instead to proceed in a piecemeal fashion, focusing first on putting 'in proper repair and good traffic order' and then clearing of obstructions verandahs in the European business quarter, namely, verandahs in Collyer Quay, Raffles Square, Prince Street, De Souza Street, and D'Almeida Street.[88] Verandahs in the 'Asiatic parts of town' would only be taken in hand 'street by street, section by section' as far as funds permitted, 'quietly and unobtrusively'.[89]

While the reform of verandahs in the European business district proceeded satisfactorily,[90] verandahs elsewhere were far more problematic. The compromise of allowing traders and shopkeepers to use the five-foot-way as long as a narrow passage was left open, a legacy of the 'verandah riots', led inevitably to 'the shopkeeper monopolising the whole space and to the police ceasing to concern themselves about the state of the colonnades'.[91] In the closing years of the nineteenth century, the municipal president reported that the 'evil' of verandah and street obstruction.was flourishing: portions of the arcades were railed off as 'private preserves'; cooking stoves, tinsmiths' stalls, and many similar erections 'equally permanent and more objectionable from a sanitary point of view' habitually lined the footways; goods constantly encumbered the public quays; and vendors of fruit, vegetables, and fresh provisions became increasingly prominent in the streets of Chinatown in particular. Even in the neighbourhood of Collyer Quay and Raffles Place, the 'nuisance' caused by itinerant food vendors was attracting regular complaints.[92] The municipal commissioners asserted that it was their duty to 'recover and retain for the public the use of the streets and pavements' and directed that the police should 'systematically compel the vendors to "move on" so that the public [might] not be deprived of the use of the streets'.[93] A circular notice in Chinese was also distributed among the hawkers, warning them that they were not allowed to 'encamp on the public street, but must move on and keep the roadway clear'.[94] As a strategy to clear verandahs and streets, the requirement for constant mobility of hawkers was temporarily successful, but 'with the least slackness' on the part of the police, 'disorder ... again prevail[ed]'.[95] It was particularly difficult to abate the 'hawker nuisance' in the Collyer Quay–Raffles Place neighbourhood as the considerable Asian and Eurasian subordinate and clerical staff employed by the large European trading firms, agency houses, and banks located here generated a high lunchtime demand for hawker food.[96]

Apart from the policy of 'moving on' hawkers, the other chief municipal strategy used in clearing verandahs was the institution of prosecutions and fines. This was 'eminently unsatisfactory', as demonstrated by the following account, which was by far the usual course of action:

A district inspector reports a case of continued obstruction after repeated warnings and threats; he is directed to apply for a summons; the summons is duly served, but the case can rarely be heard for three weeks thereafter. Meanwhile the obstruction continues unabated. One or two days before the case is called on, a show is made by the occupier of clearing the five-foot-way. In due course the offender appears before the Magistrate; probably he is one of a whole row of men similarly summoned, who all plead guilty and are dismissed with a small fine and a short exhortation from the bench. Two days thereafter the verandahs are as bad as before, while all day long the policemen stalk complacently outside and never dream of enforcing the law.[97]

Recourse to the legal process to effect the clearance of verandahs was hence tedious and ultimately ineffective. In the 1890s, fines for verandah obstruction averaged about $2.00–3.00 (Table 7.1),[98] hardly sufficient as a deterrent. During the opening years of the twentieth century, a period of increased efforts at verandah clearance, penalties for obstruction inflicted by the magistrates were almost twice as heavy but the average fine soon fell to just above the nineteenth-century level in the years immediately preceding the First World War before increasing again (Table 7.1). Table 7.2 lists a number of obstruction cases culled from the Municipal Summons Note Books for 1912–1913 which, while statistically insignificant, demonstrate that these cases were often let off lightly.[99] In a number of cases, the defendant had been convicted on several previous occasions but appeared undeterred. In the 1920s, fines for obstruction only averaged $4.00–5.00 (Table 7.1) and there were repeated complaints by municipal officers that 'prosecutions [were] a thankless business'[100] and that 'fines inflicted by the Court [were] not only ineffective but [were] so slight a deterrent as to engender in the offender a spirit of contempt both for the law and for the Officer whose duty it [was] to enforce the law'.[101] In June 1891, a Supreme Court judgement by Chief Justice Sir E. L. O'Malley laid down that the public had a full right of way over the whole of the five-foot-way and that anything that rendered the passages less commodious was an obstruction and might be removed.[102] Although this was an important legal and moral victory for the municipal authorities which helped regain some of the ground lost during the 'verandah riots', it had little consequence for the day-to-day state of the verandahs: the municipal president reported that 'prosecutions [were] still few compared to what would have been the case if the law [were] strictly enforced', verandahs continued to be 'habitually blocked'[103] and the task of clearing them virtually 'hopeless'.[104] The persistence of verandah obstruction was a consequence of the fact that most verandahs were leased to petty traders and hawkers who were loath to surrender their claim on the verandahs for what was at stake was not only a particular location but a pitch which had been paid for[105] and a clientele which had been built up over time and with effort.

Part of the ineffectiveness of municipal resolve to clear verandahs also stemmed from the fact that 'the verandah question' was strongly disputed within the Board itself. The municipal president, Alex Gentle, complained in 1892 that the efforts of the health officer and his own in

TABLE 7.1

Return of the Number of Inspections, Prosecutions,
Convictions, and Fines for the Obstruction of Verandahs,
Thoroughfares, and Drains, 1888–1925[a]

Year	Number Inspected	Number Prosecuted	Number Convicted	Total Amount of Fines ($)	Average Fine per Case ($)
1888	423	377	396	792.75	2.00
1889	–	188	185	332.50	1.80
1890	79	79	78	94.00	1.21
1891	446	439	433	1,373.75	3.17
1892	535	535	515	1,227.00	2.38
1893	202	202	167	398.00	2.38
1894	167	142	124	344.00	2.77
1895	163	163	145	274.00	1.89
1896	410	410	388	776.50	2.00
1897	161	161	147	398.50	2.71
1898	240	240	222	655.00	2.95
1899	185	185	140	554.04	3.98
1900	598	598	518	2,884.55	5.57
1901	2,673	2,673	2,382	13,241.15	5.56
1902	1,711	1,711	1,566	6,336.50	4.07
1903	–	3,070	2,757	16,058.50	5.82
1904	–	1,786	1,589	7,316.50	4.60
1905	–	1,063	988	3,926.00	3.97
1906	–	749	703	2,817.00	4.01
1907	–	964	886	3,295.00	3.72
1908	–	1,357	1,234	4,018.00	3.26
1909	–	736	684	3,099.00	4.53
1910	–	–	704	1,858.00	2.64
1911	–	593	550	1,741.00	3.17
1912	–	–	973	3,223.00	3.31
1913	–	–	764	2,820.00	3.69
1914	–	–	1,147	3,757.00	3.28
1915	–	1,526	1,438	6,249.00	4.35
1916	–	97	94	551.00	5.86
1917	–	110	103	434.00	4.21
1918	–	167	139	547.00	3.94
1919	–	215	211	1,449.00	6.87
1920	–	437	396	1,777.00	4.49
1921	–	622	570	2,615.00	3.80
1922	–	849	654	2,714.00	4.15
1923	–	475	401	1,754.00	4.38
1924	–	342	300	1,268.00	4.23
1925	–	110	16	87.00	5.44

Source: Compiled from ARSM, 1888–1925, Health Officer's Department/Health
Officer's Report/Municipal Health Officer.
[a]There were no systematic records for the number of verandah obstruction offences after
1925.

TABLE 7.2
Municipal Obstruction Cases, 1912–1913

Name/Address Where Offence Occurred	Charge	Defendant's Statement	Sentence
Kooh Kim 136 Beach Road	Goods obstructing backlane	–	Dismissed
Hak Soi 13, 14 Church Street	Goods obstructing five-foot-way	Goods were not his	Postponed
Lim Kin 7 George Street	Goods obstructing street and five-foot-way (3 separate summons)	No knowledge of obstruction	$3 + costs
Defendant	Obstruction	Pleads guilty	costs
J. Frankel 166–1 Orchard Road	Goods obstructing street and five-foot-way (several previous convictions)	Does not remember incident	$12 + costs
Wong Ah Cheng 13 Trengganu Street	Pork stall obstructing five-foot-way	–	Cautioned and discharged
Lau Ah Poh Trengganu Street	Stall obstructing five-foot-way	–	Cautioned and discharged
Chhan Sew 13 Trengganu Street	Stall obstructing five-foot-way	–	Cautioned and discharged
Tau Chin Siong Syed Alwee Road	Goods obstructing street	Not the person charged	Postponed
J. Frankel 97 Tank Road	Cases, empty cement barrel, signboard and broken marble obstructing five-foot-way (1 November 1912)	Five-foot-way was to be repaired, boxes left there to keep off cake-sellers	$5 + costs
J. Frankel 97 Tank Road (junction of Tank Road and Orchard Road)	Empty cases, signboard and marble obstructing five-foot-way (4 November 1912)	Public could pass along five-foot-way, municipal coolies did not scavenge the passage	Dismissed
J. Frankel Junction of Tank Road and Orchard Road	Packing cases, signboards and empty cement barrels obstructing five-foot-way and street (23 November 1912)	Goods were in transit and were being shifted away by *rickishas*	$5 + costs

TABLE 7.2 (*Continued*)

Name/Address Where Offence Occurred	Charge	Defendant's Statement	Sentence
Ng Ban Lane off 26 Rochor Road	Bundles of sugar-cane obstructing street (1 previous conviction)	Sugar-cane were not his but his neighbours'	$5 + costs
Go Peng (principal tenant of licensed eating house) 55 New Market Road	Tables and benches belonging to eating house obstructing five-foot-way	Permission granted by assistant municipal health officer who said he would not be summoned	$2 + costs
Haji Kader Ghani 16 Baghdad Road	Eating stall obstructing five-foot-way (1 previous conviction)	Calls witnesses to testify that he had stopped operating his eating stall before summons arrived and now hawks using a cart	$5 + costs
Chua Soon Tong 150 Orchard Road	Eating stall obstructing five-foot-way (several previous convictions)	Has stopped operating eating stall since last conviction	$10 + costs

Source: Compiled from Municipal Summons Note Books, 5 July 1912–13 December 1912 and 16 December 1912–6 May 1913.

removing obstructions from principal streets in the town met with little support from the elected members of the Board[106] and again in 1897 that 'the Commissioners still 'follow[ed] a vacillating policy, which result[ed] in most of the arcades in town being encumbered to such an extent that [made] it impossible for the public to use them as foot-ways'.[107] As already seen, the Chinese commissioners condemned verandah clearance as 'a great hardship on the Chinese and native shopkeepers'[108] and several of the other elected members (in particular Thomas Scott) were stout champions of private property rights. Whether they were actuated by public spirit alone in opposing veran-dah clearance was a moot point, for there were clearly strong personal considerations involved. Several of the commissioners themselves were large property owners and verandah leases formed part of their profits.[109] More crucially, by opposing verandah clearance, they were furthering the interest of the class of ratepayers whose vote chiefly ensured their success at municipal elections. According to the press, a 'practical walkover' was ensured to 'candidates who set their faces against too great an interference with the present license [sic] in obstructing public footways'.[110] In September 1897, the magistrates reproached the municipal commissioners for 'their haphazard way of dealing with obstructionists', pointing out that in a situation where 'almost every shop [was] guilty of the offence [of obstructing verandahs], it [was] ... useless and only calculated to cause great disorder and feeling of injustice in the minds of the Chinese to take isolated cases as scapegoats'.[111] As such, until the commissioners adopted a 'definite policy' to deal with obstructions, the acting second magistrate declined 'to inflict anything but a nominal penalty for an [obstruction] offence'.[112] This left the municipal commissioners in a highly untenable position for, on the one hand, prosecuting all offenders was impossible and, on the other, prosecuting 'a selection result[ed] in acquittal or a very small fine, because all the others in the same street [had] not been prosecuted'.[113] In vain the commissioners reminded the magistrates that their earlier attempt at complete clearance of verandahs had led to street disturbances, and as a result, they had resolved on 'partial clearance' without interfering with verandahs where 'long custom ha[d] allowed the exhibition of articles for sale'.[114] Up to the end of the nineteenth century, the spectre of another violent backlash continued to loom large in the minds of the commissioners and significantly circumscribed their policy towards verandah obstruction.

The turn of the century saw renewed efforts on the part of the municipal authorities in their campaign against verandah obstructions. Verandah clearance was extended beyond the European business district to encompass the 'Asiatic quarters' (Figure 7.1).[115] In the first few years of the twentieth century, the municipal health officer for the first time reported that although 'constant vigilance'[116] was still required, the condition of verandahs was much improved as a result of better police co-operation in keeping verandahs under surveillance[117] and also the infliction of heavier penalties by magistrates for obstruction.[118] A novel strategy was also adopted with respect to quayside verandahs and streets which were almost perpetually blocked as a result of the landing and storage of goods. In 1905, the municipal commissioners decided to capitalize on demand rather than put up a fruitless resistance. In antithesis to the Board's previous stand that quays were 'public streets' which 'could not be lawfully leased for hire',[119] public quays, wharves, and streets adjacent to waterways were partitioned into storage compartments, permits issued, and fees levied for the use of these spaces, thereby legitimizing what was originally an offence.[120] The municipal commissioners were hence not above taking a leaf out of property owners' books: while still continuing to deny property owners the right to rent out verandah spaces to tenants, the commissioners did not hesitate to equip themselves with the power to commodify, partition, and rent out certain public spaces for a fee.[121]

FIGURE 7.1
The Municipal Campaign against Verandah Obstruction:
Verandahs Cleared, 1899–1903

Source: ARSM, Health Officer's Department/Health Officer's Report, 1899–1903.

The Municipal Campaign against Street and Verandah
'Obstructionists': Hawkers and Street Traders

By the twentieth century, the municipal campaign against verandah obstruction became focused on one particular class of street and verandah 'obstructionists', the Asian street hawkers who successively 'invaded' ostensibly 'public' urban spaces. The attitude of the authorities towards street hawkers was one of ambivalence. On the one hand, hawkers were construed as major threats to public health and public order. They were considered major culprits of street, verandah, and drain obstruction,[122] agents for the spread of disease through unhygienic handling of food,[123] as well as a 'fruitful source of corruption'[124] for the lower ranks of the police and municipal service and also secret society gangs. Furthermore, by their very act of attracting crowds around them, hawkers contributed to the 'indiscipline of the road' and militated against proper policing.[125] In stigmatizing street hawkers as a community of lawbreakers and a source of obstruction, disease, and filth, the municipal authorities were variously supported by complaints from certain sectors of the public[126] and eating-house keepers who view hawkers as 'unfair competition'.[127] On the other hand, food hawkers constituted an element crucial to the reproduction of the labouring classes as they provided an indispensable service in feeding an urban working population which demanded cheap and easily accessible meals. The basic difficulty in formulating a consistent policy towards hawkers, explained C. F. J. Green, the deputy municipal president in 1921, lay in the fact that 'hawkers were so necessary to a large section of the community'.[128] As such, the municipal authorities were anxious not so much to suppress hawking entirely but to reduce hawkers to a minimum and to confine hawking to particular localities where it could be controlled and monitored. As already seen, municipal strategies used in the late nineteenth century to regulate hawkers including the institution of prosecutions and fines and the requirement for constant mobility in the case of itinerant hawkers did not meet with 'any signal success'.[129] In the early twentieth century, two other strategies were attempted: first, the registration and licensing of street traders, and second, the making of what has been termed a 'defended space' and 'defended time zone' wherein hawkers might not intrude.[130] In 1906, the Municipal Ordinance was amended by the addition of a new section empowering the municipal commissioners to set apart certain public streets and public places where hawkers might ply their trade and authorizing the commissioners to charge fees for the accommodation afforded.[131] The municipal authorities envisaged that by this new measure, they would be able 'to control and concentrate the number of hawkers of food who [did] so much to obstruct and foul the streets'.[132] Hawker by-laws for registering and controlling the occupation by night hawkers of pitches on public streets were sanctioned and became operative on the first of October 1907.[133] Stall hawkers who operated after 6 p.m. were required to acquire licences which were only granted for specific sites located in one of a list of specified streets.[134] The municipal health officer, however, bemoaned the fact that itinerant hawkers, whom he considered to be 'mostly under suspicion as the spreaders of disease', were exempted from the operation of the by-laws.[135] Some initial difficulty was experienced in the process of registration[136] but these were soon tided over and by 1908, the municipal register contained over 6,000 licensed night stall hawkers (Table 7.3), the majority of whom concentrated in over twenty streets (Figure 7.2).[137] By means of the weapon of licensing, by adding and removing from the sanctioned list of streets where night hawkers might operate and through the leverage of fee increases,[138] the municipal authorities sought to exert control over the population and spatial distribution of street hawkers in the town. As a means of establishing order in the urban environment, however, the hawker by-laws as they stood were totally inadequate, largely because day and itinerant hawkers remained beyond municipal jurisdiction. The chief police officer repeatedly pointed out to the municipal commissioners that the 'running in' of day and night itinerant hawkers for disregarding police orders was 'no good whatever'[139] and that the force was not sufficiently large to prevent hawkers from congregating on the streets in the day or to effect any 'permanent improvement'.[140] Towards the end of the first decade, the condition of verandahs and streets in the already congested 'Asiatic quarters' was further worsened by the rapid mushrooming of unauthorized fresh-produce markets.[141] Markets for fish, fowl, meat, vegetables, and fruit 'arranged in sections so that the public [knew] where to find various articles'[142] flourished at street corners, in some cases in close proximity to municipal markets with spare accommodation.[143]

Although the street hawker continued to be stigmatized as unsightly, insanitary, and obstructive, the municipal commissioners were reluctant to enforce the licensing of day and itinerant hawkers largely because of the plea that 'these men [were] a convenience to the poorer people'.[144] Municipal resolve to deal systematically with verandah encroachments was counterpoised by the awareness that hawkers were crucial to the daily working of the colonial urban economy. The exigencies of war further tipped the balance in favour of the hawker. In 1914, an attempt to clear hawkers and fresh food vendors off verandahs in Rochor Road was discontinued in deference to the wishes of the Chinese commissioners who protested that with the onset of war, any interference with hawkers and petty traders would cause undue hardship.[145] The government also directed that given the harsh circumstances dictated by war, no 'drastic measure' should be taken against street hawking[146] and consequently, the policy of clearing verandahs fell into abeyance for the duration of the war. Within a few years, the footways in town quickly reverted to their original, cluttered condition, and even verandahs in the European business quarter became 'infested'[147] with beggars and food hawkers.[148]

The 'hawker question' which had remained unresolved through the war years was again resurrected in the immediate postwar years. In

TABLE 7.3

Number of Hawker Licences Issued by the Municipality, 1907–1929

Year	Night (stall)	Day (stall)	Itinerant	Total
1907	1,587	–	–	–
1908	6,384	–	–	–
1909	7,123	–	–	–
1910	7,285	–	–	–
1911	7,399	–	–	–
1912	7,082	–	–	–
1913	6,944	–	–	–
1914	7,024	–	–	–
1915	6,385	–	–	–
1916	6,402	–	–	–
1917	6,407	–	–	–
1918	5,793	–	–	–
1919	5,551	926	4,772	11,249
1920	5,146	3,582	6,578	15,306
1921	6,448	4,311	8,612	19,371
1922	6,932	4,698	5,697	17,327
1923	6,593	4,334	3,447	14,374
1924	3,832	2,535	2,222	8,589
1925	3,094	2,117	1,316	6,527
1926	3,130	1,818	1,815	6,763
1927	923	579	2,748	4,250
1928	791	412	6,000	7,203
1929	559	258	4,696	5,513

Source: Compiled from ARSM, 1907–1929, Health Officer's Department/ Health Officer's Report/Municipal Health Office.

October 1919, by-laws for the licensing of day and itinerant hawkers were finally brought into force.[149] Like their night-time counterparts, day stall hawkers were confined to certain portions of streets during prescribed hours (5 a.m. to 3 p.m.) and prevented from intruding upon 'defended' streets where they were prohibited. To prevent 'unfair competition', stall hawkers were also banned from setting up pitches in front of municipally licensed eating houses.[150] Initially, no restrictions were imposed on the use of municipal space by itinerant hawkers but soon, this privilege was also gradually withdrawn. In 1922, itinerant hawkers were no longer allowed to trade in certain streets in the vicinity of the Esplanade[151] and in 1927, they were excluded from Raffles Place, Fullerton Road, and High Street,[152] all three major business thoroughfares. Besides attempting to license all hawkers, the municipal authorities also decided that where hawker activity was necessary or desirable, hawkers were to be transplanted from the open streets into specially erected municipal shelters. In the 1920s, a total of six shelters were opened at strategic locations: the Finlayson Green Shelter to absorb hawkers plying around the Collyer Quay–Raffles Place area in 1922;[153]

FIGURE 7.2
Major Street Concentrations of Licensed Night Stall Hawkers, 1908

Source: ARSM, Health Officer's Report, 1908.

the People's Park Shelter to clear hawkers in the Temple Street–Trengganu Street area;[154] the Balestier Road Shelter to deal with hawker trade in that vicinity in 1923;[155] the Carnie Road Shelter in 1927;[156] a shelter within the grounds of Telok Ayer Market in 1928;[157] and the Queen Street Shelter to relieve hawker congestion in the Rochor area in 1929.[158] Once a hawker shelter was opened, in order to force hawkers into the shelter, the municipal commissioners removed the streets in its vicinity from the list of specified streets where hawkers were allowed.

Despite all municipal effort, licensing by-laws and hawkers' shelters failed either to restrict the number of hawkers or relieve congestion in the streets. In the years after the by-laws came into effect, it was reported that 'there [were] as many [hawkers] without as with licences, a number of whom [plied] their trade in unauthorized streets and places'[159] and that several of the main thoroughfares were 'rendered impassable by hawkers with their paraphernalia, *rickshas* and other obstructions'.[160] The growth of the hawker population appeared to have its own inexorable logic and every decrease in the number of licensed hawkers was counterbalanced by increases in the number of unlicensed hawkers.[161] Whilst hawkers' shelters were initially successful in clearing neighbouring streets of hawkers, it quickly became apparent that even this was illusory for vacant pitches in the streets were quickly occupied by a new succession of hawkers.[162] It was also realized that the shelters failed to serve the needs of the poorest hawkers as these 'real two-basket hawkers' (as the health officer christened them) were rapidly 'shouldered out or bought out' by the large 'capitalist hawkers' who ran highly profitable businesses.[163] By the late 1920s, it had to be admitted that the municipal aim of driving hawkers off the verandahs and streets into well-regulated, specialist spaces where they could be subject to municipal control had proved to be, in the words of the superintendent of town cleansing and hawkers, 'a vain hope, conceived in a spirit of high optimism'.[164]

Restructuring Pedestrian Circulation Space: Widening Verandahs and Adding Sidewalks

By the 1920s, municipal objection to the densely packed, street-oriented habitat of the Asian communities which originated with the view that it threatened both public order and public health was further strengthened by the arrival of the motor vehicle which suddenly created a new source of danger, and of noise and fumes trapped in the street.[165] Given the rapid increase in the volume of motorized vehicular traffic, verandahs took on renewed significance as footways essential for orderly traffic and the safety of pedestrians. More so than before, it was argued, fringe economic, social, and communal activities contained in verandahs and streets had to be subordinated to the essential purpose of urban transit. Municipal architects and town planners also advocated that not only should the verandah be cleared of encumbrances, it should also be restructured to accommodate a larger width and an additional sidewalk. The existing verandah, according to a member of the newly formed Singapore Society of Architects, was 'really more of a nuisance than anything else as it complicate[d] the design of buildings and add[ed] to their instability whilst being too narrow to allow two persons to pass comfortably abreast'.[166]

In 1921, W. Sims, one of the municipal commissioners, submitted that it was 'absolutely necessary to keep pedestrians off the road' and that 'a free space of at least seven feet' wide should be provided on verandahs to accommodate pedestrians.[167] By amending building by-laws to allow for a larger width of pedestrian circulation space, it was hoped that the verandah would be given a new lease of life to serve its 'legitimate' purpose as pedestrian pathways. The commissioners referred the question to Captain E. P. Richards, the deputy chairman of the Improvement Trust (DCIT),[168] who suggested a graduated scale for arcade widths ranging from 7 to 18 feet, depending on the class and width of the road in question.[169] The proposed 'new arcading rules' laid down that a compulsory clear pathway between 5 and 15 feet was to be left either completely inside the verandah or partially provided as a narrow, open pathway between the alignment of the road and the external faces of arcade pillars.[170] The DCIT anticipated that these rules would inaugurate a 'new order of things' in street and verandah traffic and in order to help people become accustomed to them, principal streets should be 'rigorously kept clear of all obstructions' as a start.[171] The 'new arcading rules' were, however, only applicable in the case of new buildings or where a road was being widened or buildings re-erected.[172] They were generally accepted by the commissioners but were referred back to the DCIT as it was pointed out that the new rules made no allowances for cases where the building plot was very shallow.[173] In 1924, W. H. Collyer, the acting DCIT, reiterated that more effort should be made to clear verandahs of impediments and that both a wider arcade and an additional footway or sidewalk outside the verandah were increasingly vital to relieve pressure exerted by the rapid swell in both pedestrian and vehicular traffic.[174]

The question of wider verandahs and sidewalks, however, remained in abeyance and two years later, when one of the commissioners queried why sidewalks had not been provided when certain roads were remade, the municipal president had to admit that they had been 'overlooked'.[175] Although sidewalks as additional elements of public circulating space were certainly beneficial to traffic and perfectly 'possible from an engineering point of view',[176] there were several practical objections to them. Most of the existing streets in town had carriageways which were scarcely wide enough for vehicular traffic, let alone additional sidewalks. In the fully built-up parts of town, the provision of sidewalks was only possible by covering over roadside drains with grids at intervals to allow the run-off of surface water. Such a scheme would not only entail high expenses but was strongly objected to by the

acting municipal health officer, Dr W. Dawson, on the grounds that it
would hinder the cleansing of street drains.[177] Dr Dawson pointed out
that the liquid found in roadside drains 'resemble[d] sewage and
contain[ed] a large proportion of solids which [could] most conveniently be removed [from open drains] by scoops or ladles'.[178]
Covering street drains over to provide sidewalks would not only mili-
tate against efficient cleansing but also restrict ventilation and harbour
rats, lice, and other vermin.[179] It was hence decided that sidewalks
were only feasible where new roads built in suburban areas were concerned; in congested central areas, the solution still lay with keeping
existing verandahs clear for pedestrian use.[180]

Negotiating Control over 'Public' Circulation Space

From the municipal perspective, 'public' spaces such as verandahs and
streets in Asian-occupied areas with their tumult of activities, mixture
of functions and constellation of overlapping interests and timetables,
were highly chaotic and 'shockingly promiscuous'.[181] The rich anima-
tion of verandahs and streets in Singapore often moved the Western
travel writer to heights of descriptive prose in attempts to capture the
colour, activity, confusion of what one called 'a real circus',[182] and
another, 'quaint Oriental streets'.[183] The following description of 'street
scenery', culled from a contemporary tourist manual, is typical:

A stranger visiting Singapore ... beholds the pathways in front of the merchants' godowns cumbered with packages ... [from] every part of the world. If
he goes amongst the native shops he will see Klings and Chinese filling them,
all busily engaged in driving bargains. In the streets he will see the roughly-clad
Chinese coolie running madly between the shafts of his eternal *ricksha*.... On
the side-ways he will notice the Chinese cook, travelling with his portable
kitchen slung over his shoulders, walking barefooted and in every conceivable
description of dress.[184]

Such disorder in public spaces ran counter to the organic imagery of
society, prevalent among nineteenth-century civic planners and authorities, who construed streets as unimpeded arteries linking different
parts of the urban whole. In Western cities, by the nineteenth century,
public space had become a 'function of motion', without any
'independent experiential meaning of its own'.[185] Verandahs and streets
were areas to move through, not to be in. They were designed to be
what Michael Walzer calls 'single-minded space', that is, space meant
to accommodate only one type of activity as opposed to 'open-minded
space' which is designed for a variety of uses, 'including unforeseen
and unforeseeable uses'.[186]

In colonial Singapore, it was the intention of the municipal authorities that by designating one type of behaviour as appropriate to the
verandah (pedestrian movement) and forbidding other types of activities (hawking, sleeping, begging, socializing, and trading), it would be
possible to minimize the promiscuity of urban life, reduce the chances

of contamination through the close juxtaposition of desirable and un-desirable activities and instead, facilitate urban transit and enhance clarity and order in the built environment. In the municipal vision of an orderly city, fringe economic activities such as fresh-produce markets and hawking which the authorities decided to tolerate were transplanted from the verandah to specialized and more controllable places designated by the municipality and made subject to a code of by-laws and municipal surveillance. Private activities such as sleeping or the storing of goods had to be confined within the private precincts of the house, while *wayang* and other forms of Asian recreation restricted to the few public parks and playgrounds within the city. If public order were to be maintained, public behaviour and public places had to be regulated, different types of activities and landuse clearly separated, and circulation flows canalized. What the Asians regarded as their customary privilege or right through long-standing usage to occupy certain public spaces for their own purposes became increas-ingly threatened in the cause of progress. In the words of one of the municipal servants, '[w]hen custom militat[ed] against the march of progress and civilization, it [was] time to consider the justification of such custom'.[187] In this 'march of progress and civilisation', the form and function of 'public' spaces such as the verandah became increas-ingly circumscribed by a web of building regulations, hawker by-laws and other anti-obstruction restrictions. Although the promoted image of the verandah was one of freedom and unfettered movement, the reality imposed on most Asian users was one of control, discipline, and surveillance.

Municipal reforms of verandah space and attempts to confine urban street life within regulated, specialist spaces, however, did not proceed unchallenged. Stanford Anderson argues that 'the physical environment is an arena for potential actions and interpretations. The "potential environment" is reinterpreted by each user, thus yielding his or her subjective environment—the environment that is effective (or influential) for that person'.[188] In transforming a 'potential' environment into an 'effective' one, Asian users were capable of exploiting the possibilities of the verandah for their own purposes, whether or not these possibil-ities had been intended by those in authority who controlled its built form. The municipal vision of planned circulation space and well-ordered canalized streams of traffic was frustrated by the fact that verandah space was mediated through a variety of behavioural modes and appropriated for a wide-ranging and occasionally conflicting number of purposes (such as consumption, transaction, and communal and social activities) by Asian users in an almost carnivalesque fashion. The verandah was a strongly contested terrain, not simply during violent confrontations such as 'the verandah riots' but also on a daily basis between municipal inspectors, the police, and the scavenging crew on the one hand, and shopkeepers, hawkers, petty traders, fresh food vendors, and other verandah tenants on the other. Although most of the small but myriad actions in using and occupying the verandah

for their own purposes belonged to the realm of practical consciousness, in effect, these have the force not only of occasionally thwarting
municipal strategies but also ensuring that the municipal vision of an
orderly city had to be constantly modified to accommodate Asian
views, whether expressed through community leaders or 'on the
ground' through relentless, repetitive action.

Up to the 1920s, the superintendent of conservancy invariably complained in his annual reports that the work of his department was 'considerably hindered and delayed by hawker stalls, periodical *wayangs*
and vehicles of every description that [were] either permanently
or temporarily stored on the footways and streets'.[189] The obstruction
of verandahs and streets, he averred, was fast becoming 'another of
Singapore's insoluble problems' and an indisputable demonstration of
the 'utter futility of applying Western criteria to Eastern peoples'.[190] In
a similar vein, Dr P. S. Hunter, the municipal health officer, saw no
signs of the 'eternal purposeless conflict'[191] between the Health
Department and verandah users abating. Even the compulsory removal
of obstructions by municipal servants, a measure introduced in
1922,[192] proved fruitless against verandah encroachment for in most
cases, the charge of $4.00 for the return of each wagon load of goods
was promptly paid, the offending goods replaced and the obstruction
continued unabated.[193] As with many other earlier measures to clear
the streets of obstruction, what initially appeared 'promising' quickly
proved when implemented to have only a 'somewhat transient
effect'.[194] By the mid-1920s, Dr Hunter reported that his department
had been unable to stem the tide of verandah and street encroachment.
In late 1924, he represented to the municipal commissioners that 'while
the Magistrates continued to inflict small fines for [obstruction], no
improvement could be expected' in the verandahs and streets of the
town.[195] He further represented that 'the question of obstruction was
hardly a health matter' and that 'the necessity for the Health Office
conducting prosecutions [in obstruction cases] was bringing the department into disrepute' and 'dragging [the department] into quite unnecessary conflict with the people who ... look[ed] upon [the department]
as an imposition already' without the department interfering with
matters which bore little relation to public health.[196] The Health
Department was hence relieved of further responsibility for the prosecution of obstruction cases in 1925. This was not only an acknowledgement that prosecution as a strategy had failed but also a sign that
the original view of verandah 'obstructionists' as threats to public
health and order had undergone some modification and the authorities
were beginning to seek new ways of defining their relationship to those
who use verandahs.

As a place, the verandah contained a diversity of intersecting claims
and practices. It was not simply an artefact but an arena of conflict
between public definition and private use, between European ideals
and Asian reality, and between the power of the authorities and the
purpose of those who lived within the city. Whilst the authorities pos-

sessed much power in planning, producing, and controlling public spaces, this was constantly challenged by the Asian plebeian classes through their own strategies. The use and occupation of the urban environment as woven into the practice of everyday life of the Asian people constituted a dimension of colonial power relations of much significance, precisely because, as Alisdair Rogers has argued, it is a form of power most exercised by groups which cannot ordinarily exercise other forms.[197]

1. *Straits Times*, 5 July 1907.

2. Quoted in *Report of the Committee Appointed to Investigate the Hawker Question in Singapore*, 4 November 1931 (hereafter cited as *Hawker Question Report*), p. 42.

3. According to Richard Sennett, the contemporary definition of 'public' and 'private' as entirely separate and diametrically opposite domains dates from the end of the seventeenth century. 'Public' meant open to the scrutiny of everyone, whereas 'private' meant a sheltered region of life defined by one's family and friends (Richard Sennett, *The Fall of Public Man*, Cambridge: Cambridge University Press, 1974, p. 16).

4. M. Dorothy George, *London Life in the Eighteenth Century*, Harmondsworth: Penguin, 1966, p. 77.

5. Martin J. Daunton, *House and Home in the Victorian City: Working-class Housing, 1850–1914*, London: Edward Arnold, 1983, pp. 12–13.

6. Ibid., p. 12.

7. G. T. Salusbury-Jones, *Street Life in Medieval England*, Hassocks: Harvester Press, 1975 (first published in 1939).

8. François Bedarida and Anthony Sutcliffe, 'The Street in the Structure and Life of the City—Reflections on Nineteenth-century London and Paris', *Journal of Urban History*, 6 (1980): 382.

9. Sennett, *The Fall of Public Man*, p. 14.

10. M. J. D. Roberts, 'Public and Private in Early Nineteenth-century London: The Vagrant Act of 1822 and Its Enforcement', *Social History*, 13 (1988): 290.

11. David Jones, *Crime, Protest, Community and Police in Nineteenth-Century Britain*, London: Routledge and Kegan Paul, 1982, pp. 20–2.

12. Chua Beng-Huat and Norman Edwards (eds.), *Public Space: Design, Use and Management*, Singapore: Centre for Advanced Studies/Singapore University Press, 1992, p. 2.

13. The word 'verandah' was originally introduced into English from India where it is found in several of the native languages (e.g. *varanda* (Hindi), *baranda* (Bengali), *baranda* (modern Sanskrit)). However, the native form appears to be merely an adoption of the Portuguese and older Spanish *varanda*, meaning railing, balustrade, or balcony. The word also appears in Malay as *baranda*, also possibly of Portuguese origins (Henry Yule and A. C. Burnell, *Hobson-Jobson: A Glossary of Anglo-Indian Colloquial Words and Phrases and of Kindred Terms*, London: John Murray, 1903). In colonial Singapore, the term 'verandah' was used to refer both to the continuous arcade lining the front of shophouses as well as the open pillared portico around a colonial house or bungalow. This chapter focuses solely on the former.

14. Mark Francis, 'Control as a Dimension of Public-Space Quality', in Irwin Altman and Ervin H. Zube (eds.), *Public Places and Spaces*, New York: Plenum Press, 1989, p. 149.

15. The earliest verandahs in Singapore were 5 or 6 feet wide, 6 feet being the requirement for lots leased for building purposes immediately contiguous to public streets (ARSM, 1896, 'Supplement: Extracts from Municipal Minutes, Papers, & etc, Relative to the Verandah Question', p. 34).

16. Alec Dixon, *Singapore Patrol*, London: George G. Harrap & Co., 1935, p. 29.

17. Raffles to Town Committee, 4 November 1822 (reproduced in Charles B. Buckley, *An Anecdotal History of Old Times in Singapore*, Singapore: Oxford University Press, 1984 (first published in 1902), p. 84).

18. The verandah was not dissimilar in function to the sidewalk or pavement which, although almost an unknown luxury before 1800, became an increasingly common feature of nineteenth-century streets in European cities. The concept of a protected pedestrian circulation space first made its appearance with the entry of wheeled carriages: posts were driven into the roadsides to limit the area carriages could use so as to protect pedestrians and storefront property (Bedarida and Sutcliffe, 'The Street in the Structure and Life of the City', pp. 385–6; Yi-Fu Tuan, *Landscapes of Fear*, Oxford: Basil Blackwell, 1980, p. 152). As a *covered* walkway, the verandah offered the additional facility of protection from the inclemency of tropical weather. Though less common than open footways, covered or arcaded walkways could also be found in certain nineteenth-century European cities. The city of Turin possessed one of the finest arcading systems, comprising 'many miles of plesant [*sic*] and roomy arcades [which] form[ed] the footpaths of the principal city roads and streets'. A well-known Parisian example of arcading costly shopfronts could be seen along the Rue Rivoli while the streets of Bologna were 'freely arcaded with exceptionally light and beautiful arching' (SIT, 14/24, 7 May 1921: 'Memorandum on Arcaded Footpaths by Capt E. P. Richards, DCIT, 11 August 1921). Covered walkways, dating from a much earlier era, could also be found in some English towns. For example, 'Butterwalk' in Dartmouth, constructed between 1635 and 1640, was a colonnaded walk and contained part of the town's pannier market. 'Poultrywalk' and 'Butterwalk' in Totnes were similarly market sites. The well-known Chester 'Rows', dating from the fourteenth century or earlier, were long, covered passages running parallel with the streets through the first floor of buildings in the centre of the city.

19. From the early days of the town, a clause was inserted in all building leases obliging parties to make verandahs in front of their houses. By the 1840s, the minimum stipulated width of the verandah was 6 feet (Buckley, *An Anecdotal History*, p. 387). Under the Municipal Ordinance IX of 1887, by-laws were passed for securing a reserve of 7 feet for an arcade or pavement along the sides of streets but this appeared not to have been strictly complied with in the nineteenth century (ARSM, 1896, pp. 17–18). In the early twentieth century, further by-laws were passed regulating the width, construction, and maintenance of verandahs. Examples included *Municipal By-Laws with Respect to Buildings* passed by the municipal commissioners on 1 May 1908 (requiring verandahs of a minimum width of 7 feet); *Municipal By-Laws with Respect to Buildings* passed by the municipal commissioners on 27 February 1925 (sections 136–154, stipulating a minimum verandah width of 7 feet); *Municipal By-Laws with Respect to Buildings* passed by the municipal commissioners on 28 May 1926 (sections 121–135, stipulating a minimum verandah width of 7 feet); *Municipal By-Laws with Respect to Buildings* passed by the municipal commissioners on 29 November 1929 (stipulating a verandah width of at least 8 feet with a minimum of 6 feet in the clear for Chuliah Street, Raffles Place, and High Street).

20. This point was constantly reiterated in the municipal records (for example, see ARSM, 1888, p. 3; MPMC, 23 May 1888; 'Extracts from Municipal Minutes, Papers, & etc, Relative to the Verandah Question', pp. 29–43; 'Memorandum on Arcaded Footpaths by Captain E. P. Richards, DCIT, 11 August 1921'.

21. ARSM, 1892, p. 8.

22. ARSM, 1895, Health Officer's Department, p. 62.

23. *Hawker Question Report*, pp. 15–16.

24. *Straits Times*, 2 Feb 1888.

25. Richard Sennett, *The Uses of Disorder: Personal Identity and City Life*, New York: Alfred A. Knopf, 1970, pp. 56–7.

26. ARSM, 1895, Health Officer's Department, p. 62; Edwin A. Brown, *Indiscreet Memories*, London: Kelly and Walsh, 1936, p. 4.

27. ARSM, 1890, p. 6; ARSM, 1895, Health Officer's Department, p. 62; ARSM, 1899, Health Officer's Department, p. 57.

28. ARSM, 1890, 'Appendix A: Specimen Extracts from Report on Verandah Obstructions', p. 16; ARSM, 1892, 'Appendix C: Verandah Obstruction', pp. 43–5.

29. Ibid.

30. *Singapore Free Press*, 12 January 1888.

31. Sandra L. Graham, *House and Street: The Domestic World of Servants and Masters in Nineteenth-Century Rio de Janeiro*, Cambridge: Cambridge University Press, 1988, p. 16.

32. *Wayang* were street theatrical shows which were particularly popular among the Chinese. Besides the public *wayang* staged for a general audience, private *wayang* were also sponsored by wealthy Chinese traders and merchants 'just as much for [their] own glorification as for religious purposes' (SIT, 172/24, 26 November 1923: 'Minute by D. Beatty, Secretary for Chinese Affairs, 24 December 1923'.

33. *Straits Times*, 10 September 1896 and 24 September 1896.

34. MPMCOM, 26 April 1918.

35. ARSM, 1896: 'Supplement: Report on a Special Investigation into the System of Death Registration in Force in Singapore, with Some Observations on Local Conditions Adversely Affecting the Health of the Population', p. 10. For example, in March 1908, in order to prevent a 'dangerous infectious disease' from being traced to particular dwelling places, a male Chinese suffering from bubonic plague was deposited on the five-foot-way in Banda Street (MPMCOM, 27 March 1908) and another on the five-foot-way of Neil Road the following month (MPMCOM, 10 April 1908). See also Chapter 3.

36. *The Chinese Protectorate Annual Reports* contain ample evidence of secret society racketeering as well as street fighting, demarcation disputes, and struggles among rival societies for the monopoly of extortion 'rights' from hawkers, brothels, shops, music-halls, and coolie depôts within their territorial bases. See also Mak Lau Fong, *The Sociology of Secret Societies in Singapore and Peninsula Malaysia*, Kuala Lumpur: Oxford University Press, 1981, pp. 76–94; Tan Pek Leng, 'Chinese Secret Societies and Labour Control in the Nineteenth Century Straits Settlements', *Kajian Malaysia*, 1, 2 (1983): 44; Yen Ching-hwang, *A Social History of the Chinese in Singapore and Malaya 1800–1911*, Singapore: Oxford University Press, 1986, p. 123.

37. James F. Warren, *Rickshaw Coolie: A People's History of Singapore (1880–1940)*, Singapore: Oxford University Press, 1986, pp. 190–1; Katherine Yeo Lian Bee, 'Hawkers and the State in Colonial Singapore: Mid-Nineteenth Century to 1939', MA thesis, Monash University, 1989, pp. 66–70.

38. This is a Hokkien word referring to toughs, rowdies and unruly elements in society, many of whom were engaged by secret societies.

39. Meryn Llewelyn Wynne, *Triad and Tabut: A Survey of the Origin and Diffusion of Chinese and Mohamedan Secret Societies in the Malay Peninsula A.D. 1800–1935*, Singapore: Government Printing Office, 1941, p. 370.

40. Buckley, *An Anecdotal History*, p. 73.

41. John Crawfurd, *Journal of an Embassy from the Governor-General of India to the Courts of Siam and Cochin China*, London: Henry Colburn, 1828, p. 556.

42. Buckley, *An Anecdotal History*, p. 387.

43. Ibid., p. 418.

44. ARSM, 1888, p. 3.

45. Ibid., 'Extracts from Municipal Minutes, Papers, & etc, Relative to the Verandah Question', pp. 29–43.

46. *Straits Times*, 6 March 1869.

47. These rights included the right to 'hold a musical festival or ceremony' in the verandah, 'put up posts and poles for an illumination', 'place any article thereon so as to cause an obstruction', 'place any dead animal, cat or dog thereon' and 'place goods or other articles thereon for such a time as may be necessary for loading and unloading' (*Singapore Free Press*, 14 January 1888).

48. ARSM, 1888, p. 3.

49. MPMC, 6 January 1888.

50. Ibid.; for examples of the municipal notice issued, see *Straits Times*, 31 January 1888 and *Singapore Free Press*, 12 January 1888.

51. MPMC, 8 February 1888; *Straits Times*, 22 February 1888.
52. *Straits Times*, 12 January 1888.
53. Ibid.
54. *Singapore Free Press*, 14 January 1888 and 23 January 1888.
55. MPMCEM, 21 February 1888.
56. *Straits Times*, 22 February 1888.
57. CO273/151/5839: Smith to Holland, 27 February 1888, 'Inspector-General of Police to Colonial Secretary, 25 February 1888'.
58. Ibid.
59. MPMCEM, 21 February 1888.
60. *Singapore Fress Press Extra*, 21 February 1888.
61. All elected members (T. Scott, T. Shelford, Tan Kim Cheng, Lim Eng Keng, and Tan Jiak Kim) as well as one of the nominated members (T. Sohst) voted in favour of the motion. Only the president and Inspector-General of Police stood out against it (MPMCEM, 21 February 1888; *Straits Times*, 22 February 1888).
62. CO273/151/5839: Smith to Holland, 27 February 1888; MPMC, 7 March 1888.
63. The press reported that mob disorder and rioting in the Kampong Glam district soon spread to South Bridge Road, China Street, Canal Road, and Boat Quay, and later to Havelock Road, Cross Street, and New Bridge Road (*Straits Times*, 22 February 1888).
64. According to the press, most of the *samseng* who took part in the disturbances were *rikisha* coolies under the direction of the headmen of secret societies (*Straits Times*, 24 February 1888).
65. *Straits Times*, 22 February 1888; CO273/151/5839: Smith to Holland, 27 February 1888.
66. Ibid.
67. Ibid.
68. *Straits Times*, 24 February 1888.
69. CO273/51/5839: Smith to Holland, 27 February 1888.
70. *Straits Times*, 24 February 1888.
71. In September 1887, the *Singapore Free Press* commented that given the 'present cooperation of independent causes tending to produce a feeling of unsettlement and even excitement among the Chinese residents', there was every prospect of 'some form of social disorder' (*Singapore Free Press*, 12 September 1887 and 23 February 1888). One of these 'causes' was the introduction of the Burials Bill which was perceived to interfere with the sanctity of Chinese burial grounds in the city. This is discussed in detail in Chapter 8.
72. This was the Malay colloquial term for the Chinese *zhong yuan jie* or the 'Festival of the Hungry Ghosts'. The festival falls during the seventh month of the Chinese lunar calendar, usually around August. It is believed that on the first day of the seventh month, the gates of hell are opened, allowing the unborn souls to wander the world in search of food and other necessities. During the month, the Chinese offer up food, prayers, and joss papers to propitiate the 'hungry ghosts'.
73. Under section 129(3) of the Municipal Ordinance of 1887, the Asian communities could apply for a permit to occupy a space 16 feet by 3 feet along the edge of a street to put up temporary erections for religious ceremonies. In practice, a much greater width of the street was often occupied by *sembahyang* tables (ARSM, 1896, p.18). Street *wayang* required large structures built either in an open space or across a street from footway to footway (ARSM, 1924, Municipal Conservancy Department, p. H10).
74. PLCSS, 22 August 1887, p. B126.
75. The two major English dailies were divided in their opinion as to whether the government prohibition was warranted. One of the papers argued in support of the measure that 'the free use of the streets conceded to the Chinese to the exclusion of the public [was] an act of grace and favor, and they must not expect this favor till the great wrong done to the Government in the person of their Protector of Chinese ha[d] been fully expiated' (*Straits Times*, 24 August 1887). The other took the view that whilst the government's action was perfectly legal in that it had merely withheld a 'special privilege

which use and wont ha[d] ... elevated into the rank of a right', the measure was unduly harsh in that the entire community was punished for the action of one man (*Singapore Free Press*, 22 August 1887 and 3 September 1887).

76. MPMC, 1 June 1887.

77. *Singapore Free Press*, 23 August 1887.

78. Ibid.

79. *Straits Times*, 30 August 1887.

80. There had been several earlier attempts to clear verandahs of hawker obstructions but none had proved effective. The Police Act XIII of 1856 instituted fines against 'whoever sets out, or exposes for sale in or upon any stall, booth, show-board, cask, or basket, or otherwise, any meat, fish, vegetables, fruit, groceries, or any other thing whatsoever, so as to cause obstruction in any thoroughfare' (William Theobald, *The Acts of the Legislative Council of India 1834–1861*, Calcutta: Government Printing Office, 1861, pp. 22–68). This was not strictly enforced by the police and in 1872, the municipal commissioners complained to the government of the 'perfunctory manner' in which the police had carried out their duty. In late October 1872, the Inspector-General of Police, C. B. Plunket, issued an order proclaimed by beat of gong that hawkers who obstructed the streets except in the neighbourhood of the markets would be arrested. Apparently, the promulgators omitted the fact that the order applied only to stationary hawkers and that there was no intention to interfere with itinerant hawkers or 'travelling cook-shops' who sold their wares throughout the town. On 29 October, riots broke out throughout the town as 'a demonstration against the Police who had incensed the hawkers by the way they had carried out the order ... with respect to street obstruction'. Although order was restored within a few days and the anti-obstruction law orders remained in force, hawkers continued to set up stalls or ply along verandahs and streets throughout the rest of the nineteenth century (CO273/65/2013: Ord to Earl of Kimberly, 30 January 1873, 'Report of the Commission Appointed to Enquire into the Riots of October 1872').

81. MPMCOM, 11 April 1888.

82. *Straits Times*, 14 April 1888.

83. MPMCOM, 11 April 1888.

84. *Straits Times*, 14 April 1888.

85. *Straits Times*, 1 March 1888.

86. *Straits Times*, 21 April 1888.

87. Ibid.

88. The motion to commence clearing verandahs in the business district was proposed by Major H. E. McCallum, the colonial engineer, and largely supported by the official members of the Board, Scott and the Chinese commissioners declining to vote (MPMC, 23 May 1888 and 6 June 1888). When the views of the Chinese commissioners were sought, Tan Jiak Kim made it clear that while he had no objection to the total clearance of verandahs in the European business quarter, he would oppose any clearance in the native parts of the town as long as he sat on the Board as he considered this 'a great hardship on the Chinese and native shopkeepers' (MPMC, 9 January 1889).

89. ARSM, 1889, p. 2.

90. MPMC, 8 February 1889.

91. ARSM, 1890, p. 6.

92. ARSM, 1898, p. 25.

93. ARSM, 1898, 'Appendix L: Street Obstruction', p. 55.

94. Ibid.

95. ARSM, 1898, p. 25.

96. The chief police officer himself pointed out that cooked food hawkers in this area 'perform[ed] a useful function in supplying food to a multitude of clerks and office servants who otherwise would have either to go without food all day long (as there [was] not an eating house in or near this neighbourhood) or betake themselves to some houses at a distance for their food' ('Street Obstruction', p. 57).

97. ARSM, 1890, p. 6; see also ARSM, 1895, Health Officer's Department, p. 62.

98. In some cases, fines imposed by the magistrates amounted to only one cent (ARSM, 1899, Health Officer's Department, p. 57).

99. Of the 16 cases listed in Table 7.2, almost half were dismissed, postponed, or discharged with a word of caution; of the rest which incurred penalties, the majority were fined $5.00 or less.

100. ARSM, 1922, Municipal Health Office, p. 39D.

101. ARSM, 1927, Municipal Conservancy and Health Department, pp. 23J–24J. See also ARSM, 1924, Municipal Health Office, pp. 34D–35D; ARSM, 1925, Municipal Conservancy Department, p. 20J.

102. This was an appeal case where the appellant had previously been convicted by the acting senior magistrate under section 129 of the Municipal Ordinance of 1887 for causing an obstruction on 17 March by selling mutton from a stall within a verandah adjoining 251 Rochor Road and 43 Beach Road. The defendant was however only fined one dollar with costs and although the Chief Justice upheld the magistrate's conviction, he ruled that 'a nominal fine [was] all that the case demand[ed]' (ARSM, 1891, p. 6; ARSM, 1891, 'Appendix C: Report of Case of Verandah Obstruction, The Municipal Commissioners on the Prosecution of G. L. Minjoot, Respondent, vs. Manasseh and Company, Appellants', p. 24). The full report of the case is given in *Straits Law Reports*, *N.S.* 18 (1891): 18–20.

103. ARSM, 1891, p. 6.

104. ARSM, 1897, Health Officer's Department, p. 88.

105. Besides the monthly verandah rent, the tenant might also have to pay over an initial sum of money, sometimes as much as $200.00 for a particularly prominent and advantageous position on the verandahs (*Straits Times*, 5 July 1907).

106. In June 1892, an elaborate list of about ninety of 'the worst class of obstructions' was submitted to the Board for action but the president could only 'persuade' the Board to deal with about a third of them (ARSM, 1892, p. 8; ARSM, 1892, 'Verandah Obstruction', pp. 43–5).

107. ARSM, 1897, p. 16.

108. MPMC, 9 January 1889.

109. CO273/151/5839: Smith to Holland, 27 Feb 1888. Tan Jiak Kim was a partner in Kim Seng & Co., a well-known firm with large land interest whilst Lim Eng Keng was a director of the Singapore Land Co. (Song Ong Siang, *One Hundred Years' History of the Chinese in Singapore*, Singapore: Oxford University Press, 1984 (first published in 1902), pp. 188 and 194).

110. *Straits Times*, 2 February 1888.

111. ARSM, 1897, p. 16; ARSM, 1897, 'Appendix U: Obstruction of Verandahs or Arcades', p. 91.

112. ARSM, 1897, p. 16.

113. ARSM, 1911, p. 15.

114. In October 1897, instructions were issued to sanitary officers that prosecutions for verandah obstructions were only to be undertaken where obstruction was recent, increasing or complete, and where partial obstruction either led to the accumulation of filth on the five-foot-way or obstructed the adjacent street drain ('Obstruction of Verandahs or Arcades', p. 92).

115. In 1899, the commissioners passed a resolution to clear obstructions in Collyer Quay, Battery Road, Raffles Square, De Souza Street, D'Almeida Street, Malacca Street, Stanley Street, McCallum Street, Prince Street, Kling Street as far as Market Street and Bonham Street. These streets were largely in the European business quarter (ARSM, 1899, Health Officer's Department, p. 57). Verandah clearance in a number of streets located mainly in Chinatown was sanctioned in 1900 (ARSM, 1900, Health Officer's Department, p. 92). Between 1901 to 1903, roughly another 140 streets extending beyond the European business quarter and Chinatown were added to the list (ARSM, 1901, Health Officer's Department, pp. 122–3; ARSM, 1902, Health Officer's Department, p. 153; ARSM, 1903, Health Officer's Department, p. 31).

116. ARSM, 1902, Health Officer's Department, p. 153.

117. It was arranged that the police should maintain a check on verandahs after visits by sanitary inspectors and that they should furnish reports and evidence of obstruction, the actual serving of summons and prosecuting to be carried out by the Municipal

Health Department (ARSM, 1900, Health Officer's Department, p. 92; ARSM, 1901, Health Officer's Department, p. 123; ARSM, 1902, Health Officer's Department, p. 153; ARSM, 1904, Health Officer's Department, p. 27; ARSM, 1906, Health Officer's Department, p. 29).

118. ARSM, 1900, Health Officer's Department, p. 92. See Table 7.1.

119. In 1898, based on the principle that quays were meant to be 'general landing places for all', a proposal from a Chinese trader to rent a large portion of Hong Lim Quay for a fee in exchange for the right to levy charges on persons depositing goods on the Quay was turned down (ARSM, 1898, p. 25).

120. ARSM, 1905, Health Officer's Department, p. 27; MPMCSM, 24 February 1905; MPMCOM, 20 October 1905. From 1907, fees paid for about 200 licences to use public quays such as Hong Lim Quay and North Boat Quay, Raffles Quay, Sumbawa Road, and Kim Seng Road for temporary storage amounted to $4,000–5,000 per annum (for example, see ARSM, 1907, Health Officer's Report, p. 36).

121. There were several attempts to extend this principle of charging a fee for the use of public space to congested streets in other parts of the town. In 1913, in response to a petition from crockery sellers at Clyde Terrace for permission to use a portion of the roadway for temporary storage of goods, the commissioners directed that the occupiers should defray the cost of paving the verandah and that a monthly rent of $5.00 per shophouse (later reduced by half) be charged for the privilege (MMSYC, 8 August 1913; MMFGPSYC, 5 September 1913).

122. In 1895, for example, the municipal health officer complained that hawkers of fish and fruit were the 'chief offenders' of drain obstruction, and that by placing their baskets over the drains, they completely frustrated the efforts of the scavenging staff to flush and sweep them (ARSM, Health Officer's Department, p. 62).

123. There were many objections to hawkers on sanitary grounds. The municipal health officer pointed out that by being exposed in the open over drains 'containing all manner of filth, even human excreta', hawker food was 'extremely liable to contamination by emanations from the decomposing organic matters contained in the drains' (ARSM, 1895, Health Officer's Department, p. 62). The condition of hawkers' lodgings where food was prepared and stored was also conducive to food contamination. Professor Simpson, for example, observed that one of the cubicles he visited was 'overrun with cockroaches and other vermin' (W. J. Simpson, *Report on the Sanitary Conditions of Singapore*, London: Waterlow and Sons, 1907, p. 16) and food for sale was also likely to have been prepared with polluted water. Hawkers themselves might also be suffering from infectious diseases and could infect the food that they handle (ARSM, 1907, Health Officer's Report, p. 2).

124. *Hawker Question Report*, pp. 3 and 15. There is considerable evidence of corruption among police and municipal officers, particularly those on street duty. A young cadet, Frank Wilson, noted in his letters that the bribing of policemen on duty for 'small privileges such as obstructing a path, leaving a *ricksha* in the wrong place etc' was a 'prevalent custom' in Singapore (Frank Kershaw Wilson, 'Letters Home, January 1915—December 1916 while Administrative Cadet, Singapore', Mss.Ind.Ocn.s.162). Using contemporary District Court Note Books, Katherine Yeo marshals considerable evidence of police extortion among hawkers in the 1920s and 1930s (Yeo, 'Hawkers and the State in Colonial Singapore', pp. 123–8).

125. *Hawker Question Report*, p. 39.

126. For instance, Mohammedan householders complained on religious grounds of pork stalls located in front of their houses. The Chetty community in Market Street also objected to Chinese food hawkers located in their street. The majority of Asian households, however, considered the hawker a source of convenience rather than an object of offence as they often supplied food right up to the doorstep (*Hawker Question Report*, pp. 24, 33, and 49).

127. *Hawker Question Report*, pp. 23 and 27.

128. MGCM, 30 September 1921. Similarly, the 1931 report on the 'hawker question' admitted that 'so long as existing housing and social conditions prevail[ed], total abolition of hawkers in Singapore [was] not practicable' as there was 'too large a

population who [did] no cooking in the houses in which they live, and to these cooked food hawkers served an undoubted need' (*Hawker Question Report*, p. 3).

129. ARSM, 1910, p. 14; ARSM, 1895, Health Officer's Department, p. 62; ARSM, 1911, p. 15.

130. Christian M. Rogerson and K. S. O. Beavon, 'A Tradition of Repression: The Street Traders of Johannesburg', in Ray Bromley (ed.), *Planning for Small Enterprises in Third World Cities*, Oxford: Pergamon, 1985, p. 235; Christian M. Rogerson, 'The Underdevelopment of the Informal Sector: Street Hawking in Johannesburg, South Africa', *Urban Geography*, 9 (1988): 551–3.

131. ARSM, 1906, p. 17.

132. Ibid.

133. ARSM, 1907, p. 9; SSGG, 16 August 1907, pp. 1620–2.

134. The initial list contained 58 specified streets or portions of streets (SSGG, 16 August 1907, pp. 1620–2). This was increased to 82 by the following December (SSGG, 18 December 1908, pp. 2942–5).

135. ARSM, 1907, Health Officer's Report, p. 37.

136. Ibid. According to the press, there were rumours of an impending strike by Chinese hawkers but nothing came of this (*Straits Times*, 5 July 1907).

137. Compiled from ARSM, 1908, Health Officer's Report, pp. 38–9.

138. For example, in 1910, fees for hawker stalls set up in Victoria Street between Middle Road and Jalan Sultan were increased from 10 to 20 cents per foot in order to reduce the number of hawkers in that portion of the street (MMSYC, 4 February 1910 and 4 March 1910; MPMCOM, 18 March 1910). Similarly, in 1912, the president reported that the introduction of a higher scale of fees resulted in a 'desirable decrease in number of licences' (Table 7.3) and 'an equally desirable increase in [hawker] revenue' (ARSM, 1912, p. 16).

139. MMSYC, 11 January 1911.

140. MPMCOM, 25 October 1912.

141. Localities included Temple Street, Trengganu Street, Smith Street, Upper Chinchew Street, the corners of China and Pekin Streets, Rochor Canal and Selegie Roads, Rochor and North Bridge Roads, Carpenter and Lavender Streets, Albert Street and Sungei Road, and Balestier Road (ARSM, 1909, Health Officer's Report, p. 45; ARSM, 1910, Health Officer's Report, p. 44; ARSM, 1911, Health Officer's Report, p. 45; ARSM, 1912, Health Officer's Report, p. 48; ARSM, 1913, Health Officer's Report, p. 47; ARSM, 1914, Health Officer's Report, p. 38; ARSM, 1915, Health Officer's Report, p. 45).

142. ARSM, 1909, Health Officer's Report, p. 45.

143. The unauthorized market at the junction of China Street and Pekin Street for example was located only a quarter of a mile away from the Telok Ayer Market which had plenty of vacant stalls for vegetables and pork (ARSM, 1909, Health Officer's Report, p. 46).

144. MMFGPC, 12 September 1913.

145. ARSM, 1914, Health Officer's Report, p. 36.

146. MMFGPSYC, 18 December 1914.

147. MGCM, 27 September 1918.

148. In particular, attention was called by the commissioners to the obstructed condition of verandahs in certain streets, namely Collyer Quay (MGCM, 15 February 1918), Raffles Square (MGCM, 27 September 1918), Malacca Street (MGCM, 3 July 1919), Beach Road (MGCM, 12 December 1919), Rochor Canal Road (MGCM, 23 April 1920) and Queen Street (MGCM, 23 April 1920).

149. ARSM, Municipal Health Office, p. 39D; SSGG, 9 May 1919, pp. 754–8.

150. SSGG, 9 May 1919, p. 754. This form of discrimination against hawkers to protect the interest of municipally licensed places was not new. According to section 190 of the Municipal Ordinance IX of 1887 (and subsequent similar ordinances), hawkers of fresh produce were not allowed within 50 yards of a public market. A pork hawker, Tan Chye Seng, for example, was convicted in April 1891 for 'exposing for sale fresh meat' on the verandah of No. 4 Crawfurd Street within 50 yards of Rochor Market ('The

Municipal Commissioners on the Prosecution of G. L. Minjoot (Respondent) v. Tan Chye Seng (Appellant)', *Straits Law Reports, (N.S.)*, 1891: 17).

151. A by-law was passed preventing any itinerant hawker from carrying on his trade in the following streets near the Esplanade: Institution Bridge, Stamford Road (from North Bridge Road to Connaught Drive), Connaught Drive (including the approach to Anderson Bridge), Empress Place (including the approach to Cavenagh Bridge), St. Andrew's Road, and Coleman Street (from North Bridge Road to Cavenagh Bridge) (MPMCSM, 27 January 1922).

152. ARSM, 1927, p. 2.

153. ARSM, Municipal Health Office, p. 40D.

154. MMSC2, 6 December 1922.

155. ARSM, 1923, Municipal Health Office, p. 32D.

156. ARSM, 1927, Municipal Conservancy and Health Department, p. 23J.

157. ARSM, 1928, Municipal Conservancy and Health Department, pp. 17J–18J.

158. ARSM, 1929, Municipal Conservancy and Health Department, p. 17J.

159. ARSM, 1920, Municipal Engineer's Department, p. 20A. In 1931, it was estimated that there were approximately two unlicensed (and therefore illegal) hawkers to every licensed hawker (*Hawker Question Report*, p. 16). From the hawker's perspective, whilst a licence conferred certain advantages such as reduced risk of arrest, it did not entirely prevent him from running afoul of the law. Licensed hawkers were controlled by a complex arsenal of hawker legislation and contravention of any one by-law could result in prosecutions and fines. The licensed hawker was still liable to be arrested for a range of possible offences such as operating out of stipulated hours, protruding beyond the street space allocated for his stall, failing to display his licence, littering, or, in the case of the itinerant, failing to 'move on'. It was hence understandable that many hawkers preferred to run the municipal gauntlet operating without a licence. The municipal health officer admitted that 'as the same punishment [was] meted out to the licensed as to the unlicensed [hawker], the result [was] that neither [were] controlled' (ARSM, 1924, Municipal Health Office, p. 35D).

160. MPMCOM, 30 September 1921.

161. ARSM, 1924, Municipal Health Office, p. 32D.

162. ARSM, 1927, Municipal Conservancy and Health Department, p. 23J.

163. ARSM, 1924, Municipal Health Office, p. 32D; ARSM, 1925, Municipal Health Office, p. 32D.

164. ARSM, 1927, Municipal Conservancy and Health Department, p. 23J.

165. In 1919, it was estimated that there were 'between 750 to 800 carriages registered, well over 1,000 motor cars, upwards of 100 motor lorries, some 9,200 *rickishas*, 2,700 bullock carts and 2,200 handcarts, a total of say, 15,900 vehicles' on the streets of Singapore (*Straits Times*, 8 May 1919). Unlike in 'an average, high grade English town of the same size' with 'well-laid' streets 'all with broad side-walks', the pedestrian in Singapore 'must take his chance among motor cars, bullock carts, snorting lorries, and *rickishas* by the score' (*Straits Times*, 21 November 1921).

166. [S. D. Meadows], 'Impressions', *Journal of the Singapore Society of Architects*, 1, 2 (1924): 32.

167. SIT, 14/24, 7 May 1921: 'Memorandum by W. Sims, 13 May 1921'.

168. MMSC3, 8 July 1921.

169. 'Memorandum on Arcaded Footpaths by Captain E. P. Richards, DCIT, 11 August 1921'.

170. In the opinion of the DCIT, a narrow path outside the arcade pillars was essential in order to prevent danger to foot passengers stepping off the arcade to cross a street. Given the obstructed condition of the verandahs in Singapore, foot traffic was already encroaching on the road and utilizing 3–4 feet of the sides. The provision of an open footpath would simply regularize and separate foot from wheeled traffic. See Memorandum on Arcaded Footpaths by Capt E. P. Richards, DCIT, 11 August 1921.

171. Ibid.

172. SIT, 523/24, 8 July 1924: 'Memorandum by W. H. Collyer, Acting DCIT, 13 August 1924'.

173. MMSC3, 9 September 1921.

174. 'Memorandum by W. H. Collyer, Acting DCIT, 13 August 1924'.

175. MPMCOM, 27 August 1926.

176. SIT, 524/26, 24 August 1926: 'Questions by Mr Knocker Asking Why No Provisions were Made for Sidewalks'.

177. SIT, 524/26, 24 August 1926.

178. Ibid.

179. Ibid.

180. Ibid.

181. Peter Stallybrass and Allon White, *The Politics and Poetics of Transgression*, London: Methuen, 1986, p. 135.

182. Fritz Hochberg, *An Eastern Voyage: A Journal of the Travels of Count Fritz Hochberg through the British Empire in the East and Japan*, Vol. 2, London: J. M. Dent and Sons, 1910, p. 112.

183. W. Robert Foran, *Malayan Symphony*, London: Hutchinson and Co., 1935, p. 32.

184. *Souvenir of Singapore: A Descriptive and Illustrated Guide Book*, Singapore: Straits Times Press, 1905, p. 29.

185. Sennett, *The Fall of Public Man*, p. 14.

186. Michael Walzer, 'Pleasures and Costs of Urbanity', *Dissent (New York)* (Fall 1986), 470.

187. ARSM, 1927, Municipal Conservancy and Health Department, p. 24J.

188. Stanford Anderson, 'People in the Physical Environment: The Urban Ecology of Streets', in Stanford Anderson (ed.), *On Streets*, Cambridge, Mass.: MIT Press, 1978, p. 6.

189. ARSM, 1925, Municipal Conservancy Department, p. 20J; see also ARSM, 1923, Scavenging and Conservancy Department, p. 19H; ARSM, 1924, Municipal Health Department, p. H10.

190. ARSM, 1927, Municipal Conservancy and Health Department, pp. 23J–24J.

191. ARSM, 1927, Municipal Health Office, p. 37D.

192. In 1922, section 116 of the Ordinance No. 135 (Municipal) was amended to enable the municipal commissioners through their servants to remove any obstruction from certain specified streets (ARSM, 1922, p. 2; MPMCOM, 28 July 1922). By the end of 1923, only Victoria Street and Beach Road had been prescribed as streets where this amended section was applicable (ARSM, 1923, p. 9).

193. ARSM, 1927, Municipal Conservancy and Health Department, p. 24J.

194. ARSM, 1923, p. 9.

195. ARSM, 1924, Municipal Health Office, pp. 34D–35D.

196. Ibid.

197. In the context of central Los Angeles, Rogers shows that 'using and occupying streets, parks and other public places is one way that the otherwise powerless may realise alternative versions of community. In addition, they may decorate and name space independently of the ownership of property ... [and thereby] declare a claim on space or community.' Alisdair Rogers, 'Gentrification, Power and Versions of Community: A Case Study of Los Angeles', *University of Oxford School of Geography Research Paper 43*, 1989, p. 25.

8
The Control of 'Sacred' Space: Conflicts over the Chinese Burial Grounds

The cemetery stood in a place, valueless when it was chosen, which with the increase of the city's affluence was now worth a great deal of money. It had been suggested that the graves should be moved to another spot and the land sold for building, but the feeling of the community was against it. It gave the taipan a sense of satisfaction to think that their dead rested on the most valuable site on the island. It showed that there were things they cared for more than money.[1]

Singapore should adopt the modern idea of putting all future graveyards well away from the city, where they cannot interfere with free development.[2]

The 'Sacred' in the Urban Built Environment

IN traditional societies, a sense of the 'sacred' is often inherent in the form of the urban built environment, which in turn, cannot be understood apart from the 'mythical-magical concern with place'.[3] According to Mircea Eliade, the act of settlement itself is perceived as a re-enactment of the mythical creation of the world.[4] Ancient Indian cities were designed according to a mandala replicating a cosmic image of the laws governing the universe and similarly, Chinese cities were conceived as 'cosmo-magical symbols' of the universe.[5] These cities were laid out as terrestrial images of the macrocosmos, distinct spaces sacralized for habitation within a continuum of profane space.[6] As Paul Wheatley observes, 'for the ancients the "real" world transcended the pragmatic realm of textures and geometrical space, and was perceived schematically in terms of an extra-mundane, sacred experience. Only the sacred was "real", and the purely secular—if it could be said to exist at all—could never be more than trivial.'[7]

Unlike 'pure' geometric space, sacred space, 'the space of myth and magic' is non-isotropic because it is structured by rules of inherence.[8] Space and substance are 'fused' rather than distinct, and every position and direction in mythical space is endowed as it were with a particular 'accent' or 'tonality' of its own.[9] As such, in the sacred spaces of traditional cities, geographical locations, geographical directions, and other spatial properties are infused with enormous intrinsic significance. With the penetration of a capitalist order, however, space becomes viewed as separate from substances of value, as 'empty' land which could be demarcated, parcelled out, commodified, and purchased, not for its intrinsic but potential value for speculative purposes. Only under profitable circumstances do particular places and substances combine, and 'then only until a more profitable arrangement appears to dissociate and recombine them with other places and things'.[10] In colonial societies where a capitalistic order was imposed on a traditional one, 'sacred' space was often eroded away, giving way to commodified space which was more amenable to measurement, 'scientific' planning and commercial development. This process, however, did not occur without conflicts and compromises.

This chapter examines the negotiation of power over the urban landscape between the municipal authorities and the Chinese communities in colonial Singapore over a highly contentious question: whether any element of the urban built environment could be considered 'sacred' and hence inalienable and immune to changes demanded by the economic rigours of urban development. By the late nineteenth century, this debate was focused on one particular element of the city's 'sacred' geography—the burying places of the Chinese communities. Not only do the social organization and production of burial spaces and artefacts reflect the cultural norms and practices of particular social groups,[11] they should also be situated within broader developments which acknowledge the significance of the politics of power. Far from being the products of a monolithic culture, burial spaces are often construed in a plurality of ways and invested with diverse if not antithetical meanings by different individuals and social groups.[12] Within this construction of space, the burial landscape is construed as a site of control and resistance which is simultaneously drawn upon by, on the one hand, dominant groups to secure conceptual or instrumental control, and on the other, subordinate groups to resist exclusionary definitions or tactics and to advance their own claims. The focus here is on the conflicting discourses which developed over the nature and location of Chinese burial grounds, the strategies of negotiation employed by the authorities and the Chinese communities, and the resultant impact on remaking the 'sacred' geography of colonial Singapore.

Death and Cemeteries: The Western European Tradition

In western Europe, the late eighteenth century and early nineteenth century marked a profound change in people's attitude to death and their perception of cemeteries. Philippe Aries argues that until the eighteenth century, Latin Christendom was indifferent to 'the daily presence of the living among the dead'.[13] He attributes the mid-eighteenth-century movement to relocate cemeteries outside cities to profound changes in Christian attitudes towards death which engendered a strong desire among the living to detach themselves from the dead. James Riley proposes a modified view that this shift in popular attitudes towards death in Europe was led by public health arguments in the wake of the mid-eighteenth-century environmentalist campaign to modify disease-ridden environments or remove them away from human habitat.[14] In England, the campaign for cemetery reform gathered momentum in the second quarter of the nineteenth century and was inspired both by public health concerns in the wake of the cholera epidemics as well as changes in social behaviour, aesthetic taste, and moral outlook. In his 1843 report on interment in towns, the sanitarian Edwin Chadwick advocated the expulsion of the dead from the presence of the living not only on sanitary but also 'moral, religious and physical' grounds.[15] Concomitant with the new concern for urban hygiene were new sensibilities which combined to 'impart to the cemetery as well as the hospital, the slaughterhouse and the prison the taint of impure institutions that had to be banished to the periphery of the city' in order to 'assure the purity of the new Enlightenment urban milieu'.[16] In an age where the emphasis upon public hygiene was joined with a concern for moral enlightenment, the image of church graveyards as 'horrific spectacles of decomposing corpses, piles of bones and broken coffins'[17] was no longer acceptable, for graveyards not only contaminated the environment and endangered the health of the neighbourhood but also sullied the purity of the church.[18] Instead, cemeteries should present 'an ordered arrangement of monuments ... as tokens of a nation's progress in civilization' and act as "sweet breathing places" for contemplation, the indulgence of sweet melancholy, and the improvement, enlightenment and education of those whose lives had not yet run their course'.[19] It was in this mood for reform that inner-city graveyards were closed, and carefully designed and landscaped extramural cemeteries such as Liverpool's St James Cemetery, the Glasgow Necropolis, London's Kensal Green Cemetery, and the South Metropolitan Cemetery in Norwood were established. Just as the living were to live clean and godly lives, the dead were to be buried in pleasant Arcadian surroundings where they would not be disturbed and where they would not harm the living.[20]

Burial Grounds and Urban Development in Colonial Singapore

The campaign for extramural interment and properly managed cemeteries in Britain had little impact on colonial Singapore in the first half of the nineteenth century. In a plural society, the question of cemetery location was complicated by the diversity of death rituals and burial customs prevalent among different ethnic and religious communities. The government surveyor, J. T. Thomson, observed that, 'in Singapore ... native burial grounds are to be met with in all directions.... The Malays seek out sand ridges or *permatangs* in which to bury their dead. The Chinese look for round knolls and hillsides. The Hindoos burn their bodies, so that nothing may remain of what was a living soul.'[21]

The European and Eurasian Christian communities also had a separate Christian cemetery on Fort Canning Hill from the earliest days of the Settlement.[22] The provision and management of cemeteries were communal responsibilities and prior to 1857, apart from the Christian cemetery, the colonial government exerted little control over the burying places of the Asian communities. As a result, 'native' burial grounds were 'met with in unexpected places',[23] proliferating within and close to the town and often presenting spectacles of 'fresh human bones and coffins and humus sticking out of the sand by the roadsides'.[24] Of the most numerous and extensive were the Chinese burial grounds which were 'increas[ing] so fast that S'pore seem[ed] likely to become a vast Chinese cemetery'.[25] Although the Indian Conservancy Act XIV of 1856 prohibited interment in unregistered and unlicensed grounds, enforcement was slack and the Act a dead letter. Consequently, many Chinese families took full advantage of the *de facto* liberty to lay out family and private burial grounds within and outside municipal limits. Once consecrated as graves, the land was seldom given up and in effect, 'locked up in perpetuity'.[26]

By the 1880s, however, the physical expansion of the city was increasingly restricted by the scarcity of suitable land for building purposes. Building land was in short supply in the immediate area around the perimeter of the city. The northern edge of the city was surrounded by a wide expanse of low-lying swampy land ineffectively drained by the Kallang and Rochor rivers. The swamp extended through the Kallang district as far as the Geylang River and inland to Serangoon Road, with several large Malay, Boyanese, Bugis, and Orang Laut *kampung* forming a ring around the seaward edge of the swamp. The undulating high ground to the west of the city centre, the Claymore–Tanglin district, was already an exclusive residential quarter occupied by the bungalow compounds and recreational facilities of the Europeans and the wealthy Chinese. Any encroachment by working-class housing or undesirable forms of landuse would have been considered detrimental to the market value and amenity of the area. To the south, between Chinatown and New Harbour, the terrain comprised low ranges of hills interspersed with mangrove swamps and low-lying land in the valleys. Most of the high ground here had been appropriated for Chinese burial grounds, both public and private, whilst the valleys were frequently occupied by squatters. In the eyes of colonial town planners intent on relieving overcrowding in the city proper, the proximity of this area to both the southern congested area and the harbour rendered it ideal as an overspill area, but '[where] one would expect ... a network of roads, we find nothing but a space that is barren and waste, because almost all that is not swamp is given over to the dead'.[27]

From the perspective of the colonial and municipal authorities, urban building land was scarce not only because swamps around the fringe of the city rendered much of the land unsuitable for building, but also because where there were suitable plots of land, they were alienated by the Chinese who converted them into extensive public and private burial grounds.[28] The Chinese predilection for burying their dead on knolls and hillsides meant that the same places which would have made ideal residences were 'lost to the Colony as far as progress and development [were] concerned, and that in case the land should come into

TABLE 8.1

Chinese Burial Grounds within Municipal Limits in the
Late Nineteenth to Early Twentieth Century[a]

Name	Dialect/ Surname	Year Opened	Size	Location
1. Qing Shan Ting	Cantonese/ Hakka(public)	1820s	20 acres	Junction of South Bridge Road and Tanjong Pagar Road (east of Ann Siang Hill and west of Peck Seah Street, on the slopes of Mount Wallich/ Scott's Hill)
2. Heng Shan Ting	Hokkien (public)	1828		Between Silat Road and Neil Road
3. Lu Ye Ting	Cantonese/ Hakka (public)	1840	23 acres	Off Outram Road, between Havelock Road and Tiong Bahru Road
4. Tai Shan Ting	Teochew (public)	1845	70 acres	Bounded by Orchard Road, Paterson Road and Grange Road
5. Lao Yi Shan	Hainanese (public)	1862	29 acres, 3 roods	Thomson Road, 5 milestone
6. Bi Shan Ting	Cantonese (Guang-hui-zhao)	1870		Kampung San Teng, off Thomson Road
7. Xing Wang Shan/Tai Yuan Shan	Hokkien Seh Ong	1872	221 acres, 2 roods, 34 poles[b]	Near 3½ miles and north of Bukit Timah Road (Adam Road, near junction of Kheam Hock Road)

(continued)

TABLE 8.1 *(continued)*

Name	Dialect/ Surname	Year Opened	Size	Location
8. Lao Shan	Hokkien (public)		98 acres	3$\frac{1}{2}$ miles and north of Bukit Timah Rd (off Kheam Hock Rd)
9. Xie Yuan Shan	Hokkien Seh Yeo	1882	128 acres	Silat Gate (4 miles, Telok Blangah Rd)
10. Yue Shan Ting	Hakka (*Feng-shun, Yong-ding,* and *Da-bu*)	1882	143 acres	Holland Road, 7$\frac{1}{2}$ miles
11. Shuang Long Shan	Hakka (*Jia-ying*)	1887	90 acres	Holland Road, 5 miles
12. Xin Yi Shan	Hainanese (public)	1891	33 acres, 2 roods, 31 poles	Thomson Road, 5$\frac{1}{2}$ milestone
13. Guang En Shan	Teochew (public)			Kampung Chia Heng, between Balestier and Thomson Roads
14. Guang Yi Shan	Teochew (public)		44 acres	Serangoon Road, 4$\frac{1}{2}$ miles
15.	Hokkien Seh Lee		66 acres, 1 rood, 7 poles (1952 figure)	Serangoon Road, 4$\frac{3}{4}$ miles
16.	Hokkien Seh Choa			Next to the Lunatic Asylum, between Tiong Bahru and Silat Roads
17.	Hokkien Seh Khoa and Chaw			Kampong Bahru, 2$\frac{1}{2}$ miles
18. Si Jiao Ting	Hokkien (public)			Burial Ground Road (Tiong Bahru)
19.	Hokkien Seh Wee			Off Tiong Bahru Road
20.	Hokkien Seh Lim		44 acres, 3 roods, 61 poles (1952 figure)	Off Tiong Bahru Road

TABLE 8.1 *(continued)*

Name	Dialect/ Surname	Year Opened	Size	Location
21. Lin Ji Shan	Hokkien (public)			Alexandra Road, 3½ miles
22.	Hokkien (public) North-east portion set aside for Seh			Alexandra Road, 3 miles
23.	Hokkien Seh Ong			Alexandra Road, 3rd milestone, about ¾ miles off the road
24.	Hokkien (public)			Alexandra Road, 4½ miles
25. Coffee Hill Ka Fei Shan	Hokkien (public)	Licensed 1918	50 acres	Between Mount Pleasant and Whitley Roads
26. Bukit Brown Chinese Cemetery	Municipal	1922	213 acres	Junction of Kheam Hock Road and Sime Road

Sources: 'List of Existing Cemeteries, 1989' compiled by the Ministry of Environment, Singapore; PLCSS, 1905, pp. C202–8: 'List of Chinese Burial Grounds within the Municipal Limits of Singapore'; Peng Song Toh, *Directory of Associations in Singapore, 1982/1983*, Singapore: Historical Culture Publishers, 1983, pp. L17–133; '165th Anniversary Commemorative and Souvenir Publication of the Ying Fo Fui Kun, 1827–1987', Singapore, 1989, pp. 26, 30, 90–1; 'Historical Account of Hok Tek Chi and Loke Yah Teng Cemetery', Singapore, 1963, pp. 37–8; 'Souvenir Magazine of the Opening Ceremony of the Kheng Chiu Building', Singapore, 1965, p. 173; 'The Yang Clan Genealogy', Singapore, 1965, p. G16; '118th Anniversary Souvenir Magazine of the Singapore Guang Hui Zhao Bi Shan Ting', Singapore, 1988, unpaginated; 'Souvenir Magazine of the Singapore Kai Min Ong See Benevolent Association', Singapore, 1977, p. 76; Zhang Qin Yong, 'Yang Shi Jia Zu Yu Xie Yuan Shan', *Asian Culture*, 8 (1986), pp. 24–7; *Report of the Committee Appointed to Make Recommendations Regarding Burial and Burial Grounds in the Colony of Singapore*, Singapore: Government Printing Office, 1952, p. 14; and information gathered from the Singapore Hokkien Huay Kuan, the Teo Chew Poit Ip Huay Kuan, the Ying Fo Fui Kun and the Singapore Kiung Chow Hwee Kuan in September/October 1989.

[a]This list is confined to burial grounds which were opened to Chinese belonging to particular *seh* or surname groups (such as Nos. 7, 9, 15, 16, 17, 19, 20, and 23), to immigrants from particular districts in China (such as Nos. 6, 10, and 11), or more generally, to particular dialect groups (such as 1–5, 8, 12, 13, 14, 21, 24, and 25). Also included is the Chinese Municipal Cemetery (No. 26) which was opened to all Chinese regardless of affiliations. In colonial sources, these are generally referred to as 'public' burial grounds as opposed to 'private' ones reserved for family burial.

[b]About 140 acres of this estate was acquired by the municipality in 1919/1920 for the Bukit Brown Municipal Cemetery (MGCM, 8 March 1918; MPMCOM, 28 November 1919).

FIGURE 8.1
Location of Chinese Burial Grounds in and around the Municipal Area
in the Late Nineteenth to Early Twentieth Century

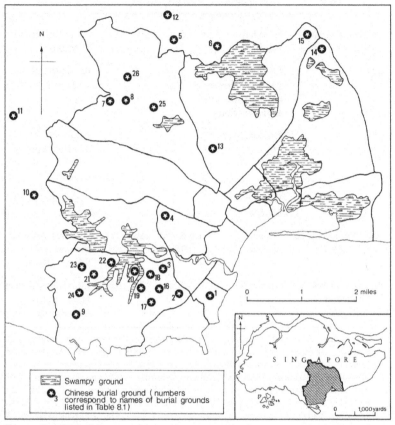

Source: See Table 8.1.

the market, its value [would] have deteriorated'.[29] Not only were there
fears that it was 'a mere question of time before the island of Singapore
would become an enormous cemetery',[30] there was also an insistence
that selecting 'the highest and most beautiful sites' for the dead while
'the living [dwelt] on the adjacent low swampy ground' was a mis-
placed priority among the Chinese.[31] To the Western mind, it was an
incomprehensible inversion that '[h]aving settled their ancestors ... on
the summits of the highest hills in the vicinity, the mourners [should]
descend to the swamps and valleys, there to live huddled together in
large numbers, [thereby becoming] an easy prey for all infective diseases, especially tuberculosis'.[32] The Chinese preference for hillside
burial places was also considered undesirable 'from a sanitary point of
view' as it was feared that springs at the base of hills from which water
for domestic purposes was obtained would become 'impregnated with

organic matter and thus generate disease among the people in the neighbourhood'.[33] By the 1880s, with the emerging pressure for sanitary reform (see Chapter 3), Chinese burial grounds located within or close to the heart of the city (for example, burial grounds numbered 1–4 and 16–20 in Table 8.1 and Figure 8.1) were increasingly perceived as hazardous to public health, particularly given the persistence of the miasmatic theory of disease which attributed epidemics to pestilential emanations arising from common graves coupled with the popular image of the tropics as a place where 'decomposition [was] rapid and where disease assume[d] its acuter forms'.[34] The emerging conception of Chinese burial grounds as both insanitary and obstructive to modern urban development soon fuelled a growing contention between the colonial authorities and the Chinese communities over the disposal of the dead, an issue which the *Singapore Free Press* considered 'the most vital question that can occupy the attention of any populous community'.[35] Just as the relocation of cemeteries in European cities which began a century before implied not simply a modification of urban topography but a general secularization and municipalization of cemetery administration and a reduction of the role of the clergy 'to one of supervision and protocol',[36] the debates surrounding the so-called 'burials question' in colonial Singapore were complicated by the crucial issue of whether burial grounds in a colonial city were to be treated as 'sacred fields of repose' governed by the religious customs of different communities or hygienic sites under municipal purview.

The Control of Chinese Burial Grounds: The Debate over the 1887 Burials Bill and Subsequent Legislation

In 1887, an attempt was made to bring Chinese burial grounds under some semblance of control through an ordinance authorizing the licensing, regulation, and inspection of burial and burying grounds.[37] The *raison d'être* for the bill appeared to be twofold. First, by licensing and controlling the siting of burial grounds, further alienation of land suitable for the construction of buildings and roads could be prevented. Second, by prescribing sanitary regulations governing the depth of graves and proper places for interment, public health would be safeguarded. The introduction of the Burials Bill during the August Legislative Council sessions generated much agitation and concern among the Chinese of all three of the Straits Settlements. Seah Liang Seah, the Chinese member of the Council requested a postponement of the second reading to allow more time for consideration as the Bill 'seriously affected the interests of the Chinese community, mostly those of the respectable class'.[38] In particular, he perceived that the Bill aimed at 'the suppression of private burial grounds', a measure which would 'much affect the much-cherished customs of the better class of the Chinese'.[39] The Colonial Secretary, in granting the postponement, assured the Council that the Bill had been drawn up 'with full regard for the feelings of the Chinese', and that there were no intentions to interfere with their 'feelings and religious sentiment'.[40] However, he made it clear that Chinese custom could only be tolerated 'within reasonable limits and without sacrificing the good of the community': 'It is not right that all other classes of the community should be sacrificed to the desires of one section [the Chinese], to secure, for instance, all the small hills, which are the only places suitable for healthy houses in these countries, and take them for ever, sometimes merely as a monument to the honour of a man's family and his own personal vanity.'[41]

From the perspective of the Chinese, the Burials Bill represented an attack on Chinese customary rituals and an erosion of Chinese control over their own sacred spaces. Whilst there were initially rumours of riots and threats of violence, the protest against the bill soon resolved itself into constitutional channels.[42] This was mainly because the bill principally affected the wealthier and more prominent members of the Chinese community who could afford private burial grounds and who were chiefly responsible for reserving large parcels of land for use as family graves. Protest against the bill was hence led by leaders of the community who were familiar with the legislative system and conversant with their own rights and privileges under such a system.[43] The means of protest adopted included letters to the press, public meetings to gather support and petitions to the Resident Councillors and the Governor. When the latter failed to arrest the progress of the bill through the Legislative Council,[44] memorials pleading their cause were sent to Sir H. T. Holland, the Secretary of State for the Colonies in London.[45]

In the main, it was argued that among the Chinese, ancestor worship, and in particular sepulchral veneration, would be incomplete if the liberty to select propitious burial sites according to the principles of geomancy were curtailed by government interference.[46] The concept of control of burial places by an external agency was an alien notion among the Chinese.[47] Feng shui,[48] or Chinese geomancy, was considered central to the Chinese faith because it was believed, so the petitioners claimed, that it was possible to site the grave in relation to the configuration of the landscape and the vicinity of watercourses in such a way that benign influences were drawn from the earth and transmitted to the descendants of the deceased. In general, a favoured burial site was one situated at the conjunction between the 'azure dragon' on the left and the 'white tiger' on the right, the former signifying boldly rising 'male' ground and the latter emblematic of softly undulating 'female' ground. The site should ideally contain three-fifths 'male' and two-fifths 'female' ground and should be open in front to breezes, shut in on the right and left, with a tortuous, winding stream running before it. Such ground contained an abundant supply of beneficial 'vital breath' which augured well for descendants. On the other hand, flat, monotonous surfaces or landscapes characterized by bold, straight lines such as the presence of a straight line of ridges, watershed, railway embankment, road, or water running off a straight course tended to concentrate malign influences and were avoided as burial sites.[49] In sum, there was a preference for tortuous or winding structures which fitted with rather than dominated the landscape, and a strong objection to straight lines and geometrical layouts which constrained and imprisoned nature rather than flowed along with it.[50] Once sited according to geomantic principles, both the tomb and its sepulchral boundaries were considered 'inviolable'[51] as any interference with them would spoil the efficacy of the *feng shui* and imperil the welfare and prosperity of living descendants. In short, burial grounds, by nature of being sacred sites, must be exempt from government control and external interference. The petitioners argued that, as in Roman law, 'the ground in which one, who had the right, buried a dead body, became *ipso facto* religious; it ceased to be private property; it would not be bought or sold, or transferred or used; it was for ever dedicated to the dead and reserved from all current usages; and it should be sacrosanct to the memory of the departed'.[52] In Chinese culture, the burial landscape itself became sacrosanct because landscape formed the basic material of architecture. As Oswald Spengler has argued, for the Chinese, 'it is the architecture of the landscape, and only that, which explains the architecture of the buildings.... The temple [or tomb] is not a self-contained building but a layout, in which hills, water, trees, flowers and stones in definite forms and dispositions are just as important as gates, walls, buildings and houses.'[53] This view stems from the basic conception of space in Chinese culture as a meandering path that wandered through the world, whereby 'the individual is conducted to his god or ancestral tomb ... by friendly Nature herself ... [by] devious ways through doors, over bridges, round hills and walls'.[54] Also basic to *feng shui* was the notion that human intervention in the natural landscape was fraught with risks and generated repercussions on society. The Chinese conception of the universe was one where 'men are bonded to the physical environment, working good or ill upon it and being done good or ill to by it'.[55] As explained by Maurice Freedman,

in the Chinese view, ... the physical universe is alive with forces that, on the one side, can be shaped and brought fruitfully to bear on a dwelling and those who live in it, and on the other side, can by oversight or mismanagement be made to react disastrously.... In principle, every act of construction disturbs a complex balance of forces within a system made up of nature and society, and it must be made to produce a new balance of forces lest evil follow.[56]

Feng shui considerations hence had to be carefully taken into account in siting a grave (or any other building) if harmony between society and the physical landscape were to be maintained. As 'a theory of location', it 'offers man the opportunity to exercise his "responsibility" to the cosmos by fixing the limits of "change" in terms of the visible landscape configuration which represents the cosmos in a living, spatial dimension'.[57] Unlike in Western science where the 'environment' as a concept is often broken down and analysed in separate categories, *feng shui* unifies the geological, atmospheric, aesthetic, and psychological qualities of the environment in one theory and code of practice which is seen as integral to the lives of the people.

On both sides, arguments were deftly marshalled to press their case home. The Chinese quoted their sages, drew out innumerable examples illustrating how deeply seated and inviolable the principles of geomancy were among the Chinese, and appealed to the Imperial Charter which provided for due respect of the religious sentiments of all races. The discourse was strategically couched in terms of religious idealism, because religion and sacred places associated with it could claim the privilege of being beyond the purview of a secular government. The Chinese were also quick to counter the charge that hillside burial grounds were insanitary and liable to contaminate the city's water supply. They argued that pollution of hill streams was impossible because unlike those of Malays and Europeans, Chinese graves were of considerable depth and lined with great quantities of quicklime. It was constantly reiterated that the grave of a wealthy Chinese was 'substantially built', 'planted round at great cost with shrubs', regularly visited and tended, and that the coffin was made of 'the hardest wood that [could] be obtained and well-lacquered' so that it was 'perfectly tight and waterproof'.[58] The memorial sent to the Secretary of State for the Colonies summarized these arguments and put forward two main requests. First, notwithstanding the assurances given by the local government that their religious customs would be respected, the memorialists feared that the Burials Ordinance opened up the possibility, and even the likely prospect, that their burial grounds would be 'seized and turned to other purposes'.[59] Their foremost request was that existing burial grounds would be protected from such 'contemplated desecration'. Second, they requested the liberty to choose their own burial sites outside a radius of two miles from the town proper in accordance with 'their religious faith in ancestor worship' and 'as directed in the teachings of their sages Confucius and Mencius'.[60] In their turn, the local government and the English language press cited land scarcity, sanitation, the public good, and attempted to counter the Chinese argument by showing that the much-vaunted religious idiosyncrasies of the Chinese were by no means respected by the mandarins in China who '[did] not scruple to appropriate with little delicacy of feeling burial grounds, shrines, etc. when required for Government purposes'.[61] The Colonial Secretary, in defending the Burials Bill, cited evidence demonstrating that in China ancestral temples were summarily demolished to make way for government projects and claimed that 'the Chinese had no real feeling against removing graves'.[62] In fact, he had been 'reliably informed' by the Consul-General of Shanghai that Chinese people 'were in the habit of placing the remains of their ancestors in urns in order that they might at any convenient season remove them from one place to another'.[63] In short, the wealthy Chinese had no 'real' grievance and were simply pandering to vested interest under the guise of religious sentiment instead of protecting public health and advancing public good. Their request to be allowed to select their own burial sites outside a 2-mile radius of the town was also dismissed by the Governor as tantamount to asking for an 'illegal privilege'.[64] The Protector of Chinese, W. A. Pickering, further reduced the issue to a question of choosing between two sets of priorities: 'Are [Chinese] customs connected with the burial and worship of the dead compatible with the sanitary welfare of the living general community, and is the practice of buying land and appropriating some of the best sites as private mausolea conducive to the interest of the Government as regards the reasonable development of the Land Revenue, and the progress and prosperity of the Colony?'[65] In his mind, the answer was unassailably clear: the claims of the living must prevail over those of the dead, customs should not be allowed to stand in the way of urban progress and land revenue, and private benefits should be subservient to 'that of the superior law, the welfare of the general public'.[66] The burials question was hence polarized into choosing between diametrically opposed priorities such as between the living and the dead, the progressive and the customary, and the public and the private.

On receiving the Chinese petitions from each of the three Settlements, Holland, the Secretary of State for the Colonies, decided that the coming into operation of the Burials Ordinance should be postponed and instructed the Governor to repeal the ordinance and re-enact it with modifications to take into account Chinese sentiments.[67] The ordinance was duly repealed but no further legislative action was taken on the burials question during the next eight years. It was left in abeyance until 1896 when the burials issue was resurrected in relation to the bill to amend the Municipal Ordinance.[68] This time, it was decided to introduce dual legislation. Outside municipal limits, the Burial Ordinance XIX of 1896 transferred the duty of licensing and controlling burial grounds to the colonial government. Within municipal limits, however, under sections 232 to 238 of the Municipal Ordinance XV of 1896,[69] the control of burial grounds was vested in the Municipal Commission rather than in the colonial government (as opposed to what was contemplated by the 1887 Burials Bill). The municipal authorities were empowered to license and inspect burial grounds, to close them if they proved 'dangerous to the health of persons living in the neighbourhood' and to impose a maximum penalty of $500.00 for the disposal of corpses in unlicensed places.

Although the new ordinance signified an important though much delayed victory for the colonial and municipal authorities, certain concessions had to be made to meet Chinese demands. Contrary to the wishes of the municipal authorities, for example, private tombs and family burial grounds were to be licensed but not prohibited.[70] The colonial government was unable to commit itself to a definite policy aimed at 'the final extinction of private burial grounds' as 'there were insufficient public burial grounds to provide for the reasonable wants of all communities' and instead preferred to leave 'the burning question of private burial grounds' to municipal discretion to be exercised through the granting and refusing of licences.[71] The introduction of licensing fees was, in the opinion of the municipal president, an

'unfortunate one' for 'a licence fee [could] not make a burial ground
sanitary or reduce the inconvenience caused by it to the general
public'.[72] Furthermore, the Chinese who could afford the luxury of
private burial grounds were unlikely to be deterred by high licence
fees but would instead regard licence fees as 'another way on the part
of Commissioners of squeezing a revenue out of them'.[73]

That the control of burial grounds was vested in the municipality
rather than the government was in itself an important concession to the
Chinese because Chinese views were far more effectively represented
on the Municipal Board than at the level of the colonial government.
Whilst Alex Gentle, the municipal president in the last decade of the
nineteenth century, had consistently opposed the extension of old burial
grounds and the multiplication of new ones, he had been frequently
outvoted by commissioners on these issues.[74] Gentle's successor, J. O.
Anthonisz, also complained that the perspective of the commissioners
was essentially myopic and far too liberal for they tended to assess what
constituted 'fit and proper places' for burial purposes on the basis of
'the effect on the public health and the avoidance of a nuisance' rather
than from the point of view of 'the effect on possible town improvements and the future development of the town'.[75] An 1897 amendment
to the Burials Ordinance of 1896 reduced the maximum penalty for
unlicensed burials outside municipal limits from $500.00 to $100.00, a
stroke described by one of the members of the Legislative Council as
being tantamount to 'an invitation to the Chinese community to bury
their dead without any licence at all, wherever they please'.[76]

Conflicting Discourses: Chinese versus Western Conceptions of the Significance of 'Sacred' Space

Chinese burial grounds remained a contentious issue primarily because
of diametrically polarized views of the significance and utility of the
space they occupied in a congested city. For the Chinese, the
significance of raising a grave went beyond the need to commemorate,
preserve, or identify, and the remains of the dead held more than emo-
tional resonance. Instead, a grave was the physical abode of the soul of
the deceased, and an essential component of ancestor worship.[77]
Chinese brick graves were normally horseshoe or omega-shaped, with
the word zu, meaning 'ancestor', and the word mu, literally 'home of
the deceased', inscribed on the top and bottom of gravestones respectively. Space set aside for the dead was sacrosanct; it was not perceived
as space which had been alienated from the living, but instead, its
preservation was believed to have a propitious and benign effect on the
welfare of the deceased's descendants. For the Chinese, the idea of the
continuity of kinship beyond death and the notion of exchange
between the living and the dead were central to their death rituals.[78]
The dead were hence linked to the living by relationships of reci-
procity: through the principles of feng shui, the living were able to use
their ancestors as the media for the attainment of worldly prosperity, and
in return, the dead received constant veneration, meticulous attention,

and offerings of food and joss paper at the ancestral tomb. It was not clear whether the Chinese believed that the remains of the dead themselves acted as the source of power and beneficence or whether these served as conductors of a power which originated in nature itself[79] but in either case, the ancestral tomb and its 'sepulchral boundaries'[80] (that is, the configuration of landscape surrounding the tomb) were considered inviolable because any form of interference could destroy the source of power and lead to a reversal of family fortunes. Geomancy was hence a means through which the Chinese not so much controlled as capitalized upon and aligned themselves to the natural environment in order to appropriate its bounty through the rituals of ancestor worship.

The classification and spatial organization of Chinese burial grounds were also related to the world of the living in another sense: the burial place and the associated rituals of death were a testimony of the social worth of the dead person as well as his family. For the immigrant Chinese, what was most prestigious was burial in an auspiciously located family-plot back in China, as Low Ngiong Ing, an early twentieth-century Hockchiu immigrant, explained:

An immigrant, if he could afford it, would return to China every few years. In his perambulations he would keep his eyes open for a desirable burial-plot, a knoll commanding a good view, and auspicious according to the laws of geomancy. For we did not mind being men of Nanyang, but that dying, we would hate to be ghost of Nanyang. If we prospered, we would pile up money in China in order to renovate the ancestral graves and the ancestral homes, to redeem the ancestral fields and add to them ... so that men might know we were somebody.[81]

For Straits-born Chinese and those who could not afford posthumous repatriation, an auspicious burial site in Singapore was the best substitute. In colonial Singapore, Chinese burial grounds were of three classes. First, there were a large number of private burial grounds belonging to wealthy families who had settled in Singapore.[82] Graves were often prepared and reserved for each member of the family well ahead of time on private estates.[83] These family graves were akin to 'substantial residences' and were often set in several acres of ground landscaped according to the requirements of *feng shui*.[84] The family name invariably dominated the burial space, and the grave itself represented not only a monument to the family's social standing but also the focus of ancestor worship, a crucial element in a social system which emphasized filial piety, the sanctity of the family name, and the perpetuation of family lineage. The importance of these notions was epitomized in the gravestone on which were inscribed the name of the deceased in green to signify death, as well as the family name in red to signify life and the hope that the family name would never pass away but would be carried on by descendants.[85] A second type of burial ground was that belonging to separate surname- or district-based associations (Plate 29). These grounds were set aside as separate burial places for 'the better class of Chinese' who in their lifetime belonged to these associations and who were guaranteed a burial place in the communal burial grounds by virtue of membership and financial subscription.[86] They signified both the complex internal divisions of Chinese society in Singapore and the solidarity of those associated by surname or provenance. A third class of burial ground was that 'presented to the Chinese community by wealthy Chinamen and by subscription among Chinamen to give decent burial to the Chinese poor'.[87] These burial grounds were opened to particular dialect groups and contained the graves of the poor who were neither able to afford a private plot nor funeral subscriptions and had to depend on charity to secure a resting place. As in life, the dead were also stratified according to a rigid socio-economic hierarchy and burial places not only symbolized but perpetuated social distinctions which death failed to eradicate. The 'social organization of death' not only reflects the social world of the living, but 'in the act of reflecting, provides a basis and a structure through which the world is sustained and reaffirmed'.[88] The principles of unity and lines of cleavage which structured Chinese society in the world of the living were hence equally inscribed and reaffirmed in the sacred spaces of the dead.

The colonial and municipal authorities, however, were unable to appreciate the sanctity of sepulchral veneration and its ramifications, and saw it as an example of the idiosyncratic prejudices of the Chinese. The Attorney-General, J. W. Bonser, for example, dismissed geomantic considerations as 'a farrago of superstitious and ignorant nonsense'.[89] Thus, as Sack has argued, the irreducible fluidity of meaning inherent in 'mythical-magical' space rendered it opaque to those in authority who subscribed to the rationality of 'scientific' planning, for the transparency of sacred symbols could only be appreciated 'through the eyes of the culture' which produced them.[90] From the colonial perspective, progress, development, and public welfare could only be advanced by a rational use of space according to the dictates of Western planning theory and sanitary science rather than the mysterious workings of geomancy. The Attorney-General further urged that while 'respect for religious superstition [was] all very well to a certain extent, ... it must not be allowed to interfere with the interest of the Community' and that it was time the Chinese learnt that the Straits Settlements was 'British and not Chinese Territory'.[91] Thus, the debate over sacred space in all its well publicized aspects was not only important in so far as its resolution had landuse and sanitary implications, but was also perceived as symbolic of the more general ideological aspect of the struggle for control over territory.

Less 'Visible' Aspects of the Conflict over Chinese Burial Grounds

The debate provoked by the introduction of burials legislation represented the better publicized, better ventilated aspects of the conflict over the rights of religious practice and the significance of burial grounds. As seen, the main protagonists on both sides employed constitutional measures to advance their claims. The concern for a suitable place of burial was, however, not confined to the wealthy Chinese but shared by a wide spectrum of Chinese immigrant society. Conflicts between the authorities and the Chinese labouring classes over burial grounds were less 'visible' but possibly more pervasive. In contrast to their wealthier compatriots, the strategies of the labouring classes in securing places of repose were less public, non-constitutional, and often clandestine.

Selecting an ideal burial site in order to procure the benefits of geomancy was to a large extent a privilege of the wealthy Chinese. The bulk of the labouring classes could not afford private sites tailored to meet personal *feng shui* esoterica but had to be content with common graveyards run by Chinese associations. *Feng shui* was not entirely disregarded in the siting of clan- or surname-based burial grounds and, as far as possible, associations selected sites in broad conformity with general geomantic principles. The Hokkien Yeo clan, for example, claimed that according to the *kan yu jia* (professional geomancer), their burial ground at Telok Blangah, Xie Yuan Shan (No. 9 in Table 8.1), had the best *feng shui* in Singapore.[92] A place of repose after death was highly important to the poorer immigrant Chinese for one of their worst fears was 'that they might die overseas, leaving their spirits wandering around without sacrifices'.[93] Without the performance of proper burial rites, the deceased would not be ensured of a safe passage through the realms of the underworld, and, unworshipped and unremembered, he would pass into and become indistinct from an undifferentiated mass of anonymous *gui* (ghosts), existing in a state of marginality.

In the first century of Indian and later colonial rule, neither the Straits government nor the municipal authorities established cemeteries to cater to the needs of the Chinese labouring classes. Corpses were sometimes dumped on five-foot-ways, backlanes, and other public places in the hope that they would be recovered by the police and given the decency of burial that the relatives themselves could not afford. As Wong Toh, an impoverished cake hawker explained: 'I had no money to bury [my brother] and no one would help me. Also the inmates would not allow the dead body to remain [in the house]. I therefore carried his body on my back to the backlane. I knew that the police would find the body and bury it, but I did not know it was against the law to put the body there.'[94] Whilst the dumping of bodies was not uncommon among the very poor or where relatives tried to prevent 'dangerous infectious diseases' from being traced to particular houses, Chinese society had its own mechanisms to confront the precariousness of life and to prepare for death. Chinese dialect, surname, and mutual benefit organizations not only supervised the various practicalities of dying,[95] but through rites and ritual, attenuated the isolation of death and imbued it with social significance. One of their chief functions was the collection of subscriptions during a member's economically productive period as a hedge against funeral expenses. Those who could not afford private burial depended on these organizations to provide a 'sacred' place of repose in communal burial grounds. The sanctity of death, the care of the burial site, and the welfare of the departed soul were also attended to by these organizations which arranged for proper funeral rites and organized visits to the graveyards to offer sacrifices to the dead during the Qing Ming and Chong Yang Festivals.[96] Chinese public burial grounds ranged in size from a few tens of acres to several hundreds (Table 8.1), and accommodated not only the resting places of the dead but also joss houses, temples, the squatters' huts, vegetable gardens, and piggeries. It was hence through their own organizations that the Chinese labouring classes were ensured the provision of sacred places for their dead, albeit in close association with the activities of the living.

From the municipal perspective, public Chinese burial grounds were more insanitary than sacred. These grounds were considered unkempt, disorderly, overgrown with brushwood, and a source of contamination to the town's water supply. By the late nineteenth century, existing public burial grounds were also overcrowded and no longer able to cope with the pressing demands of a rapidly burgeoning Chinese population. The municipal authorities passed sanitary regulations to ensure that graves were not overcrowded and were at least 5 feet deep, brushwood and jungle regularly cleared, and burial registers properly kept. Many of the public burial grounds were full and officially closed against further burial by municipal order.

Sanitary regulations were, however, little appreciated by the Chinese, nor were closing orders necessarily complied with. Even when the burial ground was considered full, 'it was the habit to dig all over the place till a vacant space was found to bury another body'.[97] Given the pressure on space, the Chinese also resorted to clandestine burials in closed or disused burial grounds as well as unused government land which frequently escaped municipal vigilance at the time of burial. A case of burials in unsanctioned grounds came to light in the early 1880s when the acting resident surgeon of the General Hospital located in Sepoy Lines lodged a 'nuisance' complaint that a piece of land adjoining the apothecaries' quarters and reserved as cantonment practice ground had been 'surreptitiously buried upon to a considerable extent' by Seh Choa Chinese who owned an adjacent burial ground which was already full.[98] In the period of uncertainty following the postponement of the 1887 Burials Bill, there was strong suspicion that in the Telok Blangah district, an area already covered by a large number of private and public burial grounds, fresh grounds were continually dug and interments carried out in unregistered and unlicensed grounds. Attempts to prosecute the trustees of various burial grounds had to be withdrawn as the present trustees denied all responsibility for past infringements of the law.[99] In 1889, Major H. E. McCallum, the Colonial Engineer and nominated member of the Municipal Board, pointed out that the question of illegal burials was becoming increasingly serious and moved that the government be requested to take immediate steps to deal with the burials question 'in the cause of sanitation and good order'.[100] It was highly desirable, urged McCallum, that 'all available building sites within easy reach of the town should be reserved for houses instead of being occupied ... by private burial grounds' for 'sickness was becoming rife in the town and the dangers from overcrowding increasing'.[101] He advocated that the municipality should be given powers to prohibit the digging of fresh graves in anticipation of future interment and the conversion of grounds presently occupied by a few graves into 'regular cemeteries'.[102] This was the first of several requests on the part of the municipal commissioners for greater powers to regulate burial grounds but until the introduction of dual legislation in 1896, no stringent measure could be introduced as 'the law [regarding burial] was unsettled as to whether the Municipality or the Government assume[d] responsibility for the supervision of this important custom'.[103]

In the meantime, the municipal authorities depended on a system of surveillance to prevent clandestine burials. An inspector of burial grounds was appointed to make periodic inspections of burial places, and to furnish regular reports and returns of burials but such a system 'signally failed' not only because the appointed inspector failed to carry out inspections 'regularly and systematically',[104] but also because the 'unsystematic' nature of Chinese burial records and the 'diffused' layout of the grounds themselves rendered the 'inspecting gaze' ineffective. In the first place, the registration, demarcation, and mapping of burial grounds were highly imperfect and as a result, encroachment of lands adjoining the boundaries of burial grounds often occurred with impunity.[105] Secondly, Chinese burial records were not systematically kept and appeared highly confusing to municipal inspection as 'it was customary among the Chinese to give another name [other than the official name] to a deceased person on the tombstone'.[106] Burial records seldom tallied with the monthly death returns for the municipality. In 1892, an attempt to collate the number of interments from an inspection of twenty burial grounds showed that whilst there had been 423 deaths during the month of May within the municipality, only 358 burials had been registered, a sign which supported the suspicion that some 15 per cent of burials had been clandestinely carried out.[107] Although the Municipal Ordinance of 1896 strengthened the hand of the authorities by requiring stricter registration and licensing of burial grounds, clandestine burials appeared to have continued unabated for several decades. The assistance of the police or the detective branch had to be enlisted to investigate certain suspected cases of illegal burials in closed grounds but even then concrete evidence could only be obtained by reopening graves to ascertain whether they contained fresh bodies, a step unpleasant enough to discourage even Dr P. S. Hunter, the municipal health officer in the 1920s.[108] It was also reported that exhumation of bodies for repatriation to China was constantly taking place without the knowledge of the authorities.[109]

The Establishment of a Municipal Cemetery for the Chinese

By the early years of the twentieth century, it became patently obvious that not only was there acute pressure for space among the living, existing burial grounds were no longer sufficient to cope with escalating numbers of the dead which inevitably accompanied a rapidly burgeoning population. Many burial grounds belonging to various communities were declared unfit for further use by the municipal authorities and officially closed against burial.[110] The shortage of burial space was further exacerbated by the compulsory acquisition of several old grounds for the purposes of modern urban development. The first of these to be cleared was the disused Cantonese public burial ground at Tanjong Pagar near Chinatown (No. 1 in Table 8.1) which was acquired in 1907/8 in order to provide filling material for the Telok Ayer Reclamation Scheme. Although the trustees of the burial ground had to bow to the Legislative Council resolution to compulsorily acquire the site, they managed to secure, through the representations of the Chinese member of Council, Tan Jiak Kim, various concessions. The government assured the Chinese that due regard would be shown for their 'superstitions and feelings', adequate arrangements made for the exhumation and reinterment of remains in other burial grounds according to Chinese rites, and compensation paid based on a 'piece-rate' according to the number of sets of bones uncovered.[111] The work of removal was entrusted to the Chinese Advisory Board which undertook to ensure that 'everything was done decently and to the satisfaction of the relatives of the deceased'.[112] According to the *Singapore Free Press*, 'the scene at the burial ground was not altogether void of pathos':

Some of the graves were opened in the presence of well-dressed male and female relatives, whilst, in other cases, two coolies with a solitary broken *changkol* [hoe or shovel] between them took it in turn to unearth all that remained of the unknown dead. The same contrast was seen in the means of transport of the unearthed remains to their new resting-places. The remains of a bygone merchant were carried under a canopy and covered with fine silk and accompanied by relatives and friends and bands of Chinese musicians while the next procession would consist of three coolies, two of whom carried the remains done up in an old 'gunny' bag slung on a carrying pole, and the third coolie preceded the procession, holding in one hand a few sprigs of bamboo with a red flag fastened to one of them and piping away on [an] old tin-whistle.[113]

A second extensive clearance of graves was carried out from the mid-1920s when 70-odd acres of land located in the Tiong Bahru area was acquired by the Singapore Improvement Trust for an improvement and housing scheme. The project involved removing over 280 huts, 2,000 squatters and their pigs as well as a large number of graves, filling in swampy ground, amalgamating irregular holdings, and re-allotting them according to a 'regular plan', and laying out roads and drains to provide an overspill area for the southern congested area.[114] The assistance of the Chinese protector was enlisted to negotiate for the removal of scattered graves and as compensation, new grave plots were offered in the municipal cemetery at Bukit Brown (see below) with a sum of money sufficient to cover the cost of removal.[115] In general, public burial grounds belonging to various *bang* or clans were allowed to remain and where acquisition was necessary, the associations were authorized to purchase a new burial site beyond municipal limits.[116]

Not only were several old burial grounds either closed or cleared by the government and the municipal authorities, conditions for the issue of new licences for private burial grounds were made increasingly stringent. After 1 July 1906, the use of any place within municipal limits was prohibited by law.[117] The Christian cemetery at Bukit Timah Road was closed at the end of 1907 and a new cemetery—the Bidadari Christian Cemetery along Upper Serangoon Road—consecrated and opened on 1 January 1908.[118] A separate section of the Bidadari estate was also purchased from the Dato Mentri of Johore by the municipal authorities, laid out as a public Mohammedan Cemetery, and opened for interment in 1910.[119] Apart from this cemetery, however, the authorities were extremely reluctant to commit themselves to a policy of providing municipal cemeteries for the various Asian communities.

On several occasions during the first two decades of the twentieth century, a section of the more 'progressive' Chinese leaders led by Dr Lim Boon Keng, a municipal commissioner in 1906 and an active campaigner for 'Straits Chinese reform',[120] petitioned the Municipal Board for the provision of a municipal cemetery for the use of the Chinese communities in view of the severe 'overcrowding' in existing burial grounds.[121] In 1906, Dr Lim urged that a Chinese municipal cemetery should be provided 'without delay' and assured the Board that 'the educated [Chinese] ... no longer believed in burying according to ideas based on geomancy [and] did not object to burying their dead under municipal regulations [similar to those] in force at the Christian cemeteries'.[122] Although geomantic complications were officially not taken into consideration, the search for a suitable piece of land for a public Chinese cemetery proved highly problematic. A minimum of 100 acres and preferably up to 400 acres was considered desirable in order for a 'reasonable number of full-sized Chinese graves' to be accommodated for a period of ten years but this was either unavailable or too expensive.[123] Although available land could be secured at Bidadari for a Chinese cemetery, the municipal commissioners rejected this option as it was felt that the burial customs of the Chinese were incompatible with the general ambience of a site already consecrated to the Christian dead. As the municipal president explained, since the burial customs of the Chinese were 'characterized by noise' and the Christians 'by silence', 'there might be clashing and inconvenience should burials be taking place in both places at one time'.[124] It was ultimately decided to use powers available under the Lands Acquisition Ordinance to compulsorily purchase a 213-acre site at Bukit Brown which formed part of an existing burial ground belonging to the

Seh Ong *kongsi*. As a result of the *kongsi*'s opposition to the compulsory acquisition of their land and 'difficulties in the way of titles and trusteeships', it was not until the end of 1919 that the site passed into municipal hands.[125]

The cemetery at Bukit Brown (Plates 29 and 30) was finally opened for interment on 1 January 1922, 18 years after the question of a municipal cemetery for the Chinese was first broached. A subcommittee comprising Commissioners See Tiong Wah and Tan Kheam Hock was set up to frame by-laws for the regulation of the cemetery in consultation with the municipal health officer, the executive engineer, and the legal adviser.[126] Initially, the municipal cemetery did not prove popular with the Chinese. Three months elapsed before the first burial took place and in August 1922, the municipal president reported that it was 'not utilized to the extent which had been anticipated'.[127] Part of the reason for the unpopularity of the municipal cemetery was to be found in its spatial layout. Grave plots were small, laid out in neat rows with one plot in ten left vacant for access, and allotted consecutively. Such a geometrical scheme produced a disciplined spatial order favoured by the municipal authorities but unpopular with those who could afford the individualizing treatment offered by geomantically prescribed burial locations. As Maurice Freedman observed, 'in geomancy there lies the inherent principle that tombs are a means of individualizing the fate of the living.... The municipal cemetery which blocks off all opportunity for grand bids for fortune by its discipline of fixed plots is hateful to the Chinese.'[128] In August 1922, after consultation with the Chinese Advisory Board, the Bukit Brown Cemetery by-laws were amended to take into account Chinese preferences: plot sizes were

TABLE 8.2
Return of Burials in the Municipal Chinese Cemetery
at Bukit Brown, 1922–1929

Year	Annual Number of Burials in Bukit Brown Chinese Cemetery			Cumulative (2)	Total Number of Chinese Burials within Municipal Limits (3)	(1) as a Percentage of (3) (4)
	General Division	Pauper Division	Total (1)			
1922	–	–	93			
1923	–	–	205	298	4,106	5.0
1924	–	–	519	817	4,422	11.7
1925	–	–	1,005	1,822	4,876	20.6
1926	–	–	1,218	3,040	5,632	21.6
1927	797	553	1,350	4,390	5,749	23.4
1928	686	668	1,354	5,744	5,322	25.4
1929	691	1,624	2,315	8,059	5,511	42.0

Source: Compiled from ARSM, 1922–1929, Health Officer's Department/Health Officer's Report/Municipal Health Officer.

increased to 20 by 10 feet in the general division and 10 by 5 feet in the paupers' division;[129] plots were oriented facing east and south,[130] and space for a path was left after every six rows of graves.[131] A temple, modelled on the Thian Hock Keng Temple in Telok Ayer, was constructed within the cemetery and farmed out to a Chinese priest who paid an annual sum to the municipality for the right to sell joss sticks and other ritual paraphernalia.[132] In subsequent years, municipal burial became increasingly accepted among the Chinese population and by 1929, Bukit Brown accounted for about 40 per cent of all officially registered Chinese burials within municipal limits (Table 8.2).

Negotiating Control over 'Sacred' Space

As already seen, the meaning of the 'sacred' spaces of death to Chinese communities was different from the significance of extramural 'eternal resting places' of Western European tradition. European cemeteries were ordered and organized as park-like expanses for family visits or monuments to illustrious persons. They served the purposes of contemplation and commemoration of those who had 'passed on' and effectively separated the dead from the living. In contrast, for the Chinese, the control and manipulation of burial places of the dead were perceived as inseparable from the fortunes of the living. Discursive elements such as ritual practices, funerary artefacts, and the epistemological precepts underlying *feng shui* which defined the very nature and meaning of mortality in Chinese society all emphasized the reciprocal and interlocking nature of relationships between the living and the dead through the medium of the landscape. 'Sacred' places devoted to the dead were hence not 'sterile' sites alienated from the living but instead formed an essential and influential part of the topography of the living. Unlike in the West where the habitations of the living and those of the dead belonged to different worlds, in the Chinese view, 'the tomb is the *yin* habitation to match the *yang* habitation of the living' within a single system.[133] It was this 'lack of rigid boundaries between the sacred and the worldly in Chinese religious culture' that Western observers found 'puzzling and at times degrading'.[134]

The attempt here is not to examine the rationality or internal consistency of Chinese geomancy as a so-called 'science' *vis-à-vis* Western conceptions of science,[135] but to demonstrate the use of *feng shui* as a strategic discourse in the encounter between the colonial authorities and the Chinese community. Through a discourse which insisted on the 'sacred' nature of burial grounds and 'mystified' landscape, the Chinese attempted to justify the immunity of certain elements of the environment from colonial or municipal control, and, in doing so, challenged Western conceptions of urban development and planning priorities. Although the authorities were quick to dismiss geomancy as mere superstition, it is ironic that its very imperviousness to Western logic made it much more difficult for the government or the municipality to regulate geomantically selected burial grounds.[136]

Whilst the existence of a separate and distinctly different Chinese discourse on death served to prolong the negotiation of control over 'sacred' space in the city between the authorities and the communities on a political level, it was the clandestine, everyday strategies of those pressed for burying space in the city which rendered municipal surveillance increasingly difficult. In the early twentieth century, the allotment of a municipally sanctified space for Chinese burials, although initially supported only by Western-educated Chinese, became an increasingly accepted compromise between the Chinese, who retained some measure of control over their burying places and the municipal authorities, who were able to ensure the gradual removal of already established burial places for the purposes of modern urban development.

1. W. S. Maugham, *Collected Short Stories*, Vol. 2, London: Pan Books, 1975, p. 310.

2. SIT, 655/24, 16 October 1924: 'Forwarding Copy of a Letter from the Secretary of Chinese Affairs on the Subject of Acquiring Land Comprised in Grant No. 49 Toah Payoh as a Burial Ground for the Seh Chua Community'.

3. Robert D. Sack, *Conceptions of Space in Social Thought: A Geographic Perspective*, London: Macmillan Press, 1980, p. 155.

4. Mircea Eliade, *The Myth of the Eternal Return*, Princeton: Princeton University Press, 1954, p. 18.

5. Paul Wheatley, *The Pivot of the Four Quarters: A Preliminary Enquiry into the Origins and Character of the Chinese City*, Chicago: Aldine, 1971, p. 411; Yi-Fu Tuan, *China*, London: Longman, 1970, p. 106. Cosmogony as a paradigmatic model for the layout of cities and the sacralizing of various urban elements were not only found in India and China but prevalent in a number of other civilizations. For further examples, see Mircea Eliade, *The Sacred and the Profane: The Nature of Religion*, trans. W. R. Trask, New York: Harvest, 1959, pp. 20–65; Amos Rapoport, *House Form and Culture*, Englewood Cliffs, NJ: Prentice-Hall, 1969, pp. 49–58.

6. Wheatley, *The Pivot of the Four Quarters*, pp. 450–1.

7. Paul Wheatley, *City as Symbol: An Inaugural Lecture Delivered at University College London, 20 Nov 1969*, London: H. K. Lewis, 1969, p. 9.

8. Sack, *Conceptions of Space*, p. 155.

9. Ibid., pp. 155–6.

10. Ibid., p. 185.

11. See, for example, F. W. Young, 'Graveyards and Social Structure', *Rural Sociology*, 25 (1960): 446–50; A. I. Ludwig, *Graven Images: New England Stonecarving and Its Symbols, 1650–1815*, Middletown: Wesleyan University Press, 1966; J. B. Jackson, 'From Monument to Place', *Landscape*, 17 (1967/68): 22–6; R. V. Francaviglia, 'The Cemetery as an Evolving Cultural Landscape', *Annals of the Association of American Geographers*, 61 (1971): 501–9; Wilbur Zelinsky, 'Unearthly Delights: Cemetery Names and the Map of the Changing American Afterworld', in David Lowenthal and M. J. Bowden (eds.), *Geographies of the Mind*, Oxford: Oxford University Press, 1975, pp. 171–95; R. G. Knapp, 'The Changing Landscape of the Chinese Cemetery', *China Geographer*, 8 (1977): 1–14; M. A. Nelson and D. H. George, 'Grinning Skulls, Smiling

Cherubs, Bitter Words', *Journal of Cultural Geography*, 15 (1982): 163–74; D. B. Knight, 'Commentary: Perceptions of Landscapes in Heaven', *Journal of Cultural Geography*, 6 (1985): 127–40; David Chuen-yan Lai, 'The Chinese Cemetery in Victoria', *BC Studies*, 95 (1987): 24–42; Wilbur Zelinsky, 'Gathering Places for America's Dead: How Many, Where, and Why?', *Professional Geographer*, 46 (1994): 29–38.

12. This interpretation draws upon recent theorizing within cultural and social geography which focuses on the plurality of cultures and the connections between cultures, human geographies, and the workings of power (see, for example, Peter Jackson, *Maps of Meaning: An Introduction to Cultural Geography*, London: Routledge, 1989; Kay J. Anderson and Fay Gale (eds.), *Inventing Places: Studies in Cultural Geography*, Melbourne: Ron Harper, 1992).

13. Philippe Aries, *In the Hour of Our Death*, trans. H. Weaver, New York: Allen Lane, 1981, pp. 92, 318ff.

14. James C. Riley, *The Eighteenth-Century Campaign to Avoid Disease*, London: Macmillan, 1987, p. 110.

15. Edwin Chadwick, 'Report on the Sanitary Condition of the Labouring Population of Great Britain: A Supplementary Report on the Results of a Special Inquiry into the Practice of Interment in Towns', *Parliamentary Papers*, 12 (1843): para. 249.

16. R. A. Etlin, *The Architecture of Death: The Transformation of the Cemetery in Eighteenth-century Paris*, Cambridge, Massachusetts: MIT Press, 1984, p. x.

17. Paul Coones, 'Kensal Green Cemetery: London's First Great Extramural Necropolis', *Transactions of the Ancient Monuments Society*, 31 (1987): 48.

18. Etlin, *The Architecture of Death*, pp. 12–17.

19. Coones, 'Kensal Green Cemetery', p. 50.

20. James Stephen Curl, *A Celebration of Death: An Introduction to Some of the Buildings, Monuments, and Settings of Funerary Architecture in the Western European Tradition*, London: Constable, 1980, p. 207.

21. John Turnbull Thomson, *Some Glimpses into Life in the Far East*, London: Richardson and Co., 1865, pp. 280–1.

22. The first cemetery was a small plot used between an undetermined date after the founding of the Settlement and late 1822 situated close to the government residence on Fort Canning Hill. This was closed and a new cemetery opened lower down the hill towards the end of 1822 (Alan Harfield, *Early Cemeteries in Singapore*, London: British Association for Cemeteries in South Asia, 1988, p. 3). The second cemetery was closed in 1865 when it became full and a new Christian cemetery, the Bukit Timah Road Cemetery, opened about 2 miles from the town.

23. *Report of the Committee Appointed to Make Recommendations Regarding Burial and Burial Grounds in the Colony of Singapore*, Singapore: Government Printing Office, 1952, p. 4.

24. Thomson, *Some Glimpses*, p. 280.

25. Ibid., p. 282.

26. *Straits Times*, 20 August 1887.

27. *Proceedings and Report of the Commission Appointed to Inquire into the Cause of the Present Housing Difficulties in Singapore and the Steps Which Should Be Taken to Remedy Such Difficulties*, 1918 (hereafter cited as *Housing Difficulties Report*), Vol. 1, p. A16.

28. According to the 1905 report of the Burials Committee, there were then in existence nineteen Chinese public burial grounds belonging to various dialect- and surname-based organizations, and fifty-nine private burial plots belonging to Chinese families widely distributed within the municipality (PLCSS, 1905, 'Report of Committee on the Question of Chinese Burial Grounds', pp. C200–13).

29. CO273/151/4517: Smith to Holland, 1 February 1888, 'Report on Chinese Memorial against the Burial Ordinance [XI] of 1887 by W. A. Pickering, Protector of Chinese, 3 January 1888'.

30. *Singapore Free Press*, 15 August 1887; *Straits Times*, 20 August 1887.

31. J. D. Vaughan, *The Manners and Customs of the Chinese of the Straits Settlements*, Singapore: Oxford University Press, 1985 (first published in 1879), p. 32; *Housing Difficulties Report*, Vol. 1, p. Al5.

32. Gordon Harrower, 'Native Medicine and Hygiene in Singapore', *British Medical Journal*, 2 (1923): 1175.

33. *Singapore Free Press*, 15 August 1887.

34. Ibid.

35. Ibid.

36. Aries, *In the Hour of Our Death*, pp. 484–91.

37. The Burials Bill of 1887 proposed to transfer the control of burial and burning grounds, hitherto vested in the municipality, to the government. It directed the Governor-in-Council to issue licences for existing burial grounds on the conditions that application for a licence was made within three months of the ordinance coming into force and also that the continued use of the burial ground would not endanger 'the health and comfort of the public'. It further specified that interment was to be allowed only in places licensed for the purpose and that burying or burning corpses in unlicensed grounds incurred a maximum penalty of $100.00. The Governor-in-Council was also authorized to frame rules and regulations to fix the depth of graves, the amount of fees charged for burials, and for registration, inspection, and management of cemeteries and burning grounds (PLCSS, 8 August 1887, pp. B91–2; *Straits Times*, 20 August 1887).

38. PLCSS, 15 August 1887, p. B101.

39. Ibid.

40. Ibid., p. B102.

41. Ibid., p. B103.

42. Among the Chinese plebeian classes, there were signs of 'much alarm', largely because of the rumour that the bill proposed to give magistrates the power to remove bodies which had already been buried (CO273/146/20985: Weld to Holland, 10 September 1887, 'Report on Ordinance [XI] of 1887 by J. W. Bonser, Attorney-General, 5 September 1887'. While widespread 'alarm' did not culminate in violence on this occasion, it was probable that ill-feeling against the government generated by the introduction of the Burials Bill contributed in part to the outbreak of the 'verandah riots' in February the following year (see Chapter 7).

43. The Penang memorialists, for example, made it clear that not only were they fully aware of the channels of protest open to them, but also from whom they had learnt their strategies: 'We admit that we have not refused to avail ourselves of constitutional means to fight our own rights and priviledges [sic] when such a vital and important matter as our time-honoured usages were threatened with interference, and probable eventual abolition.... Enlightenment, consequent on liberal education and contact with civilized people, encourages loyal subjects and citizens to act as their brethren in civilized countries, in contending for their rights and privileges when they are deemed to be encroached on, or threatened, by any Legislative measure.' (*Pinang Gazette and Straits Chronicle*, 14 October 1887.)

44. The Burials Bill was passed as Ordinance XI of 1887 on 25 August 1887, merely eighteen days after its first reading (PLCSS, 25 August 1887, p. B125).

45. Each of the Chinese communities in Penang, Malacca and Singapore sent a separate memorial to Holland. See enclosures in C0273/151/4517: Smith to Holland, 1 February 1888.

46. To the Chinese, government control of burial sites was considered oppressive because in China itself, people were neither bound by law nor custom to bury their dead in special areas. According to De Groot, everyone had 'full liberty to inter his dead wherever he [chose], provided he possess[ed] the ground, or [held] it by some legal title acquired from the legal owner'. There were also severe laws against the violation of dead bodies and the desecration of graves (Jan Jakob Mariade de Groot, *The Religious System of China*, Vol. 3, Taipei: Ch'eng-wen Publishing Co., 1969 (first published in 1892), pp. 867, 874–5, and 939). What often impressed the Western observer as unusual about the Chinese landscape was 'the ubiquity of individual and clan graves among the tilled fields' (Knapp, 'The Changing Landscape of the Chinese Cemetery', p. 1).

47. Cf. Knapp, 'The Changing Landscape', p. 7.

48. *Feng shui* literally means 'winds and waters' and is defined as 'the art of adapting residences of the living and the dead so as to cooperate and harmonise with the local currents of the cosmic breath' inherent in a particular configuration of the landscape. Although translated as 'geomancy', it is entirely different from divination methods which passed under that name in the West (Joseph Needham, *Science and Civilisation in China*, Vol. 2, Cambridge: Cambridge University Press, 1956, p. 359).

49. For more detailed discussions on the use of geomantic principles among the Chinese in Singapore and China in siting burial places, see Ernest J. Eitel, *Feng-Shui*, Singapore: Graham Brash, 1985 (first published in 1873); De Groot, *The Religious System in China*, Vol. 3, pp. 935–55; Stephan D. R. Feuchtwang, *An Anthropological Analysis of Chinese Geomancy*, Vientiane: Vithagna, 1974; Evelyn Lip, *Chinese Geomancy*, Singapore: Times Books International, 1979.

50. Needham, *Science and Civilisation*, Vol. 2, p. 361.

51. CO273/151/4517: Smith to Holland, 1 February 1888, 'The Humble Memorial from the Undersigned Chinese Merchants, Traders, Planters and other Chinese Inhabitants of Singapore to Sir H. T. Holland, 30 November 1887'.

52. Quoted by Koh Seang Tat, one of the Penang leaders in the protest, in his representation to the Governor, dated 25 August 1887 (reproduced in the *Pinang Gazette and Straits Chronicle*, 14 October 1887). The passage appears to be a loose paraphrase of paragraphs 6 and 9 in Book 2 (on 'The Law of Things') of *The Institutes of Gaius*, Part 1, trans. F. de Zulueta, Oxford: Clarendon Press, 1946, p. 67, which defends the sacrosanctity and inalienability of burial grounds.

53. Oswald Spengler, *The Decline of the West*, London: George Allen and Unwin, 1961 (first published in 1932), p. 115.

54. Ibid.

55. Maurice Freedman, 'Geomancy', in *Proceedings of the Royal Anthropological Institute of Great Britain and Ireland for 1968*, London: Royal Anthropological Institute of Great Britain and Ireland, 1969, p. 7.

56. Ibid.

57. B. Boxer, 'Space, Change and *Feng-shui* in Tsuen Wan's Urbanization', *Journal of Asian and African Studies*, 3 (1968): 235.

58. *Straits Times*, 19 August 1887; PLCSS, 22 August 1887, p. B122; 'The Humble Memorial ... 30 Nov 1887'.

59. The Humble Memorial ... 30 Nov 1887'.

60. Ibid.

61. 'Report on Chinese Memorial against the Burial Ordinance [XI] of 1887'.

62. The Colonial Secretary referred to an article published in the *Lat Pau*, the local Chinese newspaper, detailing an actual case in the Nam Hoi district in China wherein an ancestral temple and certain houses in front of the *yamen* were acquired for demolition (reproduced in PLCSS, 15 August 1887, pp. B102–3). *The Straits Times* further cited the clearance of graves to facilitate the laying of a railway line from Tientsin to Taku in China. The editor claimed that the incident was clear evidence that 'the Chinese Government set little store by any superstitious feelings among the people as regards the disturbance of graves' and urged the Straits government to 'profit from the example given by the Chinese Government' (*Straits Times*, 7 September 1887).

63. PLCSS, 15 August 1887, p. B103. In his representation to the government, Koh Seang Tat, one of the Penang Chinese leading the protest against the Burials Bill, refuted the view that storage of remains in urns was proof of lack of veneration of the dead among the Chinese. He claimed that this custom, which was only practised by the wealthy Chinese, was not contrary but in fact in accordance with their 'well-meaning and pious belief in Geomancy'. According to strict geomantic principles, the Chinese were obliged to keep the remains in urns and wait for a propitious time and place for burial (*Pinang Gazette and Straits Chronicle*, 14 October 1887). Opposing interpretations of Chinese burial customs were symptomatic of the fact that the ritual repertoire associated with death in Chinese society was extremely complex and also variable among different regions and dialect groups. For example, it has been observed that among the Cantonese, a system of 'double burial' was practised whereby the burial of a body in a coffin was followed by exhumation, the temporary storage of the bones in an urn and reburial of the

urn in a geomantically suitable tomb (James L. Watson, 'Of Flesh and Bones: The Management of Death Pollution in Cantonese Society', in Maurice Bloch and Jonathan Parry (eds.), *Death and the Regeneration of Life*, Cambridge: Cambridge University Press, 1982, p. 155). In north China, 'double burial' was uncommon and the encoffined body was directly laid to rest in substantial tombs (Susan Naquin, 'Funerals in North China: Uniformity and Variation', in James L. Watson and Evelyn S. Rawski (eds.), *Death Ritual in Late and Modern China*, Berkeley: University of California Press, 1988, p. 44). In colonial Singapore, the latter custom was the norm, it being common for wealthy Chinese to prepare elaborate family graves well ahead of time (CO273/146/20985: Weld to Holland, 10 September 1887, 'Memo. on the Burials Ordinance [XI] of 1887, Straits Settlements by D. F. A. Harvey, 25 October 1887'.

64. CO273/151/4517: 1 February 1888, Smith to Holland.

65. 'Report on Chinese Memorial Against the Burial Ordinance [XI] of 1887'.

66. Ibid.

67. PLCSS, 29 November 1887, p. B164.

68. In 1889, when the Indian Act [XIV] of 1856 and the Conservancy Ordinance [II] of 1879 (both containing clauses which hitherto governed burial grounds) were repealed by the Municipal Amendment Ordinance [XVII], no provisions were made to introduce new regulations and hence from 1889 to 1896, no laws existed for the control of burial grounds (ARSM, 1900, p. 13).

69. C. G. Garrad, *The Acts and Ordinances of the Legislative Council of the Straits Settlements from 1st April 1867 to 7th March 1898*, Vol. 2, London: Eyre and Spottiswoode, 1898, pp. 1544–6.

70. ARSM, 1896, p. 16.

71. *Straits Times*, 24 July 1896.

72. ARSM, 1900, p.13.

73. Ibid.

74. Ibid., p. 12.

75. Ibid.

76. The amendment was introduced on the orders of the Secretary of State for the Colonies, who objected to the severity of penalties prescribed by the 1896 Burials Ordinance (PLCSS, 15 April 1897, p. B29; PLCSS, 29 August 1897, pp. B41–2).

77. Jan Jakob Mariade de Groot observed that for the Chinese, 'graves ... are not a means to rid one's self of useless mortal remains in a way considered most decent; nor are they merely rendered sacred to the memory of the dead.... The grave ... is sacred especially as an abode of the soul, not only indispensable for its happiness, but also for its existence, for no disembodied spirit can long escape destruction unless the body co-exists with it to serve it as a natural support. Both the body and the soul require a grave for their preservation. Hence the grave, being the chief shelter of the soul, virtually becomes the principal altar dedicated to it and to its worship.' (De Groot, *The Religious System of China*, Vol. 3, p. 855.)

78. James L. Watson, 'The Structure of Chinese Funerary Rites: Elementary Forms, Ritual Sequence, and the Primacy of Performance', in Watson and Rawski, *Death Ritual*, p. 9; Tong Chee Kiong, 'Dangerous Blood, Refined Souls: Death Rituals among the Chinese in Singapore', Ph.D. thesis, Cornell University, 1987, pp. 340–2.

79. Rubie S. Watson, 'Remembering the Dead: Graves and Politics in Southeastern China', in Watson and Rawski, *Death Ritual*, pp. 206–7.

80. 'The Humble Memorial ... 30 Nov 1887'.

81. Low Ngiong Ing, *Recollections: Chinese Jetsam on a Tropical Shore [and] When Singapore was Syonan-to*, Singapore: Eastern Universities Press, 1983, p. 112.

82. According to municipal records, there were 162 private burial grounds belonging to all ethnic communities within municipal limits in 1900, of which 143 were still in use and 19 disused (ARSM, 1900, p. 12).

83. In 1892, Tay Geok Teat, one of the Chinese commissioners in the latter half of the year, applied for permission from the Municipal Board to make two graves, one for his recently deceased wife and one for himself, on a piece of land in Telok Blangah where nine family graves were already sited (MPMCOM, 26 October 1892). Similarly, in 1893, the wealthy landowner, Cheang Hong Lim, was buried in a grave 'already pre-

pared on his property on Alexandra Road' (MPMCOM, 15 March 1893). For the Chinese, the procurement of a grave and the preparation of grave clothes and a coffin well before death were considered symbols of longevity and status (De Groot, *The Religious System of China*, Vol. 3, p. 1031).

84. Vaughan, *The Manners and Customs of the Chinese*, p. 32; *Singapore Free Press*, 15 August 1887; 'The Humble Memorial ... 30 Nov 1887'.

85. Tong, 'Dangerous Blood, Refined Souls', p. 133.

86. 'The Humble Memorial ... 30 Nov 1887'.

87. Ibid.

88. Lindsay Prior, *The Social Organization of Death: Medical Discourse and Social Practices in Belfast*, Basingstoke: Macmillan Press, 1989, p. 111.

89. CO273/148/25479: Smith to Holland, 16 November 1887, 'Copy of a minute by J. W. Bonser, Attorney-General, 12 November 1887'.

90. Sack, *Conceptions of Space*, pp. 148, 163–4, and 189–93.

91. 'Copy of a minute by J. W. Bonser, Attorney-General, 12 November 1887'.

92. *Yang Shi Zong Pu (The Yang Clan Genealogy)*, Singapore, 1965, p. Gl6. In a different context, David Lai demonstrates retrospectively that geomantic principles were important in the siting of a late nineteenth-century burial ground belonging to the Chinese Association in Victoria, British Columbia (David C-Y. Lai, 'A Feng Shui Model as a Location Index', *Annals of the Association of American Geographers*, 64 (1974), 506–13).

93. Yen Ching-hwang, *A Social History of the Chinese in Singapore and Malaya, 1800–1911*, Singapore: Oxford University Press, 1986, p. 45.

94. Cited from *Singapore Coroners' Inquests and Inquiries* No. 229, of Wong Wan, 14 April 1931 in Katherine Yeo Lian Bee, 'Hawkers and the State in Colonial Singapore: Mid-Nineteenth Century to 1939', MA thesis, Monash University, 1989, p. 56.

95. Several clan associations and mutual benefit societies ran so-called 'dying houses' where the chronically ill or incapacitated could seek shelter and succour. In 1908, a municipal survey recorded fourteen of these dying houses (ARSM, 1908, Health Officer's Report, p. 37).

96. Qing Ming Jie (the Chinese 'All Souls' Day') and Chong Yang Jie (the 'Double Ninth' Festival) are Chinese festive periods devoted to the veneration of deceased ancestors and family members. During these festivals, it is the practice to visit the graves of forebears to make ritual offerings and to sweep the graves (*Hua Ren Li Su Jie Re Shou Ce (Chinese Customs and Festivals in Singapore)*, Singapore: Singapore Federation of Chinese Clan Associations, 1989, pp. 45–9 and 74–7.

97. SIT, 116/25, 25 November 1924: 'Unlawful Burials in the Hokkien Burial Ground, Alexandra Road'.

98. MPMC, 24 February 1881 and 2 June 1882.

99. MPMC, 3 July 1889.

100. The motion was initially opposed by the Chinese commissioners, Tan Jiak Kim and Tan Beng Wan, who claimed that the fault did not lie with the owners or trustees of burial grounds who were 'ignorant of the law' but with the authorities themselves who in former years had neglected to enforce the law. They withdrew their opposition on the assurance that any steps taken to deal with the burials question would only apply to future burial grounds and would not interfere with existing grounds (MPMC, 3 July 1889 and 31 July 1889).

101. MPMC, 31 July 1889.

102. Ibid.

103. MPMC, 27 October 1890; MPMCOM, 19 August 1891, 6 July 1892, and 20 July 1892.

104. MPMC, 31 July 1889.

105. MPMCOM, 27 October 1890.

106. SIT, 724/26, 28 October 1926: 'Tiong Bahru Improvement Scheme, Acquisition of Two Graves from Lot 185 Mukim 1'.

107. MPMCOM, 6 July 1892.

108. 'Unlawful Burials in the Hokkien Burial Ground, Alexandra Road'. The *Singapore Coroners' Inquests and Inquiries* also contain some evidence that illegal burials

were more common than discovered by municipal inspectors (for examples, see Inquests No. 185 of Unknown Female Child, 16 November 1904 and No. 99 of Tan Buan, 1 June 1905).

109. *Straits Times*, 24 July 1896.

110. ARSM, 1904, Health Officer's Department, p. 26; MPMCOM, 6 May 1904 and 18 November 1904; ARSM 1906, Health Officer's Report, p. 29; MPMCOM, 30 November 1906; ARSM 1907, Health Officer's Report, p. 38; MPMCOM, 8 May 1908.

111. Contrary to expectations, a large number of remains were unearthed despite the fact that the burial ground had been closed for forty or fifty years. As a consequence, the initial vote of $10,000 set aside in 1907 for compensation was quickly exhausted and had to be supplemented by two further votes of $10,000 each the following year (PLCSS, 17 July 1908, p. B33).

112. Song Ong Siang, *One Hundred Years' History of the Chinese in Singapore*, Singapore: Oxford University Press, 1984 (first published in 1902), p. 421.

113. Ibid.

114. ARSM, 1925, 'Appendix B: 'Singapore Improvement Trust [1925]', pp. 23–4; ARSM, 1926, 'Appendix: [Singapore Improvement Trust, 1926]', pp. 16–17; ARSM, 1927, 'Appendix: Singapore Improvement Trust [1927]', pp. 18–19.

115. 'Appendix: [Singapore Improvement Trust, 1926]', p. 17; SIT, 553/26, 23 August 1926: 'Wee Swee Teow & Co. Asks That Some Compensation May Be Included in the Award for Acquisition of Lot 15 Mukim I on Account of the Exhumation and Reburial of Their Client's Ancestors'; 'Tiong Bahru Improvement Scheme Acquisition of Two Graves from Lot 185 Mukim 1'.

116. For example, the trustees of the Hokkien Seh Choa burial ground negotiated for a new site in Toa Payoh on the outskirts of the municipality in place of their old grounds at Silat Road, Telok Blangah, which were acquired by the Singapore Improvement Trust ('Forwarding Copy of a Letter from the Secretary of Chinese Affairs on the Subject of Acquiring Land Comprised in Grant No. 49 Toah Payoh as a Burial Ground for the Seh Chua Community').

117. *Report of the Committee Appointed to Make Recommendations Regarding Burial and Burial Grounds*, p. 4.

118. MPMCOM, 26 July 1907; ARSM, 1908, Health Officer's Report, p. 40.

119. MPMCOM, 2 June 1905; ARSM, 1910, Health Officer's Report, p. 45.

120. Lim Boon Keng, 'Straits Chinese Reform', *Straits Chinese Magazine*, 3, 9 (1899): 22–5.

121. MPMCOM, 2 June 1904, 14 December 1906, and 31 August 1911.

122. Song, *One Hundred Years' History*, p. 407.

123. MMFGPSYC, 7 July 1916 and 25 May 1917; MPMCOM, 26 October 1917.

124. MMFGPSYC, 25 May 1917.

125. MGCM, 4 January 1918 and 8 March 1918; MPMCOM, 25 October 1918, 30 May 1919, 25 July 1919, 31 October 1919, and 28 November 1919.

126. MGCM, 12 March 1920; MPMCSM, 29 April 1921.

127. ARSM, 1922, Municipal Health Officer, p. 40D; MPMCSM, 25 August 1922. By the end of 1922, there had only been 98 burials in the municipal cemetery.

128. Freedman, 'Geomancy', p. 13.

129. In Chinese burial grounds, grave plots measured a minimum of about 20 by 10 feet (200 square feet) to provide space not simply for the coffin (the largest taking up about 50 square feet), but also for adequate room in front of the tombstone to provide a platform for ritual. Among the wealthy Chinese, grave plots were considerably larger. For example, in order to reinter the body of their father from a private site in Tiong Bahru to the municipal cemetery in 1926, Lim Chan Siew and Lim Chew Chye had to reserve six adjoining plots in the latter in order to accommodate the tombstone (*Report of the Committee Appointed to Make Recommendations Regarding Burial and Burial Grounds*, pp. 8–9; 'Tiong Bahru Improvement Scheme, Acquisition of Two Graves from Lot 185 Mukim 1').

130. For the beneficial power of *feng shui* to be optimized, grave alignments had to be oriented in particular directions judged auspicious for the deceased's horoscope and birth year. Each birth year is associated with particular auspicious and inauspicious compass directions, and as a result, Chinese burial grounds tended to present a helter-skelter appearance with graves oriented in different directions. The provision of plots oriented in two different directions in the municipal cemetery allowed the Chinese a limited degree of geomantic choice. In general, a southerly orientation was considered favourable as the south was traditionally considered 'the cradle of warmth, light, life and productive summer rains' (Knapp, 'The Changing Landscape', p. 7; De Groot, *The Religious System of China*, p. 942).

131. MMSC2, 9 August 1922; MPMCOM, 25 August 1922, MPMCSM, 25 August 1922. It also appears that consecutive burial was not insisted upon, particularly in the case of the wealthy. Instead, relatives could request specific plots in accordance with the geomantic requirements of the deceased. In reburying the remains of two family members from the late Cheang Hong Lim's Alexandra Road estate to Bukit Brown, the Cheang family required a site 'well above sea-level', supporting their request by reminding the authorities that 'Mr Cheang Hong Lim was one of the best known citizens in this Colony during his lifetime' (SIT, 145/28, 13 February 1928: 'Burial Grounds on Lot No. 53 Grant 63 [Tanglin] and Lot No. 4 Grant 2 [Telok Blangah]').

132. MMSC, (Special), 14 October 1921.

133. Freedman, 'Geomancy', p. 13.

134. Jean DeBernardi, 'Space and Time in Chinese Religious Culture', *History of Religions*, 31 (1992): 254.

135. Contemporary Western observers who had attempted to measure the rudimentary 'science' of *feng shui* against 'western views of physics' had often found much to disparage in the Chinese system. Rev E. J. Eitel of the London Missionary Society, for example, concludes his monograph on *feng shui* with the following words: What I have hitherto, by a stretch of charity, called Chinese physical science is, from a scientific point of view, but a conglomeration of rough guesses of nature, sublimated by fanciful play with puerile diagrams.... It is simply the blind gropings of the Chinese mind after a system of natural science, which gropings, untutored by practical observation of nature and trusting almost exclusively in the force of abstract reasoning, naturally left the Chinese mind completely in the dark. The system of *feng-shui*, therefore, based as it is on human speculation and superstition and not on [a] careful study of nature, is marked for decay and dissolution.' (Eitel, *Feng-shui*, pp. 65 and 69.) It was left to later writers in the 1950s such as Joseph Needham in his *Science and Civilisation in China* to re-evaluate the proper role of *feng shui* within the context of the development of Chinese science for a Western readership.

136. In a similar vein, Maurice Freedman observes that in the British colony of Hong Kong, the government experienced difficulty countering 'popular resistance by geomancy' largely because it was unable to 'talk back [to the people] in their own language'. He adds, 'Were the Government ... to share the belief of the people, it would be in a position to resist their consequences, for its officials would then be able to match their own *feng shui* opinions against those of the objectors, and, if necessary call in professional geomancers to argue with those retained by the people' (Freedman, 'Geomancy', pp. 8–9).

9
Conclusion: The Politics of Space in Colonial Singapore

In this account, both people who claim history as their own and the people to whom history has been denied emerge participants in the same historical trajectory.[1]

THE specific line of enquiry and the questions posed in the preceding chapters have focused on interrogating the negotiation of power between the municipal authorities and the Asian communities in shaping, representing, and using the urban built environment in colonial Singapore. This is a critical area in which the self-interest of the authorities concerned with creating a sanitary and orderly city and that of the Asian communities bent on making the built environment effective for their own purposes intersected. It is an area characterized by conflict and compromise because those vested with formal power to control the built environment and those who lived in and used it entertained differing interpretations and perspectives on health and disease, on order and disorder, and ultimately, on what constituted the effective management of life and death in an urban context.

Municipal strategies devised to control and regulate what were perceived to be pathogenic and disorderly 'Asiatic environments' were, in their main thrust, essentially spatial or environmentalist. These strategies included surveillance (Chapter 3), the modification of built form (Chapter 4), the introduction of municipal 'utilities' systems (Chapter 5), and the improvement of environmental legibility by inscribing a code of names (Chapter 6) and demarcating between public/private (Chapter 7) and obsolete/modern (Chapter 8) uses of the environment. Indeed, environmentalist thinking and strategies appeared to have been even more widely pervasive and deeply entrenched in both official circles and among the European public in colonial societies such as Singapore than in Europe itself.[2] What made 'the medicine of the environment'[3] so persuasive for so long in colonial Singapore arises, in part, from the congruence of environmentalist thought with what was perceived as the 'Asiatic misuse' of the urban environment—whether by gross overcrowding or rampant pollution—as a result of their ingrained, blatant neglect of the laws of sanitation and order. One of the tenets of environmentalism is that whilst the environment could be correctly manipulated for the people's benefits, the failure to do so would lead to disastrous consequences for their health and welfare.

While it is over-simplistic to claim that the control and manipulation of the environment was seized upon by the politically powerful in order to ensure the preservation of the colonial order or to wield a conscious policy of social pacification, it is clear that targeting the 'Asiatic environment' as the focus of colonial and municipal interventionist strategies provided a seemingly apolitical focus amenable to the language of 'improvement', without raising the more intransigent and perplexing questions relating the people's health and welfare to socio-economic critieria, the polarizations between colonizer and the colonized, or wider colonial economic policies based on safeguarding a large reserve pool of cheap labour. As Dreyfus and Rabinow have argued, 'the language of reform' was an essential component of political technologies of control and where there was resistance or failure to achieve stated aims, it was 'construed as further proof of the need to reinforce and extend the power of the experts'.[4] In reforming the 'Asiatic environment', the problems which the authorities recognized, the methods, models, and expertise they relied on in diagnosing and solving them, the theories they regarded as satisfactory in general or adequate in particular, and the evidence admitted as valid were not arrived at by abstract rationality or within a scientific vacuum but were invested with a prejudice and pragmatism typical of the colonial project as a whole. This was, however, rarely admitted, and instead, the municipal authorities attempted to legitimize their projects and schemes by claiming the support of expert advice and experience imported either from Britain or elsewhere in the British Empire.

The built spaces of the colonial city were not, however, simply shaped by dominant forces or powerful groups, but were continuously transformed by processes of conflict and negotiation involving the strategies and counter-strategies of colonial institutions of authority and the different 'colonized' groups within society. The urban built environment did not only reflect, in Henri Lefebvre's phrase, the 'representations of space' of the powerful (such as urban planners, architects, scientists, and technocrats) but also constituted the 'lived space of everyday life' which could act as 'representational spaces', or counter-spaces embodying complex symbolisms ... linked to the clandestine or underground side of social life'.[5] In other words, space 'has always been political and strategic'[6] and the clash of priorities often find resolution through a complicated process of conflict and negotiation in which individuals and groups interact to constitute the resultant landscape. Within this construction of space, the built spaces of the colonial city were construed as sites of control and resistance, simultaneously drawn upon by, on the one hand, dominant groups to secure conceptual or instrumental control, and, on the other, subordinate groups to resist exclusionary definitions or tactics and to advance their own claims. In colonial Singapore, the Asian people who sought to command spaces within the urban built environment and undermine prescribed notions and representations of space did so in a number of ways.

First, negotiation over the built environment was occasionally channelled through community leaders who were able to present in verbal or written form an alternative discourse on the management and use of the environment. English-educated leaders of the Chinese communities, for example, were able to defend the 'sacro-sanctity' of Chinese burial grounds (Chapter 8) or the traditional practice of applying night-soil as manure to vegetable gardens (Chapter 5) through petitions, letters to the press, articles in journals, and representations on various committees set up by the colonial or municipal government. By drawing on existing discourses on environmental management as embodied in Chinese geomancy or Chinese medical theory, community leaders sought to challenge Western urban planning ideals or Western sanitary science. As colonial authorities could not effectively govern without incorporating a segment of the colonized body into the ruling power structure, the views of those who acted as 'social brokers' commanded sufficient weight for them to be accommodated within decision-making bodies such as the municipal commission. Municipal policy in controlling the built environment were occasionally held back or modified to accommodate expressed views, although where policy changes were made, there was seldom any real acknowledgement of the validity of alternative Asian discourses. Instead, they were euphemistically styled as concessions made 'in deference to Chinese sentiment' and effected on pragmatic grounds in order to minimize potential conflict and increase the chances of attaining desired objectives.

More typically, however, it was at the level of the daily practices and within the spaces of everyday life that the effects of power, including their ideological aspects, were developed, sustained, renegotiated, distorted, or countered. Negotiation over the shaping of the built environment was conducted through everyday strategies of control and resistance. Sporadically, negotiation culminated in violent conflict such as during the outbreak of the so-called 'verandah riots' (Chapter 7), but in general it was through the daily practices of signification (such as the naming of urban places (Chapter 6)), of occupation and use (as in the case of the verandah (Chapter 7)), and of non-compliance to municipally prescribed formulations for ideal built forms (such as house form (Chapter 4)) and sanitary environments (Chapter 3) that a specifically Asian 'everyday discourse'[7] of the environment developed. Although such an 'everyday discourse' evolved within the realm of habitual behaviour and practical consciousness, none the less, in effect, it served to challenge the hegemony of the authorities. In more general terms, although what human agents know about what they do, and why they do it (that is, what Giddens calls their 'knowledgeability *as* agents') is largely carried out in practical consciousness and through routine, day-to-day social activity, such action 'implies power' and is capable of producing certain effects.[8] In colonial cities such as

Singapore, the 'agency' of the Asian people was further enhanced in its effectiveness because, whether tacitly or discursively, it drew upon coherent ideologies, institutional structures, and schemes of legitimation which were independent of, and largely impenetrable to the authorities in power.

The colonial urban built environment, in so far as it might be considered textual, was hence the subject of a constant contest over meaning and usage. Such a contest cannot be interpreted in solely cultural terms (though cultural perceptions were certainly influential in shaping the terms of conflict) or reduced to a question of economic competition. Instead the negotiation of power should be understood in its social materiality, its day-to-day operation, and in Foucault's terms, at the level of 'micropractices'[9] and through 'the antagonism of strategies'.[10]

Conflict over the urban built environment is an inherent part of the wider negotiation of power within the colonial process itself. While most urban landscapes can be interrogated as terrains of quotidian conflict and negotiation, the colonial city in particular lends itself to such an interpretation because the dissonance in social values, the divergence in perceptions of the environment, and the asymmetries of power between the minority in authority and the vast majority who inhabit the city are possibly more in evidence in the colonial context than elsewhere. While every society (including colonial societies) 'produces its own space',[11] it does not do so without the processes of conflict and compromise between those who control space and those who live in it, both of whom must be seen as 'participants in the same historical trajectory'.

1. Eric R. Wolf, *Europe and the People Without History*, Berkeley: University of California Press, 1982, p. 23.

2. In his study of the European campaign against disease by the prime strategy of modifying pathogenic environments, James C. Riley argues that 'most of the remedies credited with fashioning the late nineteenth century part of Europe's long mortality decline remain, despite the interim formulation of the germ theory, the remedies of environmentalism' (James C. Riley, *The Eighteenth-Century Campaign to Avoid Disease*, Basingstoke: Macmillan, 1987, p. 139). By the early twentieth century, however, Gerry Kearns argues that in tandem with (though not necessarily determined by) the shift in the epidemiologic transition from a phase dominated by epidemics to one dominated by degenerative diseases in Europe, public health strategies moved from environmentalism to the more individualist strategy of hospitalization (Gerard Kearns, 'Zivilis or Hygaeia: Urban Public Health and the Epidemiologic Transition', in Richard Lawton (ed.), *The Rise and Fall of Great Cities: Aspects of Urbanization in the Western World*, London: Belhaven Press, 1989, p. 108).

3. Riley, *The Eighteenth-Century Campaign*, p. xvi.

4. Hubert L. Dreyfus and Paul Rabinow, *Michel Foucault: Beyond Structuralism and Hermeneutics*, Brighton: Harvester Press, 1982, p. 196.

5. Henri Lefebvre, *The Production of Space*, trans. Donald Nicholson-Smith, Oxford: Basil Blackwell, 1991, p. 33.

6. Henri Lefebvre, 'Reflections on the Politics of Space', trans. M. J. Enders, in Richard Peet (ed.), *Radical Geography: Alternative Viewpoints on Contemporary Social Issues*, Chicago: Maaroufa Press, 1977, p. 341.

7. This is Lefebvre's phrase. He follows Foucault in arguing that a discourse is not merely a set of linguistic practices which reports on the world, but is composed of a whole assemblage of daily activities, events, objects, settings, and epistemological concepts (Lefebvre, *The Production of Space*, p. 25).

8. Anthony Giddens, *The Constitution of Society: Outline of the Theory of Structuration*, Cambridge: Polity Press, 1986, pp. xxiii and 5–16.

9. Dreyfus and Rabinow, *Michel Foucault*, p. 185.

10. Michel Foucault, 'The Subject and Power', in Dreyfus and Rabinow, *Michel Foucault*, p. 211.

11. Lefebvre, *The Production of Space*.

Appendix

Demographic Statistics

TABLE A.1

Population of Singapore (Settlement) by Race, 1871–1931

Year	Total	Europeans	Eurasians	Chinese	Malays	Indians	Others
1871	97,111	1,946	2,164	54,572	26,148	11,610	671
1881	139,208	2,769	3,094	86,766	33,102	12,138	1,339
1891	184,554	5,254	3,589	121,908	35,992	16,035	1,776
1901	228,555	3,824	4,120	164,041	36,080	17,823	2,667
1911	303,321	5,711	4,671	219,577	41,806	27,755	3,660
1921	418,358	6,145	5,436	315,151	53,595	32,314	5,717
1931	557,745	8,082	6,903	418,640	65,014	50,811	8,295

Sources: John F. A. MacNair, C. B. Waller, and A. Knight, *Report of the Census Officers, for the Settlement of Singapore, 1871*, Singapore: Government Printing Office, 1871, p. 7; S. Dunlop, *Report on the Census of Singapore, 1881*, Singapore: Government Printing Office, 1881, p. 306; E. M. Merewether, *Report on the Census of the Straits Settlements Taken on 5th April 1891*, Singapore: Government Printing Office, 1892, p. 47; J. R. Innes, *Report on the Census of the Straits Settlements Taken on 1st March 1901*, Singapore: Singapore Government Printers, 1901, p. 13; Hayes Marriott, 'The Peoples of Singapore', in Walter Makepeace, Gilbert E. Brooke, and Ronald St. John Braddell (eds.), *One Hundred Years of Singapore*, Vol. 1, London: John Murray, 1921, p. 360; J. E. Nathan, *The Census of British Malaya, 1921*, London: Waterlow and Sons, 1922, p. 155; C. A. Vlieland, *British Malaya: A Report on the 1931 Census and on Certain Problems of Vital Statistics*, London: Crown Agents for the Colonies, 1932, pp. 120–1; M. V. del Tufo, *Malaya, Comprising the Federation of Malaya and the Colony of Singapore: A Report on the 1947 Census of Population*, London: Crown Agents for the Colonies, 1949, pp. 158–9.

TABLE A.2

Decennial Population Change of Singapore (Settlement) by Race, 1871–1931

Decade	Total	Europeans	Eurasians	Chinese	Malays	Indians	Others
1871–81	42,097	823	930	32,194	6,954	637	559
1881–91	45,346	2,485	495	35,142	2,796	3,878	437
1891–1901	44,001	−1,430	552	42,133	88	1,788	891
1901–11	74,766	1,887	551	55,536	5,726	9,932	993
1911–21	115,037	434	765	95,574	11,789	4,559	2,057
1921–31	139,387	1,937	1,467	103,489	11,509	18,497	2,578

Source: Calculated from Table A.1.

TABLE A.3

Population of the Singapore Municipality by Race, 1871–1931

Year	Total	Europeans	Eurasians	Chinese	Malays	Indians	Others
1871	56,116[a]	3,702[c]		38,362	9,101	8,241	412[d]
	61,752[b]	(6.2)		(64.1)	(15.2)	(13.8)	(0.7)
1881	95,323[e]	1,022	2,781	63,698	16,313	10,441	1,068
		(1.1)	(2.9)	(66.8)	(17.1)	(11.0)	(1.1)
1891	153,495[f]	2,243	3,415	106,643	24,646	14,846	1,702
		(1.4)	(2.2)	(69.5)	(16.1)	(9.7)	(1.1)
1901	193,089[g]	2,748	3,982	141,865	26,230	15,646	2,618
		(1.4)	(2.1)	(73.5)	(13.6)	(8.1)	(1.4)
1911	259,610	4,981	4,427	194,016	28,000	24,494	3,692
		(1.9)	(1.7)	(74.7)	(10.8)	(9.4)	(1.4)
1921	350,355	5,170	4,620	273,393	34,105	27,777	5,290
		(1.5)	(1.3)	(78.0)	(9.7)	(7.9)	(1.5)
1931	445,719	6,518	6,134	340,614	43,373	41,356	7,724
		(1.5)	(1.4)	(76.4)	(9.7)	(9.3)	(1.7)

Sources: John F. A. MacNair, C. B. Waller, and A. Knight, *Report of the Census Officers, for the Settlement of Singapore, 1871*, Singapore: Government Printing Office, 1871, p. 11; S. Dunlop, *Report on the Census of Singapore, 1881*, Singapore: Government Printing Office, 1881, p. 314; E. M. Merewether, *Report on the Census of the Straits Settlements Taken on 5th April 1891*, Singapore: Government Printing Office, 1892, p. 88; J. R. Innes, *Report on the Census of the Straits Settlements Taken on 1st March 1901*, Singapore: Singapore Government Printers, 1901, p. 14; Hayes Marriott, *Report on the Census of the Straits Settlements Taken on 10th March 1911*, Singapore: Singapore Printing Press, 1911, pp. 13–16; J. E. Nathan, *The Census of British Malaya, 1921*, London: Waterlow and Sons, 1922, p. 155; C. A. Vlieland, *British Malaya: A Report on the 1931 Census and on Certain Problems of Vital Statistics*, London: Crown Agents for the Colonies, 1932, pp. 120–1; M. V. del Tufo, *Malaya, Comprising the Federation of Malaya and the Colony of Singapore: A Report on the 1947 Census of Population*, London: Crown Agents for the Colonies, 1949, pp. 158–9.

Note: Figures in parentheses denote percentages.

[a]Return of Native Population only, exclusive of Military and Prisoners.

[b]Return of Total Population inclusive of the Military (596 British and 415 Indian), Prisoners (923), and Floating Population.

[c]Estimated number of Europeans and Eurasians (civilian only).

[d]Exclusive of Armenians and Jews.

[e]Return of Total Population exclusive of the British Military (783). Municipal area divided into nine (A–I) Supervisor's Divisions.

[f]Return of Total Population exclusive of Floating Population and British Military. Municipal area increased by the addition of another nine Supervisor's Divisions (J–R).

[g]For 1901–31, the return is for Total Resident Population (inclusive of the Military and exclusive of the Floating Population).

TABLE A.4

Racial Composition of Census Districts A–R, 1891

Census Division	Europeans	Eurasians	Chinese	Malays	Tamils	Others	Total
A	38	21	17,680	846	1,842	53	20,480
	(0.2)	(0.1)	(86.3)	(4.1)	(9.0)	(0.3)	(100.0)
B	75	14	9,817	698	738	48	11,390
	(0.7)	(0.1)	(86.2)	(6.1)	(6.5)	(0.4)	(100.0)
C	52	36	21,704	211	1,380	42	23,425
	(0.2)	(0.2)	(92.7)	(0.9)	(5.9)	(0.2)	(100.1)
D	338	133	2,788	1,118	576	108	5,061
	(6.7)	(2.6)	(55.1)	(22.1)	(11.4)	(2.1)	(100.0)
E	671	1,246	8,370	1,000	1,412	266	12,965
	(5.2)	(9.6)	(64.6)	(7.7)	(10.9)	(2.1)	(100.1)
F	329	400	1,246	640	1,018	121	3,754
	(8.8)	(10.7)	(33.2)	(17.0)	(27.1)	(3.2)	(100.0)
G	100	973	18,862	7,376	3,747	686	31,744
	(0.3)	(3.1)	(59.4)	(23.2)	(11.8)	(2.2)	(100.0)
H	31	270	1,689	2,737	1,980	108	6,815
	(0.5)	(4.0)	(24.8)	(40.2)	(29.1)	(1.6)	(100.2)
I	10	3	1,737	3,302	100	52	5,204
	(0.2)	(0.1)	(33.4)	(63.5)	(1.9)	(1.0)	(100.1)
J	118	75	6,312	1,154	282	38	7,979
	(1.5)	(0.9)	(79.1)	(14.5)	(3.5)	(0.5)	(100.0)
K	126	29	4,118	1,499	678	62	6,512
	(1.9)	(0.4)	(63.2)	(23.0)	(10.4)	(1.0)	(99.9)
L	52	7	2,130	565	279	12	3,045
	(1.7)	(0.2)	(70.0)	(18.6)	(9.2)	(0.4)	(100.1)
M	186	42	1,647	789	101	35	2,800
	(6.6)	(1.5)	(58.8)	(28.2)	(3.6)	(1.3)	(100.0)
N	27	94	2,752	543	252	24	3,692
	(0.7)	(2.5)	(74.5)	(14.7)	(6.8)	(0.7)	(99.9)
P	26	12	1,682	887	51	8	2,666
	(1.0)	(0.5)	(63.1)	(33.3)	(1.9)	(0.3)	(100.1)
Q	50	22	1,574	418	94	27	2,185
	(2.3)	(1.0)	(72.0)	(19.1)	(4.3)	(1.2)	(99.9)
R	8	38	2,187	845	236	12	3,326
	(0.2)	(1.1)	(65.8)	(25.4)	(7.1)	(0.4)	(100.0)
Wayfarers & Vagrants	6	0	348	16	80	0	450
	(1.3)	(0.0)	(77.3)	(3.6)	(17.8)	(0.0)	(100.0)
Total	2,243	3,415	106,643	26,646	14,846	1,702	153,493

Source: Compiled from E. M. Merewether, *Report on the Census of the Straits Settlements Taken on 5th April 1891*, Singapore: Government Printing Office, 1892, pp. 52–68.

Note: Figures in parentheses denote percentages.

TABLE A.5
Racial Composition of Census Districts A–R, 1901

Census Division	Europeans	Eurasians	Chinese	Malays	Tamils	Others	Total
A	44	11	20,651	1,149	1,742	103	23,700
	(0.2)	(0.0)	(87.1)	(4.8)	(7.4)	(0.4)	(99.9)
B	103	22	14,259	706	745	74	15,909
	(0.6)	(0.1)	(89.6)	(4.4)	(4.7)	(0.5)	(99.9)
C	34	23	25,568	96	779	24	26,524
	(0.1)	(0.1)	(96.4)	(0.4)	(2.9)	(0.1)	(100.0)
D	413	220	4,106	1,249	801	120	6,819
	(6.1)	(3.2)	(58.9)	(18.3)	(11.7)	(1.8)	(100.0)
E	717	1,029	11,608	1,002	1,312	458	16,126
	(4.4)	(6.4)	(72.0)	(6.2)	(8.1)	(2.8)	(99.9)
F	340	580	2,083	798	1,176	186	5,163
	(6.6)	(11.2)	(40.3)	(15.5)	(22.8)	(3.6)	(100.0)
G	120	1,012	28,242	5,956	3,691	1,113	40,134
	(0.3)	(2.5)	(70.4)	(14.8)	(9.2)	(2.8)	(100.0)
H	55	433	2,802	2,721	2,284	208	8,503
	(0.6)	(5.1)	(33.0)	(32.0)	(26.9)	(2.4)	(100.0)
I	5	4	2,443	2,907	108	66	5,533
	(0.1)	(0.1)	(44.2)	(52.5)	(2.0)	(1.2)	(100.1)
J	117	91	8,668	2,158	912	59	12,005
	(1.0)	(0.8)	(72.2)	(18.0)	(7.6)	(0.5)	(100.1)
K	130	84	6,630	1,584	555	55	9,038
	(1.4)	(0.9)	(73.4)	(17.5)	(6.1)	(0.6)	(99.9)
L	117	45	2,219	804	439	23	3,647
	(3.2)	(1.2)	(60.8)	(22.0)	(12.0)	(0.6)	(99.8)
M	266	56	1,457	963	164	39	2,945
	(9.0)	(1.9)	(49.5)	(32.7)	(5.6)	(1.3)	(100.0)
N	113	248	4,466	1,165	493	46	6,531
	(1.7)	(3.8)	(68.4)	(17.8)	(7.5)	(0.7)	(99.9)
P	25	55	2,367	1,320	47	4	3,818
	(0.7)	(1.4)	(62.0)	(34.6)	(1.2)	(0.1)	(100.0)
Q	113	39	1,302	460	193	23	2,130
	(5.3)	(1.8)	(61.1)	(21.6)	(9.1)	(1.1)	(100.0)
R	34	30	3,084	1,192	205	19	4,564
	(0.7)	(0.7)	(67.6)	(26.1)	(4.5)	(0.4)	(100.0)
Wayfarers &	2	0	428	62	177	4	673
Vagrants	(0.3)	(0.0)	(63.6)	(9.2)	(26.3)	(0.6)	(100.0)
Total	2,748	3,982	142,293	26,292	15,823	2,624	193,762

Source: Compiled from J. R. Innes, *Report on the Census of the Straits Settlements Taken on 1st March 1901*, Singapore: Singapore Government Printers, 1901, pp. 21–37.
Note: Figures in parentheses denote percentages.

TABLE A.6

Percentage Distribution of Each Race across Census Districts A–R, 1901

Census Division	Europeans	Eurasians	Chinese	Malays	Tamils	Others	Total
A	44	11	20,651	1,149	1,742	103	23,700
	(1.6)	(0.3)	(14.5)	(4.4)	(11.0)	(3.9)	
B	103	22	14,259	706	745	74	15,909
	(3.7)	(0.6)	(10.0)	(2.7)	(4.7)	(2.8)	
C	34	23	25,568	96	779	24	26,524
	(1.2)	(0.6)	(18.0)	(0.4)	(4.9)	(0.9)	
D	413	220	4,106	1,249	801	120	6,819
	(15.0)	(5.5)	(2.8)	(4.8)	(5.1)	(4.6)	
E	717	1,029	11,608	1,002	1,312	458	16,126
	(26.1)	(25.8)	(8.2)	(3.8)	(8.3)	(17.5)	
F	340	580	2,083	798	1,176	186	5,163
	(12.4)	(14.6)	(1.5)	(3.0)	(7.4)	(7.1)	
G	120	1,012	28,242	5,956	3,691	1,113	40,134
	(4.4)	(25.4)	(19.8)	(22.7)	(23.3)	(42.4)	
H	55	433	2,802	2,721	2,284	208	8,503
	(2.0)	(10.9)	(2.0)	(10.3)	(14.4)	(7.9)	
I	5	4	2,443	2,907	108	66	5,533
	(0.2)	(0.1)	(1.7)	(11.1)	(0.7)	(2.5)	
J	117	91	8,668	2,158	912	59	12,005
	(4.3)	(2.3)	(6.1)	(8.2)	(5.8)	(2.2)	
K	130	84	6,630	1,584	555	55	9,038
	(4.7)	(2.1)	(4.7)	(6.0)	(3.5)	(2.1)	
L	117	45	2,219	804	439	23	3,647
	(4.3)	(1.1)	(1.6)	(3.1)	(2.8)	(0.9)	
M	266	56	1,457	963	164	39	2,945
	(9.7)	(1.4)	(1.0)	(3.7)	(1.0)	(1.5)	
N	113	248	4,466	1,165	493	46	6,531
	(4.1)	(6.2)	(3.1)	(4.4)	(3.1)	(1.8)	
P	25	55	2,367	1,320	47	4	3,818
	(0.9)	(1.4)	(1.7)	(5.0)	(0.3)	(0.2)	
Q	113	39	1,302	460	193	23	2,130
	(4.1)	(1.0)	(0.9)	(1.7)	(1.2)	(0.9)	
R	34	30	3,084	1,192	205	19	4,564
	(1.2)	(0.8)	(2.2)	(4.5)	(1.3)	(0.7)	
Wayfarers &	2	0	428	62	177	4	673
Vagrants	(0.1)	(0.0)	(0.3)	(0.2)	(1.1)	(0.2)	
Total	2,748	3,982	142,293	26,292	15,823	2,624	193,762
	(100.0)	(100.1)	(100.0)	(100.0)	(100.1)	(100.1)	

Source: Compiled from J. R. Innes, *Report on the Census of the Straits Settlements Taken on 1st March 1901*, Singapore: Singapore Government Printers, 1901, pp. 21–37.

Note: Figures in parentheses denote percentages.

TABLE A.7

Population Density and Average Number of Occupants per House
According to Census Divisions A–R, 1901

Census Division	Population	Area (sq. miles)	Population Density (persons/sq. mile)	Occupied Houses	Average Number of Occupants per House
A	23,700	0.28	84,600	2,105	11.3
B	15,909	0.32	49,700	1,163	13.7
C	26,524	0.16	165,800	2,039	13.0
D	6,819	0.65	10,500	958	7.1
E	16,126	0.46	35,100	1,363	11.8
F	5,163	0.37	14,000	513	10.1
G	40,134	0.34	118,000	4,267	9.4
H	8,503	0.46	18,500	1,201	7.1
I	5,533	0.14	39,500	881	6.3
J	12,005	0.72	16,700	987	12.2
K	9,038	3.34	2,700	1,455	6.2
L	3,647	1.79	2,000	578	6.3
M	2,945	2.55	1,200	375	7.9
N	6,531	4.10	1,600	1,226	5.3
P	3,818	1.22	3,100	594	6.4
Q	2,130	3.93	500	374	5.7
R	4,564	3.32	1,400	1,053	4.3
Wayfarers & Voyagers	673	–	–	–	–
Total	193,762	24.15	–	21,132	–
Average	–	–	8,000	–	9.2

Source: Calculated from J. R. Innes, *Report on the Census of the Straits Settlements Taken on 1st March 1901*, Singapore: Singapore Government Printers, 1901, p. 38.

Bibliography

Unpublished Sources

Colonial Records

(CO series deposited at the Public Record Office, Kew, London; NL series deposited at the National Archives, Singapore)
CO 273, Straits Settlements Original Correspondence, 1880–1929.
CO 275, Straits Settlements Sessional Papers, 1867–1929 (contains the Straits Settlements Proceedings of the Legislative Council and the Straits Settlements Annual Departmental Records).
CO 276, Straits Settlements Government Gazette, 1880–1929.
NL 955, Despatches from the Secretary of State to the Straits Settlements, 1887–8.

Municipal Records

(Deposited at the National Archives, Singapore)
Administrative Reports of the Singapore Municipality, 1888–1930.
Minutes of the Proceedings of the Municipal Commissioners, 1880–1929.
Municipal Building Plans, 1884–1929.
Municipal Fund: Valuation of Houses and Assessment Lists, 1880–1925.
Singapore Improvement Trust Minute Papers and Records, 1911–29.

Court Records

(Deposited at the Subordinate Courts, Singapore)
Municipal Summons Note Book, 5 July 1912–13 December 1912.
Municipal Summons Note Book, 16 December 1912–6 May 1913.
Note Book of District Court Cases No. 8, 12 December 1913–20 February 1914.
Note Book of Municipal Cases No. 1, 3 March 1921–1 April 1924.
Note Book of District Court Cases No. 2, 19 March 1923–12 July 1923.
Police Summons Note Book, 27 May 1908–12 January 1909.
Public Summons Note Book, 30 October 1908–14 June 1910.
Singapore Coroners' Inquest and Inquiries, 1904–5.

India Office Records

(Deposited at the India Office Library and Records, London)
[Council Meeting between] the Honourable Robert Fullerton, Governor, and John Prince, Esq., 8 May 1827, G/34/154.

Diary of the Proceedings of the Honourable John Prince, Esq., Resident
Councillor of Singapore, 1827, G/34/153.
Diary of the Proceedings of the Honourable John Prince, Esq., Resident
Councillor of Singapore, 1827, G/34/155.

Singapore Government Records

'List of Existing Cemeteries, 1989', compiled by the Ministry of Environment,
Singapore.

Association Papers

'Brief Account of the Kwong Wai Shui Hospital', compiled by the Secretary of
the Hospital, 1988 (deposited at the Kwong Wai Shui Hospital).
List of Hakka Medical Halls, compiled by Chen Chin Ah (deposited at the
Thong Chai Institute of Medical Research).
'Petition from Gan Eng Seng, Seong Kheng Tong, Lim Kheng Hoo and Jew
Chu Wan to the Government for a Land Grant, 1892' (deposited at the
Thong Chai Medical Institution).
'Straits Settlements Land Grant No. 3650' (deposited at the Thong Chai
Medical Institution).

Personal Papers

(Deposited at Rhodes House Library, Oxford)
Bourne, F. G., 'The Straits Settlements, a Crown Colony under the Control of
the Colonial Office (Typescript of a Talk on the Straits Settlements:
Singapore Municipality) *c.*1935', Mss.Ind.Ocn.s.239(2).
Gilman, Edward Wilmot F., 'Personal Recollections, Speeches and
Newscuttings, *c.*1920–1931 while a Civil Servant in Malaya',
Mss.Ind.Ocn.s.127.
Peel, William, 'Notes Covering His Colonial Service from 1897–1935 in
Malaya, Singapore and Hong Kong', Mss.Brit.Emp.s.208.
Wilson, Frank Kershaw, 'Letters Home, January 1915–December 1916 while
Administrative Cadet', Singapore, Mss.Ind.Ocn.s.162.

Official Publications

Reports

*Proceedings and Report of the Commission Appointed to Inquire into the Cause of
the Present Housing Difficulties in Singapore, and the Steps Which Should Be
Taken to Remedy Such Difficulties*, Vols. 1 and 2, Singapore: Government
Printing Office, 1918.
*Proceedings of the Commission Appointed to Inquire into the Administration of
Municipal Affairs in the Colony of Straits Settlements*, Vols. 1 and 2,
Singapore: Government Printing Office, 1910.
*Report of the Commission Appointed to Inquire into the Financial Position of the
Municipality*, Vols. 1 and 2, Singapore: Government Printing Office, 1921.
*Report of the Committee Appointed to Investigate the Hawker Question in
Singapore, 4 November 1931*, Singapore: Government Printing Office, 1932.
*Report of the Committee Appointed to Make Recommendations Regarding Burial
and Burial Grounds in the Colony of Singapore*, Singapore: Government
Printing Office, 1952.

Report of the Housing Committee of Singapore, 1947, Singapore: Government Printing Office, 1948.

Simpson, W. J., *Report on the Sanitary Condition of Singapore*, London: Waterlow and Sons, 1907.

Censuses

Del Tufo, M. V., *Malaya, Comprising the Federation of Malaya and the Colony of Singapore: A Report on the 1947 Census of Population*, London: Crown Agents for the Colonies, 1949.

Dunlop, S., *Report on the Census of Singapore, 1881*, Singapore: Government Printing Office, 1881.

Innes, J. R., *Report on the Census of the Straits Settlements Taken on 1st March 1901*, Singapore: Singapore Government Printers, 1901.

McNair, John F. A.; Waller, C. B.; and Knight, A., *Report of the Census Officers for the Settlement of Singapore, for 1871*, Singapore: Government Printing Office, 1871.

Marriott, Hayes, *Report on the Census of the Colony of the Straits Settlements Taken on 10th March 1911*, Singapore: Singapore Printing Press, 1911.

Merewether, E. M., *Report on the Census of the Straits Settlements Taken on 5th April 1891*, Singapore: Government Printing Office, 1892.

Nathan, J. E., *The Census of British Malaya, 1921*, London: Waterlow and Sons, 1922.

Vlieland, C. A., *British Malaya: A Report on the 1931 Census and on Certain Problems of Vital Statistics*, London: Crown Agents for the Colonies, 1932.

Acts and Ordinances

Garrad, C. G., *The Acts and Ordinances of the Legislative Council of the Straits Settlements from 1st April 1867 to 7th March 1898*, Vols. 1 and 2, London: Eyre and Spottiswoode, 1898.

Harwood, J. A., *The Acts and Ordinances of the Legislative Council of the Straits Settlements from 1st April 1867 to 1st June 1886*, Vols. 1 and 2, London: Eyre and Spottiswoode, 1886.

The Laws of the Straits Settlements Revised up to and Including the 31st Day of December 1919, but Exclusive of War and Emergency Legislation, London: Waterlow and Sons, 1920.

Municipal By-Laws with Respect to Buildings, Singapore: Government Printing Office, 1908–29.

Norton-Kyshe, James William, *Index to the Laws of the Straits Settlements from April 1867 to 28th February 1882*, Singapore: Straits Times Press, 1882.

Ordinances Enacted by the Governor of the Straits Settlements with the Advice and Consent of the Legislative Council Thereof, 1867–1929, Singapore: Government Printing Office, 1868–1930.

Straits Settlements Ordinances, Singapore: Government Printing Office, 1887–1927.

Theobald, William, *The Acts of the Legislative Council of India 1834–1861*, Calcutta: Government Printing Office, 1861.

Walters, D. K., *Index to the Municipal Ordinance of the Straits Settlements as Amended to the End of 1938*, Singapore: Government Printing Office, 1939.

_____, *Municipal Ordinance of the Straits Settlements (Annotated)*, Singapore: Government Printing Office, 1937.

Other Primary Sources

Newspapers and Periodicals

(Deposited at Colindale Newspaper Library, Colindale, London, and National University of Singapore Library, Singapore)
Pinang Gazette and Straits Chronicle, 1887–8.
Singapore Free Press and Mercantile Advertiser, 1854, 1887–8; Centenary Number, 8 October 1935.
Straits Chinese Magazine, 1897–1907.
Straits Produce, 1893–5.
Straits Times, 1846, 1862, 1869, 1883–1929.

Maps and Plans

(Deposited at the National Archives, Singapore)
'Map of Singapore', Geographical Section, General Staff, War Office, November 1906 (corrections to March 1914).
'Map of Singapore and Environs', by F. J. Pigott, Colonial Engineer and Surveyor-General, Straits Settlements, 1912.
'Map of Singapore Showing the Principal Residences and Places of Interest', Singapore: Fraser and Neave Ltd., 1923.
'Map of Singapore Town Shewing Building Allotments and Registered Numbers of Crown Leases', by Captain H. E. McCallum, Acting Colonial Engineer and Surveyor-General, Surveyor General's Office, Singapore, 1881.
'Map of the Island of Singapore and Its Dependencies', by Major H. E. McCallum, Colonial Engineer and Surveyor-General, Straits Settlements, 1885.
'Map of the Island of Singapore and Its Dependencies', Geographical Division, General Staff, War Office, 1911.
'Map of the Town and Environs', by Major J. F. A. Nair, Colonial Engineer and Surveyor-General, Straits Settlements, 1878.
'Map of the Town and Environs of Singapore', drawn by J. B. Tassin, from an Actual Survey by G. D. Coleman, Calcutta, 1836.
'Plan of Singapore Shewing the Proposed Harbour Improvements, the Routes of the Proposed Electric Tramways and Proposed Railway Extension to Wharves', first published in G. M. Reith, *1907 Handbook to Singapore*, Singapore: Fraser and Neave Ltd., 1904, facing title-page.
'Plan of Singapore Town Shewing Topographical Detail and Municipal Numbers', by Major H. E. McCallum, Colonial Engineer and Surveyor-General, Straits Settlements, 1893.
'Plan of the Town of Singapore by Lieut. [P.] Jackson [Assistant Engineer and Surveyor and Registrar of Government Lands]', first published in John Crawfurd, *Journal of an Embassy to the Courts of Siam and Cochin-China*, London: Henry Colburn, 1828, opposite p. 529.
'Singapore Residency', by J. Moniot, Surveyor-General, Straits Settlements, undated [1860s].
'Singapore Road Map 1926', F. M. S. Surveys No. 34, 1926.

Secondary Sources

Abu-Lughod, Janet L., *Cairo: 1001 Years of the City Victorious*, Princeton: Princeton University Press, 1971.

_____, 'Tale of Two Cities: The Origins of Modern Cairo', *Comparative Studies in Society and History*, 7 (1964/5), 429–57.

Agnew, John A., *Place and Politics: The Geographical Mediation of State and Society*, Boston: Allen and Unwin, 1987.

Ahern, Emily Martin, *Chinese Ritual and Politics*, Cambridge: Cambridge University Press, 1981.

Akbar, Jamel, *Crisis in the Built Environment: The Case of the Muslim City*, Singapore: Concept Media, 1988.

Allbutt, Thomas Clifford (ed.), *A System of Medicine*, Vols. 1 and 2, London: Macmillan and Co., 1896.

AlSayyad, Nezar, 'Urbanism and the Dominance Equation: Reflections on Colonialism and National Identity', in Nezar AlSayyad (ed.), *Forms of Dominance: On the Architecture and Urbanism of the Colonial Enterprise*, Aldershot: Avebury, 1992, pp. 1–26.

Anderson, Kay J., 'Cultural Hegemony and the Race-Definition Process in Chinatown, Vancouver: 1880–1980', *Environment and Planning D: Society and Space*, 6 (1988): 127–49.

_____, *Vancouver's Chinatown: Racial Discourse in Canada, 1875–1980*, Montreal: McGill–Queen's University Press, 1992.

Anderson, Kay J. and Gale, Fay (eds.), *Inventing Places: Studies in Cultural Geography*, Melbourne: Ron Harper, 1992.

Anderson, Stanford, 'People in the Physical Environment: The Urban Ecology of Streets', in Stanford Anderson (ed.), *On Streets*, Cambridge, Massachusetts: MIT Press, 1978, pp. 1–27.

Aries, Philippe, *In the Hour of Our Death*, trans. H. Weaver, New York: Allen Lane, 1981.

Armstrong, Warwick and McGee, Terence G., *Theatres of Accumulation: Studies in Asian and Latin American Urbanization*, London: Methuen, 1985.

Arnold, David, 'Cholera and Colonialism in British India', *Past and Present*, 113 (1986): 118–51.

_____, 'Introduction: Disease, Medicine and Empire', in David Arnold (ed.), *Imperial Medicine and Indigenous Societies*, Manchester: Manchester University Press, 1988, pp. 1–26.

_____, 'Smallpox and Colonial Medicine in Nineteenth Century India', in David Arnold (ed.), *Imperial Medicine and Indigenous Societies*, Manchester: Manchester University Press, 1988, pp. 45–65.

_____, 'Touching the Body: Perspectives on the Indian Plague, 1896–1900', in Ranajit Guha (ed.), *Subaltern Studies V: Writings on South Asian History and Society*, Delhi: Oxford University Press, 1987, pp. 55–90.

'Association Intelligence: Malaya Branch', *British Medical Journal*, 2 (1895): 1520.

Avetoom, T. C., 'Curious and Superstitious Native Ideas of Causation of Diseases and Their Treatment', *Journal of the Malaya Branch of the British Medical Association*, December (1905): 6–9.

'Ayurvedic Medicine: An Ancient System', *Malayan Medical Journal and Estate Sanitation*, 2, 3 (1927): 114–15.

Balandier, George, 'The Colonial Situation: A Theoretical Approach', in Immanuel Wallerstein (ed.), *Social Change: The Colonial Situation*, New York: John Wiley and Sons, 1966, pp. 34–61.

Balfour, Andrew, *The War against Tropical Disease*, London: Bailliere, Tindall and Cox, 1920.

Balfour, Andrew and Scott, Henry H., *Health Problems of the Empire: Past, Present and Future*, London: W. Collins, Sons and Co., 1924.

Bedarida, François and Sutcliffe, Anthony, 'The Street in the Structure and Life of the City—Reflections on Nineteenth-Century London and Paris', *Journal of Urban History*, 6 (1980): 379–96.

Bellam, M. E. P., 'The Colonial City: Honiara, a Pacific Islands' Case Study', *Pacific Viewpoint*, 11, 1 (1970): 66–96.

Bentley, Arthur J. M., *Beri-beri: Its Etiology, Symptoms, Treatment and Pathology*, Edinburgh: Young J. Pentland, 1893.

Boxer, B., 'Space, Change and *Feng-shui* in Tsuen Wan's Urbanization', *Journal of Asian and African Studies*, 3 (1968): 226–40.

Braddell, Ronald St. John, *The Law of the Straits Settlements: A Commentary*, Kuala Lumpur: Oxford University Press, 1982 (first published in 1915).

Brenner, Robert, 'The Origins of Capitalist Development: A Critique of Neo-Smithian Marxism', *New Left Review*, 104 (1977): 25–92.

Britton, Stephen G., 'The Evolution of a Colonial Space-Economy: The Case of Fiji', *Journal of Historical Geography*, 6 (1980): 251–74.

Brooke, Gilbert E., 'Medical Work and Institutions' in Walter Makepeace, Gilbert E. Brooke, and Ronald St. John Braddell. (eds.), *One Hundred Years of Singapore*, Vol. 1, London: John Murray, 1921, pp. 487–519.

Brown, Edwin A., *Indiscreet Memories*, London: Kelly and Walsh, 1936.

Buckley, Charles Burton, *An Anecdotal History of Old Times in Singapore*, Singapore: Oxford University Press, 1984 (first published in 1902).

Butt, John, 'Working-Class Housing in Glasgow, 1851–1914', in Stanley D. Chapman (ed.), *The History of Working-Class Housing*, Newton Abbot: David and Charles, 1971, pp. 53–92.

Carroll, Lewis, *The Philosopher's Alice: Alice's Adventures in Wonderland and Through The Looking-Glass*, annotated by Peter Heath, London: Academy Editions, 1974.

Castells, Manuel, Lee Goh, and Kwok, R. Yin-Wang, *The Shek Kip Mei Syndrome: Economic Development and Public Housing in Hong Kong and Singapore*, London: Pion, 1990.

Chadwick, Edwin, 'Report on the Sanitary Condition of the Labouring Population of Great Britain: A Supplementary Report on the Results of a Special Inquiry into the Practice of Interment in Towns', *Parliamentary Papers*, 12 (1843): 395–681.

Chalmers, A. K., *The House as a Contributory Factor in the Death-Rate*, Glasgow: Royal Philosophical Society, 1913.

Chen Jia Geng (Tan Kah Kee), 'Zhu Wu Yu Wei Sheng' [Housing and Hygiene], in Chen Jia Geng, *Nan Qiao Hui Yi Lu* [Recollections of an Overseas Chinese in South-East Asia], Hongkong: Cao Yuan Chu Ban She, 1979, pp. 384–92.

Cheng Lim-Keak, *Social Change and the Chinese in Singapore*, Singapore: Singapore University Press, 1985, pp. 28–9.

Cherry, Gordon E., 'The Town Planning Movement and the Late Victorian City', *Transactions, Institute of British Geographers*, New Series, 4 (1979): 306–19.

Chin Chan Wei, 'Short Biography of the Author', in Li Bai Gai, *Yi Hai Wen Lan* [Chinese Medical Discourses], ed. Hsu Yun-tsiao, Singapore: Lai Kai Joo, 1976, pp. 1–2.

Chinatown: An Album of a Singapore Community, Singapore: Times Books International, 1983.

'Chinese Gardeners and Disease', *Malayan Medical Journal and Estate Sanitation*, 1, 2 (1926): 33.

Chopra, Preeti, 'Pondicherry: A French Enclave in India', in Nezar AlSayyad (ed.), *Forms of Dominance: On the Architecture and Urbanism of the Colonial Enterprise*, Aldershot: Avebury, 1992, pp. 107–37.

Christopher, A. J., *The British Empire at its Zenith*, London: Croom Helm, 1988.

Chu Tee Seng, 'The Singapore Chinese Protectorate, 1900–1941', *Journal of the South Seas Society*, 26, 1 (1971): 5–45.

Chua Beng-Huat, 'Not Depoliticized but Ideologically Successful: The Public Housing Programme in Singapore', *International Journal of Urban and Regional Research*, 15 (1991): 24–41.

Chua Beng-Huat and Norman Edwards (eds.), *Public Space: Design, Use and Management*, Singapore: Centre for Advanced Studies/Singapore University Press, 1992.

Cockburn, J. A., 'Sanitary Progress during the Fifty Years, 1876–1926—Overseas Dominions and Colonial Aspect', *Journal of the Royal Sanitary Institute*, 47 (1926): 89–94.

Cohen, Anthony P., *The Symbolic Construction of Community*, London: Tavistock, 1985.

Comaroff, Jean and Comaroff, John, 'Through the Looking-Glass: Colonial Encounters of the First Kind', *Journal of Historical Sociology*, 1 (1988): 6–32.

Cook, J. A. Bethune, *Sunny Singapore*, London: Elliot Stock, 1907.

Coones, Paul, 'Kensal Green Cemetery: London's First Great Extramural Necropolis', *Transactions of the Ancient Monuments Society*, 31 (1987): 48–76.

Corbridge, Stuart, *Capitalist World Development: A Critique of Radical Development Geography*, Totowa, NJ: Rowman and Littlefield, 1986.

Corfield, P. J., 'Walking the City Streets: The Urban Odyssey in Eighteenth-Century England', *Journal of Urban History*, 16 (1990): 132–74.

Cosgrove, Denis and Daniels, Stephen, 'Fieldwork as Theatre: A Week's Performance in Venice and Its Region', *Journal of Geography in Higher Education*, 13 (1989): 169–83.

Crawfurd, John, *A Descriptive Dictionary of the Indian Islands and Adjacent Countries*, London: Bradbury and Evans, 1856.

_____, *Journal of an Embassy from the Governor-General of India to the Courts of Siam and Cochin China, etc.*, London: Henry Colburn, 1828.

Crow, Graham, 'The Use of the Concept of "Strategy" in Recent Sociological Literature', *Sociology*, 23 (1989): 1–24.

Curl, James Stevens, *A Celebration of Death: An Introduction to Some of the Buildings, Monuments, and Settings of Funerary Architecture in the Western European Tradition*, London: Constable, 1980.

Daniels, Stephen, 'Marxism, Culture, and the Duplicity of Landscape', in Richard Peet and Nigel Thrift (eds.), *New Models in Geography: The Political Economy Perspective*, Vol. 2, London: Unwin Hyman, 1989, pp. 196–220.

D'arango, B. E., *The Stranger's Guide to Singapore,* Singapore: 'Sirangoon' Press, 1890.

Darra Mair, L. W., 'Report on Back-to-back Houses', *Parliamentary Papers,* 38 (1910).

Daunton, Martin J., *House and Home in the Victorian City: Working-Class Housing, 1850–1914,* London: Edward Arnold, 1983.

_____, 'Public Place and Private Space: The Victorian City and the Working-Class Household', in Derek Fraser and Anthony Sutcliffe (eds.), *The Pursuit of Urban History*, London: Edward Arnold, 1983, pp. 212–33.

Davidoff, Leonore and Hall, Catherine, 'The Architecture of Public and

Private Life: English Middle-Class Society in a Provincial Town, 1780 to 1850', in Derek Fraser and Anthony Sutcliffe (eds.), *The Pursuit of Urban History*, London: Edward Arnold, 1983, pp. 327–45.

Davidson, G. F., *Trade and Travel in the Far East or Recollections of Twenty-one Years Passed in Java, Singapore, Australia and China*, London: Madden and Malcolm, 1846.

DeBernardi, Jean, 'Space and Time in Chinese Religious Culture', *History of Religions*, 31 (1992): 247–68.

De Groot, Jan Jakob Mariade, *The Religious System of China*, Vols. 1–6, Taipei: Ch'eng-wen Publishing Co., 1969 (first published in 1892–1910).

[Directory of the] Singapore Federation of Chinese Clan Associations, Singapore, 1988.

Dixon, Alec, *Singapore Patrol*, London: George G. Harrap and Co., 1935.

Doolittle, Justus, *Social Life in China*, Vols. 1 and 2, Singapore: Graham Brash, 1986 (first published in 1895).

Dossal, Mariam, 'Limits of Colonial Urban Planning: A Study of Mid-Nineteenth Century Bombay', *International Journal of Urban and Regional Research*, 13 (1989): 19–31.

Dream Awhile: Cartoons from Straits Produce Showing in Pictorial Form the Main Events in Local History, 1922–1933, Singapore, [1933].

Dreyfus, Hubert L. and Rabinow, Paul, *Michel Foucault: Beyond Structuralism and Hermeneutics*, Brighton: Harvester Press, 1982.

Driver, Felix, 'Moral Geographies: Social Science and the Urban Environment in Mid-Nineteenth Century England', *Transactions, Institute of British Geographers*, New Series, 13 (1988): 275–87.

———, 'Power, Space and the Body: A Critical Assessment of Foucault's *Discipline and Punish*', *Environment and Planning D: Society and Space*, 3 (1985): 425–46.

Dumarcay, Jacques, *The House in South-East Asia*, trans. Michael Smithies, Singapore: Oxford University Press, 1987.

Dumbleton, [C. E.], 'The Need of Reform in the Present System of Registration of Deaths in the Straits Settlements', *Journal of the Straits Medical Association*, 4 (1892/3): 66–75.

Duncan, James S., *The City as Text: The Politics of Landscape in the Kandyan Kingdom*, Cambridge: Cambridge University Press, 1990.

Duncan, Otis Dudley and Duncan, Beverly, 'Residential Distribution and Occupational Stratification', in Ceri Peach (ed.), *Urban Social Segregation*, London: Longman, 1975, pp. 51–66.

Eitel, Ernest J., *Feng-Shui*, Singapore: Graham Brash, 1985 (first published in 1873).

Ekwall, Eilert, *Street-Names of the City of London*, Oxford: Clarendon Press, 1954.

Eliade, Mircea, *The Myth of the Eternal Return*, Princeton: Princeton University Press, 1954.

———, *The Sacred and the Profane: The Nature of Religion*, trans. W. R. Trask, New York: Harvest, 1959.

Elliot, Adrian, 'Municipal Government in Bradford in the Mid-Nineteenth Century', in Derek Fraser (ed.), *Municipal Reform and the Industrial City*, Leicester: Leicester University Press, 1982, pp. 111–61.

Emmerson, Donald K., '"Southeast Asia": What's in a Name?', *Journal of Southeast Asian Studies*, 15 (1984): 1–21.

Etlin, R. A., *The Architecture of Death: The Transformation of the Cemetery in Eighteenth-century Paris*, Cambridge, Massachusetts: MIT Press, 1984.

Evers, Hans-Dieter, 'The Culture of Malaysian Urbanization: Malay and Chinese Conceptions of Space', in Peter S. J. Chen and Hans-Dieter Evers (eds.), *Studies in Asean Sociology: Urban Society and Social Change*, Singapore: Chopmen Enterprises, 1978, pp. 333–43.

Farr, William, 'Letter to the Registrar-General, First Annual Report of the Registrar of Births, Deaths and Marriages in England', *Parliamentary Papers*, 16 (1839): 63–71.

Feuchtwang, Stephan D. R., *An Anthropological Analysis of Chinese Geomancy*, Vientiane: Vithagna, 1974.

Firmstone, H. W., 'Chinese Names of Streets and Places in Singapore and the Malay Peninsula', *Journal of the Straits Branch of the Royal Asiatic Society*, 42 (1905): 53–208.

Fisher, J. S. (comp.), *Who's Who in Malaya: A Biographical Record of Prominent Members of Malaya's Community in Official, Professional and Commercial Circles*, Singapore: J. S. Fisher, 1925.

Foran, W. Robert, *Malayan Symphony Being the Impressions Gathered During a Six Months' Journey through the Straits Settlements, F.M.S., Siam, Sumatra, Java and Bali*, London: Hutchinson and Co., 1935.

Foucault, Michel, *Discipline and Punish: The Birth of the Prison*, trans. Alan Sheridan, London, Penguin Books, 1979.

_____, 'The Eye of Power', in Colin Gordon (ed.), *Michel Foucault: Power/Knowledge, Selected Interviews and Other Writings, 1972–1977*, Brighton: Harvester Press, 1980, pp. 146–65.

_____, 'The Politics of Health in the Eighteenth Century', in Colin Gordon (ed.), *Michel Foucault: Power/Knowledge, Selected Interviews and Other Writings, 1972–1977*, Brighton: Harvester Press, 1980, pp. 166–82.

_____, 'Power and Strategies', in Colin Gordon (ed.), *Michel Foucault: Power/Knowledge, Selected Interviews and Other Writings, 1972–1977*, Brighton: Harvester Press, 1980, pp. 134–45.

_____, 'The Subject and Power', in Hubert L. Dreyfus and Paul Rabinow, *Michel Foucault: Beyond Structuralism and Hermeneutics*, Brighton: Harvester Press, 1982, pp. 208–26.

_____, 'Truth and Power' in Paul Rabinow (ed.), *The Foucault Reader*, Harmondsworth: Penguin, 1984, pp. 51–75.

_____, 'Two Lectures', in Colin Gordon (ed.), *Michel Foucault: Power/Knowledge, Selected Interviews and Other Writings, 1972–1977*, Brighton: Harvester Press, 1980, pp. 78–108.

Francaviglia, R. V., 'The Cemetery as an Evolving Cultural Landscape', *Annals of the Association of American Geographers*, 61 (1971): 501–9.

Francis, Mark, 'Control as a Dimension of Public-Space Quality', in Irwin Altman and Ervin H. Zube (eds.), *Public Places and Spaces*, New York: Plenum Press, 1989, pp. 147–72.

Fraser, Derek, 'Introduction: Municipal Reform in Historical Perspective', in Derek Fraser (ed.), *Municipal Reform and the Industrial City*, Leicester: Leicester University Press, 1982, pp. 1–14.

_____, *Power and Authority in the Victorian City*, Oxford: Basil Blackwell, 1979.

Fraser, Henry and Stanton, A. T., 'The Cause of Beri-beri', *Lancet*, 1 (12 March 1910): 733–4.

_____, 'The Etiology of Beri-beri', *Lancet*, 2 (17 December 1910): 1755–7.

_____, *The Etiology of Beriberi*, Studies from the Institute for Medical Research No. 12, Singapore: Kelly and Walsh, 1911.

Fraser, J. M. (comp.), *The Work of the Singapore Improvement Trust, 1927–1947*, Singapore: Singapore Improvement Trust, 1948.

Freedman, Maurice F., 'Geomancy', *Proceedings of the Royal Anthropological Institute of Great Britain and Ireland for 1968*, London: Royal Anthropological Institute of Great Britain and Ireland, 1969, pp. 5–15.

_____, 'Immigrants and Associations: Chinese in Nineteenth-Century Singapore' in G. W. Skinner (ed.), *The Study of Chinese Society: Essays of M. Freedman*, California: Stanford University Press, 1979, pp. 81–3.

Friedmann, John, *Regional Development Policy: A Case Study of Venezuela*, Cambridge, Massachusetts: MIT Press, 1966.

Fu De Ci Lu Ye Ting Yan Ge Shi Te Kan [*Historical Account of Hok Tek Chi and Loke Yah Teng Cemetery*], Singapore, 1963.

Furnivall, J. S., *Colonial Policy and Practice: A Comparative Study of Burma and the Netherlands Indies*, Cambridge: Cambridge University Press, 1948.

Galloway, David J., 'Introductory Address', *Journal of the Straits Medical Association*, 1 (1890): 11–24.

_____, 'Observations on the Death Rate', *Journal of the Malaya Branch of the British Medical Association*, January (1907): 1–4.

Galloway, David J., Middleton, W. R. C., and Ritchie, John, 'Report of Committee of Malaya Branch British Medical Association Appointed to Suggest Improvements to the Medical Registration Bill', *Journal of the Malaya Branch of the British Medical Association*, January (1907): 118–23.

Gatrell, V. A. C., 'Incorporation and the Pursuit of Liberal Hegemony in Manchester 1790–1839', in Derek Fraser (ed.), *Municipal Reform and the Industrial City*, Leicester: Leicester University Press, 1982, pp. 15–60.

Geertz, Clifford, *Peddlars and Princes: Social Change and Economic Modernization in Two Indonesian Towns*, Chicago: University of Chicago Press, 1963.

George, M. Dorothy, *London Life in the Eighteenth Century*, Harmondsworth: Penguin, 1966.

Gibson, R. M., 'Beri-beri in Hong Kong, with Special Reference to the Records of the Alice Memorial and Nethersole Hospitals, and Some Notes on Two Years' Experience of the Disease', *Journal of Tropical Medicine*, 4 (1901): 96–9.

Giddens, Anthony, *The Constitution of Society: Outline of the Theory of Structuration*, Cambridge: Polity Press, 1986.

_____, *A Contemporary Critique of Historical Materialism, Volume One: Power, Property and the State*, Basingstoke: Macmillan, 1981.

_____, *The Nation-State and Violence, Volume Two of A Contemporary Critique of Historical Materialism*, Cambridge: Polity Press, 1987.

Ginsburg, Norton S., 'Urban Geography and "Non-Western" Areas', in Philip M. Hauser and Leo F. Schnore (eds.), *The Study of Urbanization*, New York: John Wiley and Sons, 1965, pp. 311–46.

Godley, Michael R., *The Mandarin-Capitalist from Nanyang: Overseas Chinese Enterprise in the Modernisation of China, 1893–1911*, Cambridge: Cambridge University Press, 1981.

Gordon, Colin, 'Afterword', in Colin Gordon (ed.), *Michel Foucault: Power/Knowledge, Selected Interviews and Other Writings, 1972–1977*, Brighton: Harvester Press, 1980, pp. 229–59.

Goss, Jon, 'The Built Environment and Social Theory: Towards an Architectural Geography', *Professional Geographer*, 40 (1988): 392–403.

Gould-Martin, K., 'Hot Cold Clean Poison and Dirt: Chinese Folk Medical Categories', *Social Science and Medicine*, 12, 1B (1978): 39–46.

Graham, A., *Medical Topography of Singapore and Sarawak*, Edinburgh: Murray and Gibb, 1852.

Graham, Sandra Lauderdale, *House and Street: The Domestic World of Servants and Masters in Nineteenth-Century Rio de Janeiro*, Cambridge: Cambridge University Press, 1988.

Gregory, Derek, *Regional Transformation and Industrial Revolution: A Geography of the Yorkshire Woollen Industry*, London: Macmillan Press, 1982.

_____, *Geographical Imaginations*, Oxford: Blackwell, 1994.

Hallifax, F. J., 'Municipal Government', in Walter Makepeace, Gilbert E. Brooke, and Ronald St. John Braddell (eds.), *One Hundred Years of Singapore*, Vol. 1, London: John Murray, 1921, pp. 315–40.

Hamadeh, Shirine, 'Creating the Traditional City: A French Project', in Nezar AlSayyad (ed.), *Forms of Dominance: On the Architecture and Urbanism of the Colonial Enterprise*, Aldershot: Avebury, 1992, pp. 241–59.

A Handbook of Information Presented by the Rotary Club and the Municipal Commissioners of the Town of Singapore, Singapore: Publicity Committee of the Rotary Club of Singapore, 1933.

Harfield, Alan, *Early Cemeteries in Singapore*, London: British Association for Cemeteries in South Asia, 1988.

Harley, J. B., 'Historical Geography and Its Evidence: Reflections on Modelling Sources', in Alan R. H. Baker and Mark Billinge (eds.), *Period and Place: Research Methods in Historical Geography*, Cambridge: Cambridge University Press, 1982, pp. 261–73.

Harries, J. W., 'The Singapore Municipal Commission', *British Malaya*, 4, 4 (1929): 113–15.

Harrison, Cuthbert Woodville, *Some Notes on the Government Services in British Malaya*, London: Malayan Information Agency, 1929.

Harrison, Gordon, *Mosquitoes, Malaria and Man*, London: John Murray, 1978.

Harrower, Gordon, 'Native Medicine and Hygiene in Singapore', *British Medical Journal*, 2 (1923): 1175–6.

Hart, Ernest, *The Nurseries of Cholera: Its Diffusion and Its Extinction*, An Address Delivered before the Section of Public Medicine of the British Medical Association at Newcastle, August 1893, reprinted from *British Medical Journal*, London: Smith, Elder and Co., 1894.

Harvey, David, 'Labour, Capital, and Class Struggle around the Built Environment in Advanced Capitalist Societies', in Anthony Giddens and David Held (eds.), *Classes, Power, and Conflict: Classical and Contemporary Debates*, Basingstoke: Macmillan, 1982, pp. 545–61.

_____, 'On the History and Present Condition of Geography: An Historical Materialist Manifesto', *Professional Geographer*, 36 (1984): 1–11.

Hassan, J. A., 'The Growth and Impact of the British Water Industry in the Nineteenth Century', *Economic History Review*, 38 (1985): 531–47.

Hassan, Riaz, *Families in Flats: A Study of Low Income Families in Public Housing*, Singapore: Singapore University Press, 1977.

Haughton, H. T., 'Native Names of Streets in Singapore', in Mubin Sheppard (ed.), *Singapore 150 Years*, Singapore: Times Books International, 1982, pp. 208–19 (first published in 1891).

_____, 'Notes on the Names of Places in the Island of Singapore and Its Vicintity', *Journal of the Straits Branch of the Royal Asiatic Society*, 20 (1889): 75–82.

Headrick, Daniel R., *The Tools of Empire: Technology and European Imperialism in the Nineteenth Century*, Oxford: Oxford University Press, 1981.

Hennock, E. P., *Fit and Proper Persons: Ideal and Reality in Nineteenth-Century Urban Government*, London: Edward Arnold, 1973.

Hillier, S. M. and Jewell, J. A., 'Chinese Traditional Medicine and Modern Western Medicine: Integration and Separation in China', in S. M. Hillier and J. A. Jewell, *Health Care and Traditional Medicine in China 1800–1982*, London: Routledge and Kegan Paul, 1983, pp. 306–35.

Hintze, K., 'Sanitäre Verhältnisse und Einrichtungen in den Straits Settlements und Federated Malay States (Hinterindien)', *Archiv für Schiffs-und Tropen-Hygiene*, 10 (1906): 523–36.

Ho, Suzanne Chan; Kwok Chan Lun; and Ng, W. K. Cheng Hin, 'The Role of Chinese Traditional Medical Practice as a Form of Health Care on Singapore—II. Some Characteristics of Providers', *American Journal of Chinese Medicine*, 11 (1983): 16–23.

_____, 'The Role of Chinese Traditional Medical Practice as a Form of Health Care on Singapore—III. Conditions, Illness Behaviour and Medical Preferences of Patients of Institutional Clinics', *Social Science and Medicine*, 18 (1984): 745–52.

Ho Kong Chong and Lim, Valerie Nyuk Eun, 'Backlanes as Contested Regions: Construction and Control of Physical Space', in Chua Beng-Huat and Norman Edwards (eds.), *Public Space: Design, Use and Management*, Singapore: Centre for Advanced Studies/Singapore University Press, pp. 40–54.

Hobsbawm, E. J., 'History from Below: Some Reflections', in Frederick Krantz (ed.), *History from Below: Studies in Popular Protest and Popular Ideology*, Oxford: Basil Blackwell, 1988, pp. 13–27.

Hochberg, Fritz, *An Eastern Voyage: A Journal of the Travels of Count Fritz Hochberg through the British Empire in the East and Japan*, Vols. 1 and 2, London: J. M. Dent and Sons, 1910.

Hodder, B. W., 'Racial Groupings in Singapore', *Malayan Journal of Tropical Geography*, 1 (1953): 25–36.

Holmer, N. M., 'Indian Placenames in South America and the Antilles', *Names*, 8 (1960): 133–49.

Hoops, A. L., 'The Prevention of Disease in the Tropics', in A. L. Hoops and J. W. Scharff (eds.), *Transactions of the Fifth Biennial Congress Held at Singapore, 1923, Far Eastern Association of Tropical Medicine*, London: John Bale, Sons and Danielsson, 1924, pp. 4–14.

Horvath, Ronald J., 'In Search of a Theory of Urbanization: Notes on the Colonial City', *East Lakes Geographer*, 5 (1969): 69–82.

Hsieh Chiao-Min, 'Sequent Occupance and Place Names', in Ronald G. Knapp (ed.), *China's Island Frontier: Studies in the Historical Geography of Taiwan*, Honolulu: University Press of Hawaii, 1980, pp. 107–14.

Hua Ren Li Su Jie Re Shou Ce [*Chinese Customs and Festivals in Singapore*], Singapore: Singapore Federation of Chinese Clan Associations, 1989.

Hutchins, Francis G., *The Illusion of Permanence: British Imperialism*, Princeton: Princeton University Press, 1967.

The Institutes of Gaius, trans. Francis de Zulueta, Oxford: Clarendon Press, 1946.

Jackson, J. B., 'From Monument to Place', *Landscape*, 17 (1967/8): 22–6.

Jackson, Peter, *Maps of Meaning: An Introduction to Cultural Geography*, London: Routledge, 1989.

Jewell, J. A., 'Theoretical Basis of Chinese Traditional Medicine', in S. M.

Hillier and J. A. Jewell, *Health Care and Traditional Medicine in China 1800–1982*, London: Routledge and Kegan Paul, 1983, pp. 221–41.

Johansson, S. Ryan and Mosk, Carl, 'Exposure, Resistance and Life Expectancy: Disease and Death during the Economic Development of Japan, 1900–1960', *Population Studies*, 41 (1987), 207–35.

Jones, David, *Crime, Protest, Community and Police in Nineteenth-Century Britain*, London: Routledge and Kegan Paul, 1982.

J. T. L., 'William Geoffrey's Profit', reprinted from the *Straits Times Annual*, December 1906, Singapore: Straits Times Press, 1906.

Jyoti, Hosagrahar, 'City as Durbar: Theater and Power in Imperial Delhi', in Nezar AlSayyad (ed.), *Forms of Dominance: On the Architecture and Urbanism of the Colonial Enterprise*, Aldershot: Avebury, 1992, pp. 83–105.

Kaye, Barrington, *Upper Nankin Street, Singapore: A Sociological Study of Chinese Households Living in a Densely Populated Area*, Singapore: University of Malaya Press, 1960.

Kearns, Gerard, 'Zivilis or Hygaeia: Urban Public Health and the Epidemiologic Transition', in Richard Lawton (ed.), *The Rise and Fall of Great Cities: Aspects of Urbanization in the Western World*, London: Belhaven Press, 1989, pp. 96–124.

Khoo Kay Kim, 'The Municipal Government of Singapore 1887–1940', Academic Exercise, University of Malaya, Singapore, 1960.

King, Anthony D., 'Colonial Cities: Global Pivots of Change', in Robert Ross and Gerard J. Telkamp (eds.), *Colonial Cities: Essays on Urbanism in a Colonial Context*, Dordrecht: Martinus Nijhoff, 1985, pp. 7–32.

_____, 'Colonialism, Urbanism and the Capitalist World Economy', *International Journal of Urban and Regional Research*, 13 (1989): 1–18.

_____, *Colonial Urban Development: Culture, Social Power and Environment*, London, Routledge and Kegan Paul, 1976.

_____, 'Exporting Planning: The Colonial and Neo-Colonial Experience', in Gordon E. Cherry, *Shaping an Urban World: Planning in the Twentieth Century*, London: Mansell, 1980, pp. 204–26.

_____, *Global Cities: Post-Imperialism and the Internationalization of London*, London: Routledge, 1990.

_____, 'The Impress of Empire', *Journal of Historical Geography*, 15 (1989): 193–5.

_____, 'Rethinking Colonialism: An Epilogue', in Nezar AlSayyad (ed.), *Forms of Dominance: On the Architecture and Urbanism of the Colonial Enterprise*, Aldershot: Avebury, 1992, pp. 339–55.

_____, *Urbanism, Colonialism and the World Economy: Cultural and Spatial Foundations of the World Urban System*, London: Routledge, 1990.

Kirk, James, 'Analysis of One Hundred and Fifty Cases of Local Fever', *Journal of the Malaya Branch of the British Medical Association*, January (1904): 49–57.

Knapp, Ronald G., 'The Changing Landscape of the Chinese Cemetery', *China Geographer*, 8 (1977): 1–14.

_____, *China's Vernacular Architecture: House Form and Culture*, Honolulu: University of Hawaii Press, 1989.

Knight, A., 'Chinese Names of Streets', *Journal of the Straits Branch of the Royal Asiatic Society*, 45 (1906): 287–8.

Knight, D. B., 'Commentary: Perceptions of Landscapes in Heaven', *Journal of Cultural Geography*, 6 (1985): 127–40.

Kohl, David G., *Chinese Architecture in the Straits Settlements and Western*

Malaya: Temples, Kongsis and Houses, Kuala Lumpur: Heinemann Educational Books, 1984.

Kosambi, Meera and Brush, John E., 'Three Colonial Port Cities in India', *Geographical Review*, 78 (1988): 32–47.

Krantz, Frederick (ed.), *History from Below: Studies in Popular Protest and Popular Ideology*, Oxford: Basil Blackwell, 1988.

Lai, Chuen-yan David, 'The Chinese Cemetery in Victoria', *BC Studies*, 95 (1987): 24–42.

_____, 'A *Feng Shui* Model as a Locational Index', *Annals of the Association of American Geographers*, 64 (1974): 506–13.

Lang, Jon, 'Cultural Implications of Housing Design Policy in India', in Setha M. Low and Evre Chambers (eds.), *Housing, Culture and Design: A Comparative Perspective*, Philadelphia: University of Pennsylvannia Press, 1989, pp. 375–91.

Lee Kip Lin, *The Singapore House, 1819–1942*, Singapore: Times Editions, 1988.

Lee Yong Kiat, 'The Grand Jury in Early Singapore (1819–1873)', *Journal of the Malayan Branch of the Royal Asiatic Society*, 46 (1973): 55–150.

Leeming, Frank, *Street Studies in Hong Kong: Localities in a Chinese City*, Hong Kong: Oxford University Press, 1977.

Lefebvre, Henri, *The Production of Space*, trans. Donald Nicholson-Smith, Oxford: Basil Blackwell, 1991.

_____, 'Reflections on the Politics of Space', trans. M. J. Enders, in Richard Peet (ed.), *Radical Geography: Alternative Viewpoints on Contemporary Social Issues*, Chicago: Maaroufa Press, 1977, pp. 339–52.

Lewandowski, Susan J., 'Urban Growth and Municipal Development in the Colonial City of Madras, 1860–1900', *Journal of Asian Studies*, 34 (1975): 341–60.

_____, 'The Built Environment and Cultural Symbolism in Post-Colonial Madras', in John A. Agnew, John Mercer, and David E. Sopher (eds.), *The City in Cultural Context*, Boston: Allen and Unwin, 1984, pp. 237–54.

Lewes, F. M. M., 'Dr Marc D'Espine's Statistical Nosology', *Medical History*, 32 (1988): 301–13.

Ley, David, 'Styles of the Times: Liberal and Neo-Conservative Landscapes in Inner Vancouver, 1968–1986', *Journal of Historical Geography*, 13 (1987): 40–56.

_____, 'From Urban Structure to Urban Landscape', *Urban Geography*, 9 (1988): 98–105.

Li Bai Gai (Li Peh Khai), *Yi Hai Wen Lan* [Chinese Medical Discourses], ed. Hsu Yun-tsiao, Singapore: Lai Kai Joo, 1976.

_____, 'Yuan Zhen Ji Su' [A Brief Description of Diagnosis at the Thong Chai Medical Institution] in *Tong Ji Yi Yuan Da Sha Luo Cheng Ji Nian Te Kan* [Souvenir Magazine of the Opening Ceremony of the Newly Completed Thong Chai Medical Institution], Singapore, 1979, pp. 155–6.

Li Chung Chu, 'A Description of Singapore in 1887', trans. Chang Chin Chiang, *China Society of Singapore 25th Anniversary Journal*, 1975: 20–9 (first published in 1895).

Li Song, 'Xin Jia Bo Zhong Yi Yao De Fa Zhan 1349–1983' [The Development of Chinese Medicine in Singapore 1349–1983], *Xin Jia Bo Zhong Yi Xue Bao* [Singapore Journal of Traditional Chinese Medicine], 10 (1983): 46–60.

Lim Boon Keng, 'Straits Chinese Reform', *Straits Chinese Magazine*, 3, 9 (1899): 22–5.

_____, 'Tuberculosis among the Singapore Chinese', *Journal of the Malaya Branch of the British Medical Association*, January (1904): 16–23.

Lip, Evelyn, *Chinese Geomancy*, Singapore: Times Books International, 1979.

Little, Robert, 'On the Medical Topography of Singapore Particularly in Its Marshes and Malaria', *Journal of the Indian Archipelago and Eastern Asia*, 2 (1848): 449–94.

Liu, Gretchen, *Singapore Historical Postcards from the National Archives Collection*, Singapore: Times Editions, 1986.

Liu Yangchi, *The Essential Book of Traditional Chinese Medicine*, Vols. 1 and 2, New York: Columbia University Press, 1988.

Low Ngiong Ing, *Recollections: Chinese Jetsam on a Tropical Shore [and] When Singapore was Syonan-to*, Singapore: Eastern Universities Press, 1983.

Ludwig, A. I., *Graven Images: New England Stonecarving and Its Symbols, 1650–1815*, Middletown: Wesleyan University Press, 1966.

Lyons, Maryinez, 'Sleeping Sickness Epidemics and Public Health in the Belgian Congo', in David Arnold (ed.), *Imperial Medicine and Indigenous Societies*, Manchester: Manchester University Press, 1988, pp. 105–24.

McDowell, Linda, 'Towards an Understanding of the Gender Division of Urban Space', *Environment and Planning D: Society and Space*, 1 (1983): 59–72.

McGee, Terence G., 'The Changing Cities', in R. D. Hill (ed.), *South-east Asia: A Systematic Geography*, Kuala Lumpur: Oxford University Press, 1979, pp. 180–91.

_____, *The Southeast Asian City: A Social Geography of the Primate Cities of Southeast Asia*, London: G. Bell and Sons, 1967.

_____, *The Urbanization Process in the Third World: Explorations in Search of a Theory*, London: G. Bell and Sons, 1971.

McGregor, William, 'An Address on Some Problems of Tropical Medicine', Address Delivered at the London School of Tropical Medicine on 3 October 1900, *Journal of Tropical Medicine*, October (1900): 63.

Mackenzie, Suzanne, 'Women in the City', in Richard Peet and Nigel Thrift (eds.), *New Models in Geography*, Vol. 2, London: Unwin Hyman, 1989, pp. 109–26.

Macleod, Kenneth, 'The Scope of the Section's Work', Presidential Address Delivered at the Opening of the Section of Tropical Diseases at the Annual Meeting of the British Medical Association, Ipswich, July–August 1900, *Journal of Tropical Medicine*, August (1900): 17–19.

Macpherson, Kerrie L., *A Wilderness of Marshes: Origins of Public Health in Shanghai, 1843–1893*, Hong Kong: Oxford University Press, 1987.

Mak Lau Fong, *The Sociology of Secret Societies in Singapore and Peninsula Malaysia*, Kuala Lumpur: Oxford University Press, 1981.

Makepeace, Walter, 'Concerning Known Persons', in Walter Makepeace, Gilbert E. Brooke, and Ronald St. John Braddell (eds.), *One Hundred Years of Singapore*, Vol. 2, London: John Murray, 1921, pp. 416–64.

Manderson, Lenore, 'Race, Colonial Mentality and Public Health in Early Twentieth Century Malaya', in Peter J. Rimmer and Lisa M. Allen (eds.), *The Underside of Malaysian History: Pullers, Prostitutes, Plantation Workers ...*, Singapore: Singapore University Press, 1990, pp. 193–213.

Manson, Patrick, *Tropical Diseases*, London: Cassell and Co., 1898.

Marriott, Hayes, 'The Peoples of Singapore', in Walter Makepeace, Gilbert E.

Brooke, and Ronald St. John Braddell (eds.), *One Hundred Years of Singapore*, Vol. 1, London: John Murray, 1921, pp. 341–62.

Maugham, W. Somerset, *Collected Short Stories*, Vol. 2, London: Pan Books, 1975.

[Meadows, S. D.], 'Impressions', *Journal of the Singapore Society of Architects*, 1, 2 (1924): 32–3.

'Medical Progress in the Straits Settlements: An Editorial Note', *Journal of the Malaya Branch of the British Medical Association*, December (1905), unpaginated.

Middleton, W. R. C., 'The Sanitation of Singapore', *Journal of State Medicine*, 8 (1900): 696–707.

_____, 'The Working of the Births and Deaths Registration Ordinance', *Malaya Medical Journal*, 9, 3 (1911): 33–50.

Mitchell, Timothy, *Colonising Egypt*, Cairo: American University of Cairo Press, 1989.

Mugliston, T. C., 'Unqualified Practice in Singapore', *Journal of the Straits Medical Association*, 5 (1893/4): 68–85.

'Municipal Government in the Straits Settlements', *Colonial Office Journal*, 4, 3 (1911): 220–3.

Nair, 'Etiology and Early Diagnosis of Pulmonary Tuberculosis', *Malayan Medical Journal and Estate Sanitation*, 1, 1 (1926): 15–16.

Naquin, Susan, 'Funerals in North China: Uniformity and Variation', in James L. Watson and Evelyn S. Rawski (eds.), *Death Ritual in Late and Modern China*, Berkeley: University of California Press, 1988, pp. 37–70.

Needham, Joseph, *Science and Civilisation in China*, Vol. 2, Cambridge: Cambridge University Press, 1956.

Nelson, M. A. and George, D. H., 'Grinning Skulls, Smiling Cherubs, Bitter Words', *Journal of Cultural Geography*, 15 (1982): 163–74.

The New Atlas and Commercial Gazetteer of the Straits Settlements and the Federated Malay States, Shanghai: Far Eastern Geographical Establishment, 1917.

Ng Siew Yoong, 'The Chinese Protectorate in Singapore, 1877–1900', *Journal of South East Asian History*, 2 (1961): 76–97.

Nicholson, Malcolm, 'Medicine and Racial Politics: Changing Images of the New Zealand Maori in the Nineteenth Century', in David Arnold (ed.), *Imperial Medicine and Indigenous Societies*, Manchester: Manchester University Press, 1988, pp. 66–104.

'Obituary: William Robert Colvin Middleton', *British Medical Journal*, 2 (1921): 1135.

Oldenburg, Veena Talwar, *The Making of Colonial Lucknow, 1856–1877*, Princeton: Princeton University Press, 1984.

Oman, W. C., 'A Plea for the Registration of Architects in Singapore', *Journal of the Singapore Society of Architects*, 1, 4 (1924): 32–4.

Ooi Giok Ling, 'Conservation-Dissolution: A Case-Study of Chinese Medicine in Peninsular Malaysia', Ph.D. thesis, Australian National University, 1982.

Owen, Norman G., 'Towards a History of Health in Southeast Asia', in Norman G. Owen (ed.), *Death and Disease in South-East Asia: Explorations in Social, Medical and Demographic History*, Singapore: Oxford University Press, 1897, pp. 3–30.

Papers Relating to the Investigation of Malaria and Other Tropical Diseases and the Establishment of Schools of Tropical Medicine, London: HMSO, 1903.

Pearson, H. F., 'Lt. Jackson's Plan of Singapore', in Mubin Sheppard (ed.), *Singapore: 150 Years*, Singapore: Times Books International, 1982, pp. 150–4.

Peet, George L., *Rickshaw Reporter*, Singapore: Eastern Universities Press, 1985.

Pelling, Margaret, *Cholera, Fever and English Medicine 1825–1865*, Oxford: Oxford University Press, 1978.

Peng Song Toh, *Directory of Associations in Singapore, 1982/1983*, Singapore: Historical Culture Publishers, 1983.

Pirie, Gordon H., 'Letters, Words, Worlds: The Naming of Soweto', *African Studies*, 43 (1984): 43–51.

'Powers of Colonial Governments', *Colonial Magazine*, 3 (1840): 177–81.

Prior, Lindsay, *The Social Organisation of Death: Medical Discourse and Social Practices in Belfast*, Basingstoke: Macmillan Press, 1989.

Rabinow, Paul, 'Governing Morocco: Modernity and Difference', *International Journal of Urban and Regional Research*, 13 (1989), 32–46.

Raja-Singam, S. Durai, *Malayan Street Names: What They Mean and Whom They Commemorate*, Ipoh: Mercantile Press, 1939.

Ranger, Terence, 'The Influenza Pandemic in Southern Rhodesia: A Crisis of Comprehension', in David Arnold (ed.), *Imperial Medicine and Indigenous Societies*, Manchester: Manchester University Press, 1988, pp. 172–88.

Rapoport, Amos, 'Culture and the Urban Order', in John A. Agnew, John Mercer, and David E. Sopher (eds.), *The City in Cultural Context*, Boston: Allen and Unwin, 1984, pp. 50–75.

_____, *House Form and Culture*, Englewood Cliffs, NJ: Prentice-Hall, 1969.

_____, *Human Aspects of Urban Form: Towards a Man–Environment Approach to Urban Form and Design*, Oxford: Pergamon Press, 1977.

Rathborne, Ambrose B., *Camping and Tramping in Malaya*, London: Swan, Sonnenschein and Co., 1898.

[Read, William Henry Macleod], *Play and Politics, Recollections of Malaya by an Old Resident*, London: Wells Gardner, Darton and Co., 1901.

Redfield, R. and Singer, M. S., 'The Culture Role of Cities', *Economic Development and Cultural Change*, 3 (1954): 53–73.

Reed, Robert R., *Colonial Manila: The Context of Hispanic Urbanism and the Process of Morphogenesis*, Berkeley: University of California Press, 1978.

_____, 'From Suprabarangay to Colonial Capital: Reflections on the Hispanic Foundation of Manila', in Nezar AlSayyad (ed.), *Forms of Dominance: On the Architecture and Urbanism of the Colonial Enterprise*, Aldershot: Avebury, 1992, pp. 45–81.

_____, *Hispanic Urbanism in the Philippines: A Study of the Impact of Church and State*, Manila: University of Manila, 1967.

Reith, G. M., *1907 Handbook to Singapore*, revised by Walter Makepeace, Singapore: Oxford University Press, 1986 (first published in 1892).

'Report on the Diseases of Singapore', *Madras Quarterly Medical Journal*, 1 (1839): 59–77.

Rex, John, 'Racism and the Structure of Colonial Societies', in Robert Ross (ed.), *Racism and Colonialism. Essays on Ideology and Social Structure*, The Hague: Martinus Nijhoff, 1982, pp. 199–218.

Riberio, Luiz Cesar de Queiroz, 'The Constitution of Real-Estate Capital and Production of Built-up Space in Rio de Janeiro, 1870–1930', *International Journal of Urban and Regional Research*, 13 (1989): 47–67.

Richman, Joel, *Traffic Wardens: An Ethnography of Street Administration*, Manchester: Manchester University Press, 1983.

Riley, James C., *The Eighteenth-Century Campaign to Avoid Disease*, Basingstoke: Macmillan, 1987.

Rimmer, P. J. and Forbes, D. K., 'Underdevelopment Theory: A Geographical Review', *Australian Geographer*, 15 (1982): 197–211.

Rimmer, Peter J.; Manderson, Lenore; and Barlow, Colin, 'The Underside of Malaysian History', in Peter J. Rimmer and Lisa M. Allen (eds.), *The Underside of Malaysian History: Pullers, Prostitutes, Plantation Workers ...*, Singapore: Singapore University Press, 1990, pp. 3–22.

Roberts, M. J. D., 'Public and Private in Early Nineteenth-Century London: The Vagrant Act of 1822 and Its Enforcement', *Social History*, 13 (1988): 273–94.

Robinson, David J., 'The Language and Significance of Place in Latin America', in John A. Agnew and James S. Duncan (eds.), *The Power of Place*, Boston: Unwin Hyman, 1989, pp. 157–84.

Rodney, Walter, *History of the Guyanese Working People, 1881–1905*, Kingston: Heinemann, 1981.

Rogers, Alisdair, 'Cinco de Mayo and the 15th January: Contrasting Situations in a Mixed Ethnic Neighbourhood', in Alisdair Rogers and Steven Vertovec (eds.), *The Urban Context: Ethnicity, Social Networks and Situational Analysis*, London: Berg, 1995, pp. 117–40.

_____, *Gentrification, Power and Versions of Community: A Case Study of Los Angeles*, University of Oxford School of Geography Research Paper 43, Oxford, 1989.

Rogerson, Christian M., 'The Underdevelopment of the Informal Sector: Street Hawking in Johannesburg, South Africa', *Urban Geography*, 9 (1988): 549–57.

Rogerson, Christian M. and Beavon, K. S. O., 'A Tradition of Repression: The Street Traders of Johannesburg', in Ray Bromley (ed.), *Planning for Small Enterprises in Third World Cities*, Oxford: Pergamon Press, 1985, pp. 233–45.

Ross, Robert and Telkamp, Gerard J., 'Introduction', in Robert Ross and Gerard J. Telkamp (eds.), *Colonial Cities: Essays on Urbanism in a Colonial Context*, Dordrecht: Martinus Nijhoff, 1985, pp. 1–6.

Sack, Robert D., *Conceptions of Space in Social Thought: A Geographic Perspective*, London, Macmillan Press, 1980.

Sadka, Emily, *The Protected Malay States 1874–1895*, Kuala Lumpur: University of Malaya Press, 1968.

Said, Edward W., 'Forward to Subaltern Studies', in Ranajit Guha and Gayatri Chakrovorty Spivak (eds.), *Selected Subaltern Studies*, New York: Oxford University Press, 1988, pp. v–x.

_____, *Orientalism*, New York: Vintage, 1979.

Salusbury-Jones, Goronwy Tidy, *Street Life in Medieval England*, Hassocks: Harvester Press, 1975 (first published in 1939).

Sandhu, Kernial Singh, 'Some Aspects of Indian Settlement in Singapore, 1819–1969', *Journal of Southeast Asian History*, 10 (1970): 193–201.

Saw Swee Hock, 'The Changing Population Structure of Singapore during 1824–1962', *Malayan Economic Review*, 9 (1964): 90–101.

_____, 'Population Trends in Singapore, 1819–1967', *Journal of Southeast Asian History*, 10 (1969): 36–49.

Schlegel, G., 'A Singapore Streetscene', *Internationales Archiv für Ethnographie*, 1 (1888): 121–9.

Scott, James C., *Weapons of the Weak: Everyday Forms of Peasant Resistance*, New Haven: Yale University Press, 1985.

Sennett, Richard, *The Fall of Public Man*, Cambridge: Cambridge University Press, 1974.

_____, *The Uses of Disorder: Personal Identity and City Life*, New York: Alfred A. Knopf, 1970.

Simmons, James S.; Whayne, T. F.; Anderson, G. W.; and H. M. Horack, *Global Epidemiology, A Geography of Disease and Sanitation*, Vol. 1., London: William Heineman, 1944.

Simon, David, 'Colonial Cities, Postcolonial Africa and the World Economy', *International Journal of Urban and Regional Research*, 13, 1 (1989): 68–91.

_____, 'Third World Colonial Cities in Context: Conceptual and Theoretical Approaches with Particular Reference to Africa', *Progress in Human Geography*, 8 (1984): 493–514.

Simon, Max F., 'Some Remarks on the Nature of So-called "Beri-beri" or Peripheral Neuritis, in the Tropics, and on Its Place in the "Nomenclature of Diseases"', *Journal of the Straits Medical Association*, 3 (1891/2): 54–60.

Simpson, W. J. R. (and discussants), 'Discussion on Sanitation of Villages and Small Towns, with Special Reference to Efficiency and Cheapness', *British Medical Journal*, 2 (1911): 1273–6.

Singapore Housing and Development Board, *Homes for the People: A Review of Public Housing*, Singapore: Straits Times Press, 1965/6.

A Singapore Merchant, *How to Govern a Colony*, London, A. H. Bailey and Co., 1869.

Singapore Municipal Commissioners, *Singapore Water Works: Water Supply from Johore*, Presented on the Occasion of the Completion of the Reservoir at Pontian Kechil, Singapore: Printers Ltd., 1932.

Singapore Municipality, Waterworks: Opening of New Works, 26 March 1912, Glasgow: Aird and Coghill, 1912.

Singapore Street Directory and Sectional Maps, Singapore: Ministry of Culture, 1954.

Sjoberg, Gideon, 'Cities in Developing and in Industrial Societies: A Cross-Cultural Analysis', in Philip M. Hauser and Leo F. Schnore (eds.), *The Study of Urbanization*, New York: John Wiley and Sons, 1965, pp. 213–63.

_____, *The Pre-Industrial City*, New York: Free Press, 1960.

Slater, David, 'Capitalism and Urbanisation at the Periphery: Problems of Interpretation and Analysis with Reference to Latin America', in David Drakakis-Smith (ed.), *Urbanisation in the Developing World* (London, Routledge, 1988), pp. 17–21.

_____, 'Peripheral Capitalism and the Regional Problematic', in Richard Peet and Nigel Thrift (eds.), *New Models in Geography: The Political Economy Perspective*, Vol. 2, London: Unwin Hyman, 1989, pp. 267–94.

Smith, D. A. and Nemeth, R. J., 'Urban Development in Southeast Asia: An Historical Structural Analysis', in David Drakakis-Smith (ed.), *Urbanisation in the Developing World*, London: Routledge, 1988, pp. 121–39.

Smith, F., 'Municipal Sewerage: Part I', *The Journal of Tropical Medicine*, 6 (1903): 285–91.

Smith, F. B., *The People's Health 1830–1910*, London: Croom Helm, 1979.

Smith, Michael Peter and Tardanico, Richard, 'Urban Theory Reconsidered: Production, Reproduction and Collective Action', in Michael Peter Smith and Joe R. Feagin (eds.), *The Capitalist City: Global Restructuring of Community Politics*, Oxford: Basil Blackwell, 1987, pp. 87–110.

Snow, John, *On the Mode of Communiciation of Cholera*, London: J. Churchill, 1855.

'Some Points in the Municipal Administration of Singapore', *Straits Chinese Magazine*, 8, 4 (1904): 163–6.

Song Ong Siang, *One Hundred Years' History of the Chinese in Singapore*, Singapore: Oxford University Press, 1984 (first published in 1902).

Souvenir of Singapore: A Descriptive and Illustrative Guide Book of Singapore, Singapore: Straits Times Press, 1905.

Spencer, Joseph E., 'Chinese Place Names and the Appreciation of Geographic Realities', *Geographical Review*, 31 (1941): 71–94.

Spengler, Oswald, *The Decline of the West*, London: George Allen and Unwin, 1961 (first published in 1932).

Stallybrass, Peter and White, Allon, *The Politics and Poetics of Transgression*, London: Methuen, 1986.

Stedman Jones, Gareth, *Outcast London: A Study of the Relationship between Classes in Victorian Society*, London: Peregrine Books, 1984.

Stump, Roger W., 'Toponymic Commemoration of National Figures: The Case of Kennedy and King', *Names*, 36 (1988): 203–16.

Su Xiao Xian, 'Tong Ji Yi Yuan Yan Ge Shi Lue' [A Brief History of the Thong Chai Medical Institution], in *Tong Ji Yi Yuan Da Sha Luo Cheng Ji Nian Te Kan* [Souvenir Magazine of the Opening Ceremony of the Newly Completed Thong Chai Medical Institution Building], Singapore, 1979.

Sundaram, Jomo Kwame, *A Question of Class: Capital, the State and Uneven Development in Malaya*, Singapore: Oxford University Press, 1986.

'Supreme Court: Sir E. O'Malley, C.J., Singapore, 9 June 1891, The Municipal Commissioners on the Prosecution of G. L. Minjoot (Respondent) v. Tan Chye Seng (Appellant)', *Straits Law Reports (N.S.)* (1891): 17.

'Supreme Court: Sir E. O'Malley, C.J., Singapore, 9 and 10 June 1891, The Municipal Commissioners on the Prosecution of G. L. Minjoot (Respondent) v. Manasseh and Co. (Appellant)', *Straits Law Reports (N.S.)*, 1891: 18–20.

Swanson, Maynard W., 'The Sanitation Syndrome: Bubonic Plague and Urban Native Policy in the Cape Colony, 1900–1909', *Journal of African History*, 18 (1977): 387–410.

Tan Pek Leng, 'Chinese Secret Societies and Labour Control in the Nineteenth Century Straits Settlements', *Kajian Malaysia*, 1, 2 (1983): 14–48.

Tan, Thomas Tsu Wee, 'Singapore Modernisation: A Study of Traditional Voluntary Associations in Social Change', Ph.D. thesis, University of Virginia, 1983.

Teo Siew Eng and Savage, Victor R., 'Singapore Landscape: A Historical Overview of Housing Change', *Singapore Journal of Tropical Geography*, 6 (1985): 48–63.

Thio, Eunice, *British Policy in the Malay Peninsula 1880–1910*, Vol. 1, Singapore: University of Malaya Press, 1969.

_____, 'The Singapore Chinese Protectorate: Events and Conditions Leading to Its Establishment', *Journal of the South Seas Society*, 16 (1960): 40–80.

Thomas, Nicholas, 'Sanitation and Seeing: The Creation of State Power in Early Colonial Fiji', *Comparative Studies in Society and History*, 32 (1990): 149–70.

Thompson, E. P., *The Making of the English Working Class*, Harmondsworth: Penguin, 1968.

Thomson, John Turnbull, *Sequel to Life in the Far East*, London: Richardson and Co., 1865.

_____, *Some Glimpses into Life in the Far East*, London: Richardson and Co., 1864.

Thrift, Nigel, 'Flies and Germs: A Geography of Knowledge', in Derek Gregory and John Urry (eds.), *Social Relations and Spatial Structure*, Basingstoke: Macmillan, 1985.

Tilly, Charles, 'Notes on Urban Images of Historians', in Lloyd Rodwin and Robert M. Hollister (eds.), *Cities of the Mind: Images and Themes of the City in the Social Sciences*, New York: Plenum Press, 1984, pp. 119–32.

Timberlake, Michael, 'World-System Theory and the Study of Comparative Urbanization', in Michael Peter Smith and Joe R. Feagin (eds.), *The Capitalist City: Global Restructuring of Community Politics*, Oxford: Basil Blackwell, 1987, pp. 37–65.

Tolkien, J. R. R., *The Lord of the Rings*, London: Unwin Hyman, 1978 (first published in 1954).

Tong Chee Kiong, 'Dangerous Blood, Refined Souls: Death Rituals among the Chinese in Singapore', Ph.D. thesis, Cornell University, 1987.

Trexler, R., *Public Life in Renaissance Florence*, New York: Academic Press, 1980.

Tuan Yi-Fu, *China*, London: Longman, 1970.

_____, *Landscapes of Fear*, Oxford: Basil Blackwell, 1980.

Turnbull, Constance Mary, 'Communal Disturbances in the Straits Settlements in 1857', *Journal of the Malayan Branch of the Royal Asiatic Society*, 31 (1958): 94–144.

_____, *A History of Singapore, 1819–1975*, Kuala Lumpur: Oxford University Press, 1977.

_____, *The Straits Settlements, 1826–67: Indian Presidency to Crown Colony*, Singapore: Oxford University Press, 1972.

Vaughan, James Daniel, *The Manners and Customs of the Chinese of the Straits Settlements*, Singapore: Oxford University Press, 1985 (first published in 1879).

Vidler, Anthony, 'The Scenes of the Street: Transformation in Ideal and Reality, 1750–1871', in Stanford Anderson (ed.), *On Streets*, Cambridge, Massachusetts: MIT Press, 1978, pp. 28–111.

Voon Phin-Keong, 'The Origins of Chinese Place Names', *Geographica*, 5 (1969): 34–47.

Walzer, Michael, 'Pleasures and Costs of Urbanity', *Dissent (New York)*, Fall (1986): 470–5.

Ward, T. M. and Grant, J. P., *Official Papers on the Medical Statistics and Topography of Malacca and the Prince of Wales' Island, and on the Prevailing Diseases of the Tenasserim Coast*, Pinang: Government Press, 1830.

Warren, James Francis, *Ah Ku and Karayuki-san: Prostitution in Singapore, 1870–1940*, Singapore: Oxford University Press, 1993.

_____, 'Prostitution and the Politics of Venereal Disease: Singapore, 1870–98', *Journal of Southeast Asian Studies*, 21 (1990): 360–83.

_____, *Rickshaw Coolie: A People's History of Singapore (1880–1940)*, Singapore: Oxford University Press, 1986.

Washbrook, D. A., 'Ethnicity and Racialism in Colonial Indian Society', in Robert Ross (ed.), *Racism and Colonialism: Essays on Ideology and Social Structure*, The Hague: Martinus Nijhoff, 1982, pp. 143–81.

Watson, James L., 'Of Flesh and Bones: The Management of Death Pollution in Cantonese Society', in Maurice Bloch and Jonathan Parry (eds.), *Death and the Regeneration of Life*, Cambridge: Cambridge University Press, 1982, pp. 155–86.

_____, 'The Structure of Chinese Funerary Rites: Elementary Forms, Ritual Sequence, and the Primacy of Performance', in James L. Watson and Evelyn S. Rawski (eds.), *Death Ritual in Late Imperial and Modern China*, Berkeley: University of California Press, 1988.

Watson, Rubie S., 'Remembering the Dead: Graves and Politics in Southeastern China', in James L. Watson and Evelyn S. Rawski (eds.), *Death Ritual in Late Imperial and Modern China*, Berkeley: University of California Press, 1988.

Weld, Frederick A., 'The Straits Settlements and British Malaya [Speech Delivered on 10 June 1884]', in Paul H. Kratoska (ed.), *Honourable Intentions: Talks on the British Empire in South-East Asia Delivered at the Royal Colonial Institute 1874–1928*, Singapore: Oxford University Press, 1983, pp. 43–90.

Wheatley, Paul, *City as Symbol: An Inaugural Lecture Delivered at University College London, 20 Nov 1969*, London: H. K. Lewis, 1969.

_____, *The Pivot of the Four Quarters: A Preliminary Enquiry into the Origins and Character of the Chinese City*, Chicago: Aldine, 1971.

Williams, Lea E., 'Chinese Leadership in Early British Singapore', *Asian Studies*, 2 (1964): 170–9.

Wohl, Anthony S., *Endangered Lives: Public Health in Victorian Britain*, London: J. M. Dent and Sons, 1983.

_____, 'The Housing of the Working Classes in London', in Stanley D. Chapman (ed.), *The History of Working-class Housing*, Newton Abbot: David and Charles, 1971, pp. 13–54.

Wolf, Eric R., *Europe and the People without History*, Berkeley: University of California Press, 1982.

Wong, Aline K. and Yeh, Stephen H. K. (eds.), *Housing a Nation*, Singapore: Maruzen and Housing Development Board, 1985.

Wong K. Chimin and Wu Lien-Teh, *History of Chinese Medicine Being a Chronicle of Medical Happenings in China from Ancient Times to the Present Period*, Tientsin: Tientsin Press, 1932.

Wood, R., 'Mortality and Sanitary Conditions in the "Best Governed City" in the World—Birmingham, 1870–1910', *Journal of Historical Geography*, 4 (1978): 35–56.

Worboys, Michael, 'British Colonial Medicine and Tropical Imperialism: A Comparative Perspective', in G. M. van Heteren, A. de Kneckt-van Eekelen, and M. J. D. Poulissen (eds.), *Medicine in the Malay Archipelago*, Amsterdam: Atlanta GA, 1989, pp. 153–67.

Wynne, Meryn Llewelyn, *Triad and Tabut: A Survey of the Origin and Diffusion of Chinese and Mohamedan Secret Societies in the Malay Peninsula A.D. 1800–1935*, Singapore: Government Printing Office, 1941.

Xin Jia Bo Cha Yang Hui Guan Bai Nian Ji Nian Kan [Souvenir Magazine of the Centenary Celebration of the Singapore Char Yong Association], Singapore, 1958.

Xin Jia Bo Guang Hui Zhao Bi Shan Ting Qing Zhu 118 Zhou Nian Ji Nian Te Kan [118th Anniversary Souvenir Magazine of the Singapore Guang Hui Zhao Bi Shan Ting], Singapore, 1988.

Xin Jia Bo Qiong Zhou Hui Guan Da Sha Luo Cheng Ji Nian Te Kan [Souvenir Magazine of the Opening Ceremony of the Kheng Chiu Building], Singapore, 1965.

Xin Jia Bo Wang Shi Ci Shan Kai Min Gong Si Te Kan Ji Nian [Souvenir Magazine of the Singapore Kai Min Ong See Benevolent Association], Singapore, 1977.

Xin Jia Bo Ying Fo Hui Guan Yi Bai Liu Shi Wu Zhou Nian Ji Nian Te Kan, 1827–1987 [165th Anniversary Commemorative and Souvenir Publication of the Ying Fo Fui Kun, 1827–1987], Singapore, 1989.

Yang Shi Zong Pu [The Yang Clan Genealogy], Singapore, 1965.

Yeh, Stephen H. K. (ed.), *Public Housing in Singapore*, Singapore: Singapore University Press, 1975.

Yen Ching-hwang, *A Social History of the Chinese in Singapore and Malaya, 1800–1911*, Singapore: Oxford University Press, 1986.

Yeo Lian Bee, Katherine, 'Hawkers and the State in Colonial Singapore: Mid-Nineteenth Century to 1939', MA thesis, Monash University, 1989.

Yeoh Saw Ai, Brenda, 'The Decline of a Community—The Babas (A Socio-Geographical Perspective)', BA dissertation, Cambridge University, 1985.

———, 'Municipal Control, Asian Agency and the Urban Built Environment in Colonial Singapore, 1880–1929', D. Phil. thesis, Oxford University, 1991.

Yeoh, Brenda S. A. and Kong, Lily, 'Reading Landscape Meanings: State Constructions and Lived Experiences in Singapore's Chinatown', *Habitat International*, 18, 4 (1994): 17–35.

Yong Ching Fatt, 'Chinese Leadership in Nineteenth Century Singapore', *Journal of the Island Society*, 1, 1 (1967): 1–18.

———, 'Emergence of Chinese Community Leaders in Singapore, 1890–1941', *Journal of the South Seas Society*, 30, 2 (1975): 1–18.

———, 'A Preliminary Study of Chinese Leadership in Singapore, 1900–1941', *Journal of South East Asian History*, 9 (1968): 258–85.

Young, F. W., 'Graveyards and Social Structure', *Rural Sociology*, 25 (1960): 446–50.

Yule, Henry and Burnell, A. C., *Hobson-Jobson: A Glossary of Anglo-Indian Colloquial Words and Phrases and of Kindred Terms*, London: John Murray, 1903.

Zelinsky, Wilbur, 'Gathering Places for America's Dead: How Many, Where, and Why?', *Professional Geographer*, 46 (1994): 29–38

———, 'Nationalism in the American Place-Name Cover', *Names*, 31 (1983): 1–28.

———, 'Unearthly Delights: Cemetery Names and the Map of the Changing American Afterworld', in David Lowenthal and M. J. Bowden (eds.), *Geographies of the Mind*, Oxford: Oxford University Press, 1975, pp. 171–95.

Zhuang Qin Yong (David K. Y. Chng), 'Yang Shi Jia Zu Yu Xie Yuan Shan' [Hiap Guan Sun Cemetery Stone Tablet and the Yeo Families], *Asian Culture*, 8 (1986): 24–7.

Index